Lecture Notes in Control and Information Sciences

Volume 456

About this Series

This series aims to report new developments in the fields of control and information sciences—quickly, informally and at a high level. The type of material considered for publication includes:

1. Preliminary drafts of monographs and advanced textbooks
2. Lectures on a new field, or presenting a new angle on a classical field
3. Research reports
4. Reports of meetings, provided they are
   (a) of exceptional interest and
   (b) devoted to a specific topic. The timeliness of subject material is very important.

More information about this series at http://www.springer.com/series/642

Jan H. van Schuppen · Tiziano Villa
Editors

Coordination Control of Distributed Systems

Springer

Editors
Jan H. van Schuppen
Delft Institute of Applied Mathematics
Delft University of Technology
Delft
The Netherlands

Tiziano Villa
Dipartimento d'Informatica
Università degli studi di Verona
Verona
Italy

ISSN 0170-8643 ISSN 1610-7411 (electronic)
ISBN 978-3-319-10406-5 ISBN 978-3-319-10407-2 (eBook)
DOI 10.1007/978-3-319-10407-2

Library of Congress Control Number: 2014947683

Springer Cham Heidelberg New York Dordrecht London

Printed on acid-free paper

Springer is part of Springer Science+Business Media (www.springer.com)

Preface

Aim of Book

The aim of this book is to provide to its readers an exposition and a summary of research results of control of distributed and of multilevel/hierarchical systems. The chapters are based on research for the C4C Project which was sponsored by the European Commission. The chapters of the book cover case studies and theory.

The case studies of the project include the following distributed systems: control of underwater vehicles as available at the University of Porto, control of aerial vehicles, control of road networks with a traffic control center, control of straddle carriers transporting containers on the floor of a container terminal, and control of a high-speed printer with many local sensors and actuators.

In regard to theory, the book focusses attention on the integration of control, information and communication, computation, verification, and related aspects of engineering and of computer science.

The book originated from an extended technical report written for the project officer of the C4C Project and for the three anonymous reviewers of the project.

The authors of the report are researchers who have been involved in the C4C Project or were affiliated with the teams of the C4C Project.

Style of Parts and Chapters

The book is structured into parts with chapters each having its own research topic. The titles of the parts reflect the organization of the project and the various control architectures considered.

The chapters of the book have the character of essays of at most eight pages. An essay is a short text with a focus on a problem, concepts, the main body of the relevant theory, a discussion of research issues, and finally suggestions for further reading. Due to their special character, several chapters are longer than eight pages.

The motivation for the choice of chapters in the form of essays follows. Long chapters quickly make a book unreadable for most readers. But an essay can be read by a knowledgeable reader of engineering or of mathematics in about 15 minutes. After reading one essay, the reader can go on to other essays. With the table of contents, each reader can chart her or his own route through the book.

C4C Project

The book also describes partly the results of the Project *Control for Coordination of Distributed Systems* (CON4COORD and C4C, both acronyms are used) which was sponsored by the European Commission via Grant Agreement INFSO-ICT-223844. The lifetime of the project was 1 May 2008 till 1 September 2011. The participants of the project are the following organizations:

- Centrum Wiskunde & Informatica (CWI), in Amsterdam, The Netherlands.
- The research center CERETETH at the University of Thessaly, Volos, Greece.
- The Faculty of Technology, Policy, and Management of the Delft University of Technology in Delft, The Netherlands.
- The Faculty of Mechanical Engineering of the Eindhoven University of Technology in Eindhoven, The Netherlands.
- The Department of Electrical and Computer Engineering of the University of Cyprus in Nicosia, Cyprus.
- The Department of Electrical Energy, Systems, and Automation, of Ghent University in Ghent, Belgium.
- The Faculty of Engineering of the University of Porto in Porto, Portugal.
- The Department of Computer Science of the University of Verona in Verona, Italy.
- The company PSA Antwerp (formerly Hesse-Noord Natie) in Antwerp, Belgium.
- The company Ocean Scan–Marine Systems Technology in Porto, Portugal.
- The company Océ Technologies in Venlo, The Netherlands.
- The company Trinité Automation B.V. in Uithoorn, The Netherlands.

Acknowledgments

The teams of the C4C Consortium herewith gratefully acknowledge the financial support in part provided by the European Commission for the project *Control for Coordination of Distributed Systems* via Grant Agreement INFSO-ICT-223844.

Amsterdam, May 2014 Jan H. van Schuppen
Verona Tiziano Villa

Acknowledgments

The Editors of the book are very grateful to all the members of the book Editorial Board for their dedicated efforts to improve the quality of the chapters. The task of a member of the board was to handle the review procedures of several chapters of the book. The members of the Editorial Board include: René K. Boel, Charalambos D. Charalambous, Chris Hadjicostis, Marios Polycarpou, João Bourges de Sousa, Bert (D.A.) van Beek, Jan H. van Schuppen, Tiziano Villa, and Jos L.M. Vrancken.

In addition, the members of the Editorial Board are very grateful to the following researchers who reviewed the chapters of the book and provided careful comments to help the authors in improving their essays:

Alessandro Abate, Tamer Başar, Luca Benvenuti, Rene Boel, Davide Bresolin, Marta Capiluppi, Charalambos D. Charalambous, Themistoklis Charalambous, Manolis Christodoulakis, Ferdinando Cicalese, Milan Derpich, Alina I. Doban, Tony Ephremides, Francesco Fabris, Lorenzo Farina, Paolo Fiorini, Federica Garin, Luca Geretti, Christoforos Hadjicostis, Bruce Hajek, Pia L. Kempker, Georgios Kladis, Panayiotis Kolios, Christos Kourtellaris, Ioannis Krikidis, P.R. Kumar, Alexander B. Kurzhanski, Stephane Lafortune, Aditya Mahajan, Jasen Markovski, Petros Mavroeidis, Manuel Mazo, Lyudmila Mihaylova, Riccardo Muradore, Christos Panayiotou, Mihály Petreczky, Alexander Pogromsky, Marios Polycarpou, Davide Quaglia, Nicanor Quijano, Andre Ran, Stefan Ratschan, Michel Reniers, Vasso Reppa, Michael Rotkowitz, Carsten Scherer, KlausWerner Schmidt, Thomas Schreiter, Roberto Segala, Stephen L. Smith, Lou Somers, Photios A. Stavrou, Rong Su, Lennart Swartjes, Demos Teneketzis, Stelios Timotheou, Frits Vaandrager, Alan van Hulst, Jan H. van Schuppen, Dirk van Zwieten, Alexander Verbraeck, Tiziano Villa, Stelios Vrachimis, Jos L.M. Vrancken, Rafal Wisniewski, Serdar Yüksel.

The editors are grateful to the secretary Dorothée W.M. Engering (Department of Applied Mathematics, Delft University of Technology) for secretarial assistance.

Acknowledgments

Contents

Part I Case Studies in Control of Distributed Systems

1 **C4C Case Studies** 3
Jan H. van Schuppen

2 **A Model Predictive Control Approach to AUVs Motion Coordination** 9
Fernando Lobo Pereira, J. Borges de Sousa, R. Gomes and P. Calado

3 **Dynamic Optimization Techniques for the Motion Coordination of Autonomous Vehicles** 19
Jorge Estrela da Silva, João Borges de Sousa and Fernando Lobo Pereira

4 **Coordination Challenges in Networked Vehicle Systems: Are We Missing Something?** 27
J. Borges de Sousa and Fernando Lobo Pereira

5 **Leader–Follower Coordination Control for Urban Traffic** 35
René Boel and Nicolae Marinică

6 **Prediction of Traffic Flow in a Road Network** 45
Yubin Wang, Jan H. van Schuppen and Jos L.M.Vrancken

7 **Zone-Control-Based Traffic Control of Automated Guided Vehicles** 53
Qin Li, Jan Tijmen Udding and Alexander Pogromsky

8 **Coordination Control of Complex Machines** 61
Jos C.M. Baeten, Bert van Beek, Jasen Markovski and Lou J.A.M. Somers

9 **Coordinated Model-Predictive Control for Avoiding Voltage Collapse in an Electric Power Transmission Net** 69
René Boel, Mohammad Moradzadeh and Lieven Vandevelde

Part II Architectures of Distributed Systems and Their Control

10 **System Architectures of Distributed and of Multilevel Systems** . . . 81
Jan H. van Schuppen

11 **Control Architectures** 89
Jan H. van Schuppen

Part III Coordination Control

12 **What is Coordination Control?** 99
Jan H. van Schuppen

13 **Introduction to Coordinated Linear Systems** 107
Pia L. Kempker

14 **Coordinated Linear Systems** 113
André C.M. Ran and Jan H. van Schuppen

15 **LQ Optimal Control for Coordinated Linear Systems** 123
Pia L. Kempker

16 **Supervisory Control of Discrete-Event Systems** 129
Jan Komenda and Tomáš Masopust

17 **Coordination Control of Distributed Discrete-Event Systems**. 137
Jan Komenda and Tomáš Masopust

Part IV Distributed Control

18 **What Is Team Theory?**. 147
Jan H. van Schuppen

19 **Team Theory and Information Structures of Stochastic Dynamic Decentralized Decision** 155
C.D. Charalambous and N.U. Ahmed

20 Signaling of Information 165
César A. Uribe and Jan H. van Schuppen

21 Distributed Control of Manufacturing Networks: Analysis of Performance 173
Konstantin K. Starkov, Alexander Y. Pogromsky and Jacobus E. Rooda

Part V Distributed Control with Communication

22 What Is Distributed Control with Direct Communication Between Controllers? 183
Jan H. van Schuppen

23 Communication Constraints in Control and Observation of Distributed Systems 191
A.Yu. Pogromsky and A.S. Matveev

24 Information Structures 197
Jan H. van Schuppen

25 Control Theory with Information Structures 205
Jan H. van Schuppen

26 Common, Correlated, and Private Information in Control of Decentralized Systems 215
Jan H. van Schuppen

27 Distributed State Estimation with Communication of Observations 223
André C.M. Ran and Jan H. van Schuppen

Part VI Multilevel Control

28 Complexity and Scalability of Hierarchical Control for Networked Infrastructures 231
Jos Vrancken

29 Hierarchical Control for Road Traffic Management 239
Jos Vrancken

Part VII Communication and Control of Distributed Systems

30 Layered Backpressure Scheduling for Delay-Aware Routing in Ad Hoc Networks 251
Dimitrios Katsaros

31 Cache Control Issues in Pub–Sub Networks and Wireless Sensor Networks 259
Dimitrios Katsaros

32 Detecting Influential Nodes in Complex Networks with Range Probabilistic Control Centrality 265
Dimitrios Katsaros and Pavlos Basaras

33 Model Predictive Controllers over Differentiated Services Packet Networks 273
Riccardo Muradore, Davide Quaglia and Paolo Fiorini

34 A SystemC/MATLAB Co-simulation Tool for Networked Control Systems 283
Davide Quaglia, Riccardo Muradore and Paolo Fiorini

35 On Shannon's Duality of a Source and a Channel and Nonanticipative Communication and Communication for Control 291
Charalambos D. Charalambous, Christos K. Kourtellaris and Photios A. Stavrou

36 Directed Information on Abstract Spaces: Properties and Extremum Problems 307
Charalambos D. Charalambous, Photios A. Stavrou and Christos K. Kourtellaris

37 Information Nonanticipative Rate Distortion Function and Its Applications 317
Photios A. Stavrou, Christos K. Kourtellaris and Charalambos D. Charalambous

38 Nonanticipative Duality of Sources and Channels with Memory and Feedback 325
Christos K. Kourtellaris, Charalambos D. Charalambous and Photios A. Stavrou

Part VIII Informatics of Distributed Systems: Modeling and Verification

39 An Introduction to the Verification of Hybrid Systems Using ARIADNE 339
Davide Bresolin, Luca Geretti, Tiziano Villa and Pieter Collins

40 Formal Verification Applied to Robotic Surgery 347
Davide Bresolin, Luca Geretti, Riccardo Muradore, Paolo Fiorini and Tiziano Villa

41 Computing Reachable Sets of Differential Inclusions 357
Sanja Živanović Gonzalez and Pieter Collins

42 Modeling Objects Moving in a Complex Environment with World Automata 367
Marta Capiluppi and Roberto Segala

43 Distributed Average Consensus in Digraphs 375
Christoforos N. Hadjicostis and Alejandro D. Domínguez-García

Part IX Research Program

44 Research Program for Control of Distributed and of Multilevel Systems 385
Jan H. van Schuppen

Index 393

Acronyms

C4C	Project Control for Coordination of Distributed Systems. Project financed by the European Commission via the ICT Program INFSO—223844.
CER	Research Group Communication Networks of the CEnter for REsearch and TEchnology (CERETETH) established at the University of Thessaly, Volos, Greece.
CON4COORD	Same as C4C Project, this is an earlier version of the acronym.
CWI	Centrum Wiskunde & Informatica, a research institute in Amsterdam, The Netherlands.
HNN	The company Hessen-Noord Natie, Antwerp, Belgium. In the year 2010 the name of the company was changed to PSA Antwerp. A port operator with a container terminal.
MST	Ocean Scan–Marine Systems Technology, Porto, Portugal. A company which produces underwater vehicles.
OCE	Océ Technologies, Venlo, The Netherlands. The company produces copy machines and high-speed printers.
TRI	Trinité Automatisering B.V., Uithoorn, The Netherlands. A software company specializing in control of road networks. In 2010 the name of the company was changed to Trinité Automation B.V.
TUD	Systems Engineering Section, Faculty of Technology, Policy, and Management, Delft University of Technology, Delft, The Netherlands.
TUE	Systems Engineering Group, Faculty of Mechanical Engineering, Eindhoven University of Technology, Eindhoven, The Netherlands.
UCY	Department of Electrical and Computer Engineering, University of Cyprus, Nicosia, Cyprus.
UGE	the SYSTeMS Research Group, Faculty of Engineering, Ghent University, Ghent, Belgium.

UPO	Decision and Control Engineering Research Group of the Department of Engineering, of the University of Porto, Porto, Portugal.
UVR	Department of Computer Science (Dipartimento d'Informatica) of the University of Verona, Verona, Italy.

Contributors

N.U. Ahmed University of Ottawa, School of Engineering and Computer Science, Ottawa, Canada

Jos C.M. Baeten Eindhoven University of Technology, Eindhoven, The Netherlands

Pavlos Basaras Electrical and Computer Engineering Department, University of Thessaly, Volos, Greece

René Boel SYSTeMS Research Group, Ghent University, Gent, Belgium

Davide Bresolin Dipartimento di Informatica, Università di Verona, Verona, Italy; Department of Computer Science, University of Verona, Verona, Italy

P. Calado Faculdade de Engenharia, Universidade do Porto, Porto, Portugal

Marta Capiluppi Dipartimento di Informatica, Università di Verona, Verona, Italy

Charalambos D. Charalambous Department of Electrical and Computer Engineering (ECE), University of Cyprus, Nicosia, Cyprus

Pieter Collins Department of Knowledge Engineering, Maastricht University, Maastricht, The Netherlands

Jorge Estrela da Silva School of Engineering, Polytechnic Institute of Porto, Porto, Portugal

J. Borges de Sousa Faculdade de Engenharia, Universidade do Porto, Porto, Portugal

Alejandro D. Domínguez-García Department of Electrical and Computer Engineering, University of Illinois at Urbana–Champaign, Urbana, USA

Paolo Fiorini Department of Computer Science, University of Verona, Verona, Italy

Luca Geretti Dipartimento di Informatica, Università di Verona, Verona, Italy

R. Gomes Faculdade de Engenharia, Universidade do Porto, Porto, Portugal

Christoforos N. Hadjicostis Department of Electrical and Computer Engineering, University of Cyprus, Nicosia, Cyprus; Department of Electrical and Computer Engineering, University of Illinois at Urbana–Champaign, Urbana, USA

Dimitrios Katsaros Electrical and Computer Engineering Department, University of Thessaly, Volos, Greece

Pia L. Kempker TNO, Delft, The Netherlands

Jan Komenda Institute of Mathematics, Academy of Sciences of the Czech Republic, Brno, Czech Republic

Christos K. Kourtellaris Department of Electrical and Computer Engineering (ECE), University of Cyprus, Nicosia, Cyprus

Qin Li Statoil Research Centre, Porsgrunn, Norway

Fernando Lobo Pereira Departamento de Engenharia Electrotécnica e Computadores, Faculdade de Engenharia, Universidade do Porto, Porto, Portugal

Nicolae Marinică SYSTeMS Research Group, Ghent University, Gent, Belgium

Jasen Markovski Eindhoven University of Technology, Eindhoven, The Netherlands

Tomáš Masopust Institute of Mathematics, Academy of Sciences of the Czech Republic, Brno, Czech Republic

A.S. Matveev St. Petersburg State University, St. Petersburg, Russia

Mohammad Moradzadeh Electrical Energy Laboratory, Ghent University, Gent, Belgium

Riccardo Muradore Department of Computer Science, University of Verona, Verona, Italy

Alexander Pogromsky Eindhoven University of Technology, Eindhoven, The Netherlands

Davide Quaglia Department of Computer Science, University of Verona, Verona, Italy

André C.M. Ran Department of Mathematics, VU University Amsterdam, Amsterdam, The Netherlands; Unit for BMI, North-West University, Potchefstroom, South Africa

Jacobus E. Rooda Eindhoven University of Technology, Eindhoven, The Netherlands

Roberto Segala Dipartimento di Informatica, Università di Verona, Verona, Italy

Lou J.A. M. Somers Eindhoven University of Technology, Eindhoven, The Netherlands

Konstantin K. Starkov NSPYRE B.V., Utrecht, The Netherlands

Photios A. Stavrou Department of Electrical and Computer Engineering (ECE), University of Cyprus, Nicosia, Cyprus

Jan Tijmen Udding Eindhoven University of Technology, Eindhoven, The Netherlands

César A. Uribe Coordinated Science Laboratory, University of Illinois at Urbana–Champaign, Urbana, IL, USA

Bert van Beek Eindhoven University of Technology, Eindhoven, The Netherlands

Jan H. van Schuppen Van Schuppen Control Research, Amsterdam, The Netherlands

Lieven Vandevelde Electrical Energy Laboratory, Ghent University, Gent, Belgium

Tiziano Villa Dipartimento di Informatica, Università di Verona, Verona, Italy

Jos L.M. Vrancken Delft University of Technology, GA, Delft, The Netherlands

Yubin Wang Delft University of Technology, Delft, GA, The Netherlands

Sanja Živanović Gonzalez Department of Mathematics and Computer Science, Barry University, Miami Shores, FL, USA

Part I
Case Studies in Control of Distributed Systems

Each Part page provides suggestions to the readers by the editors on the chapters included in the Part.

The chapters of every part are distinguished into *research chapters and introductory chapters*. A research chapter presents an essay on a topic recently investigated by its authors. An introductory chapter provides to the reader an introduction and a tutorial of a research area which will help the reader to better understand the research chapters of the same part.

Chapter 1 is introductory.

Chapters 2-4 treat control of distributed underwater vehicles. Control of road networks is treated in the Chaps. 5 and 6. Control of vehicles on the floor of a container terminal is treated in Chap. 7. Chapter 8 deals with the control software design process and illustrates it when designing the controller of a printer. Finally, there is an application to coordinated model predictive control of an electric power transmission net in Chap. 9.

Chapter 1
C4C Case Studies

Jan H. van Schuppen

1.1 Motivation

This chapter provides an overview of the five case studies of the C4C Project.

The case studies of control of distributed systems were selected for the C4C Project and are listed below. Each of the case studies represents a problem of control engineering which is currently the focus of research at the related companies or government agencies. The case studies are far from trivial and the research for the case studies has continued after the lifetime of the project. The strength of the C4C Project was partly in the technological advanced character of its case studies.

Almost all case studies related to all of the four theoretical workpackages on control, communication, informatics, and tools are also listed below. Per case study are discussed the aim of the research, the owner of the problem, the motivation, the main research issues, and the C4C Teams involved in the investigation.

1.2 Purposes of Case Studies

The European Commission provides financial support for research. The research is expected to benefit the societies of the European Union countries. Therefore, the focus is on joint research of industrial firms, government agencies, and academic research institutes.

The interactions between these organizations is always a two-way process. The concrete technological problems of the firms or of government agencies are formulated, transferred to academic circles, and transformed to academic research problems. Conversely, theoretical approaches are detailed, adapted to the concrete

J.H. van Schuppen (✉)
Van Schuppen Control Research, Gouden Leeuw 143, 1103 KB Amsterdam, The Netherlands
e-mail: jan.h.van.schuppen@xs4all.nl

J.H. van Schuppen and T. Villa (eds.), *Coordination Control of Distributed Systems*,
Lecture Notes in Control and Information Sciences 456,
DOI 10.1007/978-3-319-10407-2_1

engineering problems, converted into algorithms, and evaluated in regard to their usefulness for the companies. Both sides therefore benefit, companies in the form of solutions to problems and in access to scientific research. Academic institutes benefit because of the experience with concrete engineering problems and the need to develop concepts, theories, and algorithms for motivated research.

Most research proposals financed by the European Commission for applied research now include case studies and involve companies and/or government agencies. The evaluation process of research proposals includes criteria about the usefulness of the project to problems of the European Union societies in broad terms and to science. The evaluation criteria of research proposals related to science refer to novelty, to substance of the proposed research, and to the expertises and experiences of the research teams.

For a particular project, the term *case study* is used to describe the joint research activity of a company and one or more academic research institutes of the consortium. The research efforts for a case study mostly take the full lifetime of the project. The communication process for a case study between all parties proceeds via the following process. First, there is a discussion about the problem formulation. Secondly, there are discussions about possible approaches. Next, there are discussed one or several theoretical solutions. Discussions of the solutions in regard to practical constraints of the company take much time. Finally, there are discussions about the possible implementations. During the lifetime of the project, all parties experience a demanding learning effort.

What after the project remains with the companies is a solution to an engineering problem. Often, these solutions need further development in the companies. What after the project remains with the academic researchers is the knowledge about a concrete engineering problem, the scientific knowledge about the theory for such problems, the expertise of modeling realistic engineering problems, and the experience of communication with companies. These effects may not be realized by all parties during the process, but the long-term effect may more reflect this.

Since the C4C Project has ended, many of the team leaders of the academic institutes have continued the cooperation with companies of the project either informally or by engaging with several parties in a research proposal for a subsequent project.

1.3 WP2 Underwater Vehicles

Work Package 2 autonomous underwater vehicles (AUVs) aimed at the development of control algorithms and at the demonstration of their effectiveness. The focus of the case study was to develop control algorithms for the vehicles to operate autonomously and to coordinate the activities of two or more vehicles.

The main research and demonstration tasks included the following: (1) coordination control of multiple AUVs, (2) autonomous operation of the AUV individually,

(3) communication between the AUV and surface vessels, and (4) a demonstration of the AUVs and other vehicles at the C4C Review Meeting 2011 in Porto.

The leader of Work Package 2 Underwater Vehicles (WP2) of the C4C Project for this case study was the Department FEUP of the University of Porto. There is a laboratory at this university with both underwater vehicles and aerial vehicles. The list of all C4C Teams involved in this case study was primarily the team UPO of the University of Porto with Fernando Lobo Pereira and João Sousa as its leaders, the team MST of the company Oceanscan-Marine Systems Technology with Alexandre Sousa as its leader, and, to a minor extent, the teams CWI, UCY, UGE, and UVR.

In this description and those below are mentioned the main research teams active for the work package and by name their leaders. Teams whose relative effort to the work package is minor are mentioned only by their acronym.

1.4 WP3 Aerial Vehicles

The aim of the case study was to develop control theory for coordination of uninhabited aerial vehicles (UAVs). The motivation of the case study was the use of aerial vehicles for environmental monitoring like for forest fires or for oil spills. This motivation required coordination between the vehicles during monitoring missions to organize searches in such a way that the possible object, an oil spill or a forest fire, is located as fast as possible.

The main research issues of the work package included the following: (1) control of search missions and (2) coordination of multiple vehicles during search missions.

The leader of Work Package 3 Aerial Vehicles was the Department Electrical and Computer Engineering of the University of Cyprus. The research leader of the package was Marios Polycarpou. The department does not operate a laboratoria with vehicles. The required background in aerial vehicles was obtained by Polycarpou before the start of the project. The list of all C4C Teams involved in the WP3 was UCY and UPO, and, to a minor extent, several other teams.

1.5 WP4 Road Networks

The aim of the case study was to develop control engineering and control theory for the control architecture and for the control of road traffic in a hierarchically structured road network. The motivation of the case study was to develop measures to counter the negative effects of traffic and of transportation, primarily the loss of human lives, the damage to the environment, and the economic costs. In the Netherlands, there are now five traffic control centers for monitoring and control of motorway road networks. There are similar networks for provincial and urban roads. A provincial–urban network in Belgium was also part of the case study.

The main research issues are as follows: (1) How to structure the system and control multilevel architecture of these networks? (2) How to develop coordination controllers of the many levels of these hierarchical systems? (3) How to predict traffic flow in a large-scale road network? (4) How to model and how to control a regional urban road network?

The leader of Work Package 4 Road Networks (WP4) was the Department Technology, Policy, and Management of the Delft University of Technology. The leader of that group, Jos L.M. Vrancken, was involved in the computer architecture of the online monitoring and control of the traffic control centers for motorways in the Netherlands. The company Trinité Automation B.V. (TRI) is a software house which up to recently developed almost all of the software of the traffic control centers for motorways. Its leader was and is Frank Ottenhof. The team of the University of Gent with its leader René Boel developed modeling and control of a regional urban road network. The list of C4C Teams involved in WP4 was TUD, TRI, UGE, and CWI.

1.6 WP5 Automated Guided Vehicles

The aim of the case study was to develop algorithms for control of automated guided vehicles on a container terminal. The C4C Team HNN operates a container terminal in the harbor of Antwerp, Belgium. The motivation was the automatic transportation of containers from the quai to a yard and from there to a truck for further transportation outside the container terminal, and conversely. Currently, the transportation with straddle carriers is carried out by human drivers. The objectives were to organize the movements of the carriers such that they operate autonomously, interact with the other vehicles, and adjust their routes if necessary.

The main research issues were the following: (1) coordination control to avoid blockingness and (2) coordination control of the driving of vehicles, including the interaction at intersections.

The leader of Work Package 5 Automated Guided Vehicles (WP5) was the Department of Mechanical Engineering of the Eindhoven University of Technology. The person directing the research was Jan Tijmen Udding. The company with the container terminal is Hessen Noord Natie in Antwerp whose official name changed in 2010 to PSA Antwerp. The list of all teams involved in WP5 is TUE, UCY, UGE, and HNN.

1.7 WP6 Complex Machines

The aim of the case study was to develop coordination control of complex machines consisting of many different sensors, actuators, and local control computers. High-speed printers are a typical example of such machines, but other technological machines are also relevant.

The main research issues included the following: (1) multilevel/hierarchical modeling of the operation of the complex machine as an automaton to be controlled and (2) development of control synthesis of the machines, using supervisory control of discrete-event systems and control of hybrid systems.

The leader of Work Package 5 Automated Guided Vehicles (WP5) was the Department Mechanical Engineering of the Eindhoven University of Technology. The research was directed by Koos (J.E.) Rooda, later by Jos (J.) Baeten, and by Bert van Beek. The company involved was Océ Technologies B.V. established in Venlo, the Netherlands. The leader of the team OCE was the company researcher Lou Somers. The list of all teams involved in WP6 is TUE, OCE, and, to a minor extent, CWI and UVR.

1.8 Communalities and Differences of the C4C Case Studies

The C4C case studies were intentionally chosen by the coordinator of the project because they address the coordination of distributed systems. The character of the case studies leads to the conclusion that no single method of control of distributed systems is an answer to all cases studies; hence, the theoretical investigations had to have a broad focus.

The main aim of the project was to advance research on coordination control of distributed systems. All five case studies required algorithms and theory of coordination control. In particular, the project required a joint research effort of researchers active in the research areas of control, communication, and informatics (computer science). This broad view was as essential for the project as it was and is for control engineering and for control theory. Thus, teams were selected with expertise not only in control theory, but also in the research areas of communication and informatics.

The fact that five case studies were considered led during the project meetings during the lifetime of the project to constant comparisons of the approaches for each of the case studies. The differences were considerable. Vehicles are treated in the case studies of underwater vehicles, aerial vehicles, and automated guided vehicles on a container terminal. These case studies were exemplary for the coordination control of distributed systems. The case study of road networks was a typical example of a control of a multilevel system. The coordination aspects showed up at every level. Yet, the formalization of the multilevel approach as initiated requires more research than it has so far received in the C4C project. The case study of control of complex machines had in principle the distributed systems as a model. The focus in this case study was primarily on the software engineering process and on supervisory control. The differences were useful for the scientific aspects of the process, and in that sense, the variety of the case studies was useful.

1.9 Further Reading

The reader may find information about the case studies in Part I of the book in which this chapter appears, in particular in the Chaps. 2–4 on the underwater vehicles, Chaps. 5, 6, 28, and 29 on control of road networks, Chap. 7 on control of automated guided vehicles, and Chap. 8 on control of complex machines. An additional case study, not mentioned in the C4C Project proposal but carried out within the project, was control of electric power networks, see Chap. 9.

Further details are also available in the C4C Deliverables. These are available at the C4C Web site with address http://www.c4c-project.eu. The Web site is planned to remain online till about 2020 or even longer.

Chapter 2
A Model Predictive Control Approach to AUVs Motion Coordination

Fernando Lobo Pereira, J. Borges de Sousa, R. Gomes and P. Calado

2.1 Motivation

This chapter concerns the decentralized coordinated control of a formation of autonomous underwater vehicles (AUVs) subject to a given set of constraints. The need of AUV motion coordination is due to observation and actuation requirements, such as, spatial and temporal distribution, persistence, event detection and monitoring, etc., which are critical to address a wide range of applications, and can only be achieved by distributing sensors and actuators by a number of distributed fixed and mobile platforms. Examples of application areas are climate change, environment sustainability, natural resources management, surveillance, and security. A selected sample of a vast literature is [1, 4, 9, 14, 20, 24, 26, 27, 31].

Thus, the vast research effort undertaken to design systems for the coordinated control of multiple autonomous vehicles is not surprising. The cooperative control of a team of distributed agents with decoupled nonlinear dynamics and exchanging delayed information has been addressed in a number of works, notably, [2, 6, 7, 10, 11, 16, 19, 22, 23, 29, 32]. The last reference is a chapter of the recently published book edited by Lunze referred to in Sect. 2.5 in which multiple issues pertinent to networked control are considered. The schemes proposed in the above references

F. Lobo Pereira (✉) · J. Borges de Sousa · R. Gomes · P. Calado
Faculdade de Engenharia, Universidade do Porto, Rua Dr. Roberto Frias, s/n, 4200-465 Porto, Portugal
e-mail: flp@fe.up.pt

J. Borges de Sousa
e-mail: jtasso@fe.up.pt

R. Gomes
e-mail: rgomes@fe.up.pt

P. Calado
e-mail: pcalado@fe.up.pt

J.H. van Schuppen and T. Villa (eds.), *Coordination Control of Distributed Systems*, Lecture Notes in Control and Information Sciences 456,
DOI 10.1007/978-3-319-10407-2_2

are decentralized in that each agent computes its control law locally by exchanging, possibly delayed, state information with neighboring agents.

Model predictive control (MPC)-like schemes have been widely adopted to formulate decentralized cooperative control problems. The seminal work of Mayne and co-workers reported in the two Automatica articles cited in Sect. 2.5 address fundamental MPC stability, optimality, and robustness issues that lay down important foundations for further research effort on decentralized coordinated control. Typically, in the approaches to decentralized control, control laws depend on the local state variables and on, possibly delayed, information from neighboring agents. Information exchange strategies that improve the formation stability and performance and, at the same time, are robust to changes in the communication topology are considered in [3]. The sensed and communicated information flow is modeled by a graph, and stability conditions are obtained in terms of the eigenvalues of the graph Laplacian. The problem of unreliable communication channels between the MPC controller output and the actuator input has been addressed in, among others, [8]. Here, the mechanism for compensation of packet dropouts has been incorporated in the MPC scheme for discrete time problems. This article also includes some stability and sub-optimality analysis under an asymptotic controllability assumption. In order to show stability, the authors prove that, under the considered assumptions, the value function associated with the underlying optimal control problem exhibits Lyapunov properties.

Although very significant to motion coordinated control challenges, these approaches are not tailored for the specific requirements arising in the marine environment. Highly nonlinear and complex dynamics due to hydrodynamic effects, [5], huge variability of underwater phenomena, severe communication constraints, and scarcity of onboard resources compound to make the networked AUV formation control problem a formidable one, [27]. Due to the fact that radio waves are strongly attenuated in the underwater milieu, acoustics are the most common form of communication but, unfortunately, not only exhibits low bandwidth, high-noise level, and low reliability, but also requires relatively high-power consumption, [25].

2.2 The AUV Formation Control Problem

The AUV formation control problem considered here is based on a MPC scheme and targets field demonstrations with NAUV vehicles from LSTS—The Laboratory for Underwater Systems and Technologies of Porto University—(http://lsts.fe.up.pt) and consists in tracking a given trajectory while maintaining a given formation pattern and satisfying state, control, and communications constraints. The key reason to choose an MPC scheme relies on the fact that it enables to combine the highly desired optimization of scarce onboard resources with the feedback control nature of the scheme that allows to cope with the significant perturbations and with the wide variability of the underwater milieu.

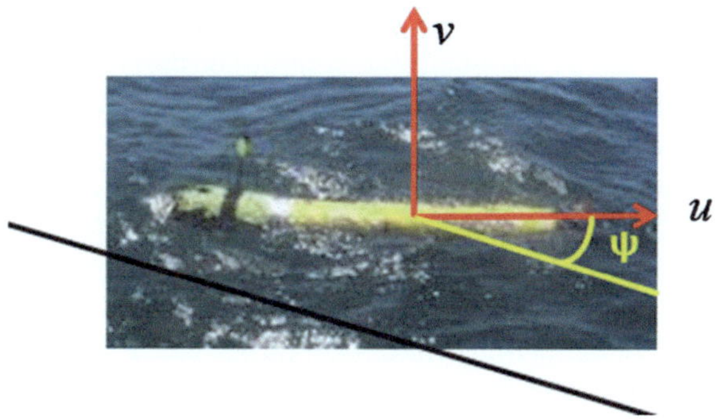

Fig. 2.1 The AUV pose and velocity coordinates are, respectively, in external and in body-fixed reference frameworks. The *line in black* is the reference trajectory

The NAUV is a small torpedo-shaped vehicle with one propeller and four control fins. It is equipped with an advanced miniaturized onboard computer system with a real-time Linux kernel, a Benthos acoustic modem, and an accurate positioning system comprising an ADCP and an IMU,[1] [28]. The model of the AUV NAUV for the motion in the horizontal plane, depicted in Fig. 2.1, is given by (2.1), [5].

The value of the model coefficients was extracted from elaborated identification procedures combining data from [21] coupled with data from LSTS field experiments. The AUV state $x^T = [\eta^T, \nu^T]$[2] satisfies

$$\dot{\eta} = \begin{bmatrix} u\cos(\psi) - v\sin(\psi) \\ u\sin(\psi) + v\cos(\psi) \\ r \end{bmatrix}, \quad \dot{\nu} = \begin{bmatrix} \frac{\tau_u - (m - Y_{\dot{v}})vr - X_{u|u|}u|u|}{m - X_{\dot{u}}} \\ \frac{(m - X_{\dot{u}})ur - Y_{v|v|}v|v|}{m - Y_{\dot{v}}} \\ \frac{\tau_r - (Y_{\dot{v}} - X_{\dot{u}})uv - N_{r|r|}r|r|}{I_{zz} - N_{\dot{r}}} \end{bmatrix}, \tag{2.1}$$

where $\eta = [x, y, \psi]^T \in \mathbf{R}^3$, $\nu = [u, v, r]^T \in \mathbf{R}^3$, $\tau = [\tau_u, \tau_r] \in \mathbf{R}^2$, and m are, respectively, the vehicle's pose (position and yaw), velocity (surge, sway, and yaw rate), input forces (surge and yaw), and mass. In these equations, the parameter I_{zz} is the rotational mass, and while the triple $(X_{\dot{u}}, Y_{\dot{v}}, N_{\dot{r}})$, represents the surge, sway, and yaw hydrodynamic added mass, the triple $(X_{u|u|}, Y_{v|v|}, N_{r|r|})$, are the surge, sway, and yaw hydrodynamic quadratic drag coefficients.

The control strategy for AUV i, $i = 1, \ldots, n_v$, should minimize, over a given time interval, a cost functional penalizing the tracking error relative to the reference trajectory, η^i_{ref}, and the control effort, i.e.,

$$\int_t^{t+T} \left[(\eta^i(s) - \eta^i_{\text{ref}}(s))^T \mathbf{Q} (\eta^i(s) - \eta^i_{\text{ref}}(s)) + \tau^{iT}(s) \mathbf{R} \tau^i(s) \right] ds, \tag{2.2}$$

---

[1] ADCP and IMU stand by acoustic Doppler current profiler, and inertial measurement unit, respectively. While the former provides water current velocity measurements, the former measures position, velocity, and orientation.

[2] From now on, "T" in upper script will denote transposed.

subject to the vehicle dynamics (2.1), (i) position endpoints constraints, $\eta^i(t+T) \in C_{t+T}$, (ii) pointwise control constraints, $\tau^i(s) \in \mathscr{U}^i$, (iii) state constraints, $(\eta^i(s), \nu^i(s)) \in \mathscr{S}^i$, and (iv) two graph constraints specifying, respectively, the communication links $g^c_{i,j}(\eta^i(s), \eta^j(s)) \in C^c_{i,j}$, $\forall j \in \mathscr{G}^c$, and the formation pattern, $g^f_{i,j}(\eta^i(s), \eta^j(s)) \in C^f_{i,j}$, $\forall j \in \mathscr{G}^f$. While control constraints reflect saturations, state constraints incorporate safety, and communication constraints ensure the AUVs connectivity. The severe power constraints impose the need of each AUV to communicate only with its neighbors and thus imposes the need of decentralization. The communications structure is described by a triple $(g^c, C^c, \mathscr{G}^c)$, where $g^c : \mathbf{R}^n \times \mathbf{R}^n \to \mathbf{R}^M$, $C^c \in \mathbf{R}^M$, and $\mathscr{G}^c$ is a graph specifying the communication links among AUVs. The formation constraints specify the AUVs relative positions and are described by triple $(g^f, C^f, \mathscr{G}^f)$ where $g^f : \mathbf{R}^n \times \mathbf{R}^n \to \mathbf{R}^M$, $C^f \in \mathbf{R}^M$, and $\mathscr{G}^f$ is a graph representing the vehicles' formation relations.

2.3 The Approach

Outline. Our approach is based on a MPC scheme, being the information exchanged over acoustic communication channels. To deal with the bandwidth limitation that precludes closing low-level (fast) feedback loops over acoustic communications, the following two-layer control framework distributed over the AUVs in formation is considered: The lower layer deals with the fast low-level control in each vehicle. The upper layer deals with acoustic communications and provides control corrections to the lower layer. Each vehicle has a fast low-level formation controller. This is a feedback controller for the whole formation. We use a model-based approach to close the control loop around state estimates from the vehicle and from models of the other vehicles. This is done without communications with the other AUVs. We use MPC for the high-level controller, which runs in each vehicle. The model state value is reset when a message with the true state data of other AUVs is received. The MPC is run with the model updates to generate a sequence of control inputs for the AUVs in the formation. These control inputs are sent to the other AUVs for coordination. The MPC cost function is targeted at balancing the control effort and the quadratic error to the given formation reference trajectory and to the given formation pattern. While control constraints reflect control saturations and other model features, state constraints preclude collisions with obstacles.

Implementation. The main features of the implemented discrete time overall MPC controller of the AUVs formation are as follows:

- Decentralization. Each vehicle runs its own MPC scheme (which are identical for all vehicles) and communicates only with its neighbors;
- Computational efficiency. The MPC optimal control problem is approximated by a LQ optimization problem involving: (i) quadratic cost functionals, (ii) AUV

dynamics approximated by a discrete time linear model, and (iii) state and control constraints given by linear inequalities;

- Easy incorporation of communication delays and packet dropouts; and
- Accommodation of noise and disturbances in the vehicles-simulated motion.

Let N_p, n_v, and T be, respectively, the prediction horizon, the number of vehicles, and the sampling period. Then, the optimal formation control problem to be solved in AUV i involves data from all its neighbors as defined by the formation graph and, for some reference time t, can be stated as follows:

$$\text{Minimize} \quad \sum_{k=1}^{N_p} \left[\|y_k^{\text{ref},i} - y_k^i\|_{\mathbf{Q}^i}^2 + \|\tau_{k-1}^i\|_{\mathbf{R}^i}^2 + \sum_{j \in \mathcal{G}(i)} \|\mathbf{D}^{ij}(y_k^i - y_k^j) - d^{ij}\|_{\mathbf{L}^{ij}}^2 \right] \tag{2.3}$$

$$\text{subject to} \quad x_{k+1}^j = \Phi^j(T)x_k^j + \Psi^j(T)\tau_k^j, \quad x_0^j = x_t^j \tag{2.4}$$

$$y_k^j = \mathbf{C}^j x_k^j \tag{2.5}$$

$$x_k^j \in [x_{LB,t}^j, x_{UB,t}^j], \quad \tau_k^j \in [\tau_{LB}^j, \tau_{UB}^j]. \tag{2.6}$$

We observe that the cost functional consists in a weighted sum of three terms: trajectory error, control effort, and a penalization of the deviation form the formation configuration. We observe that (2.4) represents not only the discrete time linearized dynamics of vehicle i,[3] but also those of all the vehicles with which the vehicle i is linked through the graph $\mathcal{G}(i)$. Notice that the constraints hold for $j \in \{i\} \cup \mathcal{G}(i)$, being, for each time k, $\mathcal{G}(i)$ the set of nodes of the graph specifying the vehicles linked to AUV i.

Here, y_k^i and $y_k^{\text{ref},i}$ are, respectively, the vector of outputs of vehicle i and its reference, x_t^j is the initial state of vehicle j at the initial time t, $\mathbf{D}^{ij}$ is the adjacency matrix reflecting the formation relation between vehicles i and j, d^{ij} is a vector parameter specifying distances between vehicles i and j, and $x_{LB,t}^j$, $x_{UB,t}^j$, τ_{LB}^j, and τ_{UB}^j are bounds for state and control at time t, respectively. The matrices $\mathbf{Q}^i$, $\mathbf{R}^i$, and $\mathbf{L}^{ij}$ are the chosen performance weights for AUV i.

Now, we describe the MPC scheme running onboard each AUV. If communication packets dropout or arrive late, then the vehicles will not share the same data and there will be differences in the control strategies computed by the various vehicles. This issue is mitigated by replacing the missing sampled data by simulated data. The MPC scheme for AUV i is as follows:

1. Initialization: prediction and control horizons, and other optimal control problem parameters that depend on specific mission requirements, such as, level of perturbations, existence of obstacles, relative weight of trajectory tracking, and formation pattern errors.

[3] The matrices $\Phi^j(T)$ and $\Psi^j(T)$ are obtained by integrating the piecewise constant linear system in (x, u) approximating the original system over the sampling period T.

2. Sample the state variable, compute its estimate, and communicate it to its neighbors via acoustic modem.
3. Obtain the state variable of its neighbors via acoustic modem.
   – If data are available go to step 4.
   – Otherwise, generate estimates of the neighbors' state by running their models.
4. Solve the linear quadratic optimization problem (LQP^i) at the current time t, for the current prediction horizon and for the given reference output trajectory. This yields the optimal control for vehicle i.
5. Apply the control τ^i for the current control horizon.
6. Slide time for the optimization problem and adjust parameters as needed.
7. Let time elapse until the end of the current control horizon and go to step 2.

Results. We evaluated the MPC controller by taking into account conditions which are representative of field operations. We introduce the following four metrics for performance evaluation: trajectory tracking (*TM*), formation tracking (*FM*), control effort (*CM*), and total cost (*C*). While the first two are the L_2 norm of trajectory and formation tracking errors, the third one is the total control fuel consumption. In this assessment, three different scenarios were considered for a side-by-side formation of two vehicles along a sine wave trajectory with a nominal velocity of 1 m/sec: no communication, communication without delays, and communication with a 0.1 sec that corresponds to a 150 m distance between vehicles. In this last scenario, a prediction model was used to mitigate the impact of the delay. For each scenario, Gaussian noise with mean and variance values (0, 0.1), (0, 0.25), and (0.1, 0.02) is considered. In the case of no noise and no delay, the values $TM = 0.7$, $FM = 0.2$, $CM = 0.2$, and C = 34.4 were obtained. In Table 2.1, it is shown how our MPC controller performed in the various situations. Its entries were obtained by averaging the performances of 10 runs with independent realizations of the random variables.

From the data in the table, some conclusions emerge as follows:

- The value of the cost function and performance measures of the formation controller degrades as the noise level increases whatever simulation scenario, being the impact of the noise mean far greater than that of its variance.
- The performance of the controller improves significantly with enabled communications relatively to open-loop case.

Table 2.1 MPC performance table

Noise (Mean,Var.)	(0, 0.1)	(0, 0.25)	(0.1, 0.02)
Criteria	TM FM CM C	TM FM CM C	TM FM CM C
Comms disabled	11.8 2.8 40.6 524.9	33.5 4.8 48.2 1158.0	211.7 39.6 57.7 8197.0
Comms enabled No delay	0.8 0.3 14.7 48.4	1.0 0.4 25.9 70.3	1.1 0.4 17.6 81.3
Comms enabled Delay = 1 s	0.9 0.3 24.5 52.5	1.2 0.4 34.7 74.9	1.6 0.8 18.3 105.5

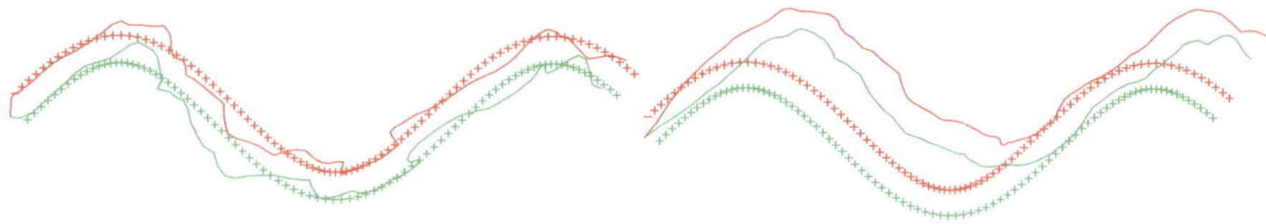

Fig. 2.2 Side-by-side AUV formation trajectories with communications and noise: (Mean = 0, Variance = 0.25), (Mean = 0.1, Variance = 0.02)

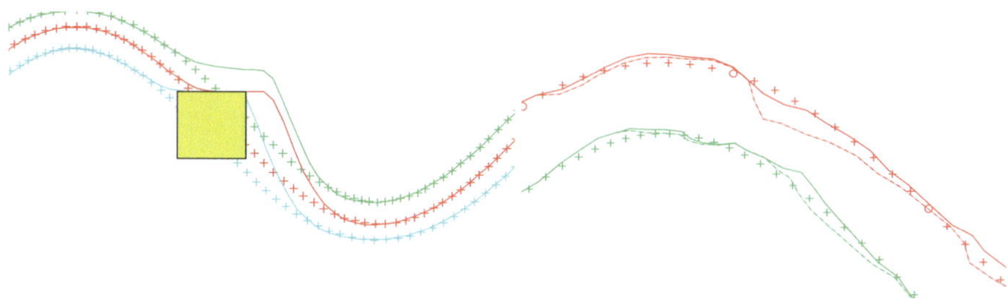

Fig. 2.3 *In the left* Obstacle avoidance of a three AUV triangle formation control and obstacle avoidance. *In the right* Effect of communications dropouts in two AUV formation control

- The use of prediction mitigates the impact of delay, as it significantly prevents performance degradation.

These conclusions are backed by a cursory inspection of the trajectories obtained with simulation runs shown in Fig. 2.2 where solid lines represent actual trajectories and the "+" the reference trajectory waypoints. Figure 2.3 shows the flexibility and robustness of our MPC approach. In the left, three AUVs moving in a triangle formation are able to avoid collision with an obstacle whose emergence can be regarded as perturbation forcing the vehicles to deviate from their originally nominal trajectories. In the right, the impact of random communication dropouts, are marked with "o," of the red AUV in receiving messages from the green AUV in the controller performance is shown. It is clear from the trajectory with communication dropouts, represented by the dash-dot line that the MPC controller is able to recover after a certain transient.

2.4 The Reach Set MPC Research Challenges

Although the conclusions in the previous section are extremely relevant for control design, there is still plenty of room to improve the control performance. One consists in improving state estimates relatively to the ones provided by the linear approximation of the AUV dynamics by taking into account the nonlinear nature of the system. Unfortunately, this will imply a much higher computational complexity. In order to address both these issues, we propose a new formulation of the MPC scheme, [15], that relies in the observation[4] that the optimal control problem in Sect. 2.2 is equivalent to

$$\text{Minimize } \{V(t+\Delta, \bar{x}(t+\Delta)) \colon \bar{x}(t+\Delta) \in \mathscr{R}(t+\Delta, (t, \bar{x}(t)))\}\,, \tag{2.7}$$

where $\mathscr{R}(t_2, (t_1, x_1))$ is the Reach set of the extended system $\dot{\bar{x}} = [l(x,\tau), f(x,\tau)]^T$, where l is the integrand in (2.2) and f is specified by (2.1), i.e., set of points that the extended system can reach at t_2 when x departs from x_1 at $t_1 \leq t_2$, [12], and $V(t,z)$ is the Value function, i.e., the minimum cost from the point (t,z) onward, [13, 30]. Under appropriate assumptions, $V(t,z)$ can be computed as a solution to the following Hamilton–Jacobi–Bellman equation with an appropriate boundary condition, [13, 30],

$$\frac{\partial}{\partial t}V(t,\bar{x}) + \min_{\tau \in \mathscr{U}} \left\{ \left\langle \frac{\partial}{\partial \bar{x}} V(t,\bar{x}), (f(x,\tau), l(x,\tau)) \right\rangle \right\} = 0. \tag{2.8}$$

For time invariant systems, both Reach set and Value functions can be computed off-line, being, with respect to the former, the online computational burden reduced to (i) rotations and translations of the Reach set to take into account the pose of the vehicle at t, and (ii) the computation of the optimal control in $[t, t+\Delta]$. Moreover, both computational complexity and amount of information to be shared among the vehicles can be further reduced by considering polyhedral approximations to the Reach sets. In spite of powerful tools available, [17, 18], solving (2.8) with state constraints, even off-line and for systems with a moderate dimension, remains a huge challenge. The book by Stanley Osher and Ronald Fedkiw mentioned in Sect. 2.5 provides a good overview on level set methods to generate pertinent computational schemes. Another challenge concerns the "online" update of $V(\cdot,\cdot)$ which depends strongly on the types of perturbations. We note that, for the case of obstacle emergence, $V(\cdot,\cdot)$ has to be updated only in the region encompassing all the possible paths joining the current pose and the best one for which the obstacle is overcome.

2.5 Most Relevant Literature

- D. Mayne, J. Rawlings, C. Rao, and P. Scokaert. Constrained model predictive control: Stability and optimality. *Automatica*, vol. 36, 2000, pp. 789–814.
- Stanley Osher and Ronald Fedkiw. *Level Set Methods and Dynamic Implicit Surfaces*, Springer-Verlag, 2002.
- D. Mayne, S. Rakovic, R. Findeisen, and F. Allgower. Robust output feedback model predictive control of constrained linear systems: Time varying case. *Automatica*, vol. 45, 2009, pp. 2082–2087.
- J. Lunze (ed.), *Control Theory of Digitally Networked Dynamic Systems*, Springer-Verlag, 2014.

[4] State constraints are omitted to facilitate the exposition.

References

1. Curtin T, Bellingham J, Catipovic J, Webb D (1993) Autonomous Ocean Sampling Networks. Oceanography 6(3):86–94
2. P. Deshpande, P. Menon, and C. Edwards. Delayed static output feedback control of a network of double integrator agents. Automatica, 49(11):3498-Â3501, 2013.
3. Fax J, Murray M (2004) Information flow and cooperative control of vehicle formations. IEEE Trans. Automat. Control 49:1465–1476
4. Fiorelli E, Leonard N, Bhatta P, Paley D, Bachmayer R, Fratantoni D (2006) Multi-AUV Control and Adaptive Sampling in Monterey Bay. IEEE J. of Oceanic Eng. 31(4):935–948
5. T. Fossen. Guidance and Control of Ocean Vehicles. John Wiley & Sons Ltd., 1994.
6. Franco E, Magni L, Parisini T, Polycarpou M, Raimondo D (2008) Cooperative constrained control of distributed agents with nonlinear dynamics and delayed information exchange: A stabilizing receding-horizon approach. IEEE Trans. Automatic Control 53:324–338
7. L. Grüne, F. Allgöwer, R. Findeisen, J. Fischer, D. Groß, U. Hanebeck, B. Kern, M. Müller, J. Pannek, M. Reble, O. Stursberg, and P. Varuttiand K. Worthmann. chapter Distributed and Networked Model Predictive Control, pages 111–167. Springer-Verlag, 2014.
8. L. Gruene, J. Pannek, and K. Worthmann. A networked constrained nonlinear mpc scheme. In Procs European Control Conference, Budapest, Hungary, 2009.
9. Jones M, Miller L, Woodruff D, Ewert D (2007) Mapping of Submerged Aquatic Vegetation Using Autonomous Underwater Vehicles in Nearshore Regions. Proc. Oceans 2007:1–7
10. Keviczky T, Borrelli F, Fregene K, Godbole D, Balas G (2008) Decentralized receding horizon control and coordination of autonomous vehicle formations. IEEE Trans. Control Systems Technology 16:19–33
11. G. Kladis, P. Menon, and C. C. Edwards. Cooperative tracking for a swarm of unmanned aerial vehicles: A distributed tagaki-sugeno fuzzy framework design. In 50th IEEE Conference on Decision and Control and European Control Conference, Orlando, FL, USA December 12–15, 2011.
12. A. Kurzhanski and P. Varaiya. Dynamic optimization for reachability problems. J. Optimization Theory and Applications, 108, 2001.
13. A. Kurzhanski and P. Varaiya. Analysis and Design of Nonlinear Control Systems, chapter The Hamilton-Jacobi Type Equations for Nonlinear Target Control and Their Approximation, pages 77–90. Springer-Verlag, 2008.
14. Leonard N, Paley D, Lekien F, Sepulchre R, Fratantoni D, Davis R (2007) Collective Motion, Sensor Networks, and Ocean Sampling. Procs of the IEEE 95(1):48–74
15. F. Lobo Pereira, J. Sousa, R. Gomes, and P. Calado. Reach set formulation of a model predictive control scheme. In MTNS 2012, 20th Int. Symp. on Mathematical Theory of Network of Systems, Melbourne, Australia, July 9–13, 2012.
16. Menon P, Edwards C (2009) Decentralised static output feedback stabilisation and synchronisation of networks. Automatica 45(12):2910–2916
17. I. Michel. The flexible, extensible and efficient toolbox of level set methods. J. of Scientific Computing, 35.
18. Michel I, Bayen A, Tomlin C (2005) Computing reachable sets for continuous dynamics games using level sets methods. IEEE Trans. on Automatic Control 50:980–1001
19. P. Ogren, M. Egerstedt, and X. Hu. A control lyapunov function approach to multi-agent coordination. In 40th IEEE Conference on Decision and Control, Orlando, Florida USA, December, volume I, pages 1150–1155, 2001.
20. D. Popa, A. Sanderson, R. Komerska, S. Mupparapu, D. Blidberg, and S. Chappel. Adaptive sampling algorithms for multiple autonomous underwater vehicles. In IEEE/OES Autonomous Underwater Vehicles, pages 108–118, 2004.
21. T. Prestero. Verification of a six-degree of freedom simulation model for the remus autonomous underwater vehicle. Master's thesis, MIT, WHOI, Cambridge, MA, 2001.
22. Ren W, Beard R (2008) chapter Distributed Consensus in Multi-vehicle Cooperative Control. Springer-Verlag, London

23. Ren W, Cao Y (2011) chapter Distributed Coordination of Multi-agent Networks. Springer-Verlag, London
24. Rigby P, Williams S, Pizarro O, Colquhoun J (2007) Effective Benthic Surveying with Autonomous Underwater Vehicles. Oceans 2007:1–6
25. H Riksfjord, O. Haug, and J. Hovem. Underwater acoustic networks - survey on communication challenges with transmission simulations. In Procs of SENSORCOMM '09, pages 300–305, 2009.
26. H. Schmidt, J. Bellingham, M. Johnson, D. Herold, D. Farmer, and R. Pawlowicz. Real-time frontal mapping with AUVs in a coastal environment. In Proc. MTS/IEEE OCEANS '96, pages 1094–1098, 1996.
27. J. Sousa, B. Maciel, and F. Pereira. Sensor systems and networked vehicles. Networks and Heterogeneous Media, 4, 2009.
28. J. Sousa and R. Martins. Control architecture and software tool set for networked operations of ocean vehicles. In Proc. 9th IFAC MCMC, 2012.
29. Stilwell D, Bishop B (2000) Platoons of underwater vehicles. IEEE Control Systems Magazine 20(6):45–52
30. R. Vinter. Optimal Control. Birkhauser, 2000.
31. Willcox J, Bellingham J, Zhang Y, Baggeroer A (2001) Performance Metrics for Oceanographic Surveys with Autonomous Underwater Vehicles. IEEE J. of Oceanic Eng. 26(4):711–725
32. Y. Zhang and H. Mehrjerdi. A survey on multiple unmanned vehicles formation control and coordination: normal and fault situations. In ICUAS - International Conference on Unmanned Aircraft Systems, Atlanta, GA, USA, May 28–31, pages 1087–1096, 2013.

Chapter 3
Dynamic Optimization Techniques for the Motion Coordination of Autonomous Vehicles

Jorge Estrela da Silva, João Borges de Sousa and Fernando Lobo Pereira

3.1 Problem

Consider the problem of formation control (see, e.g., [3, 6]). Informally, the objective in a formation control problem is to maintain the relative positions of a set of agents such that the shape of the formation follows a given reference. Several approaches have been proposed for the formation control problem (e.g., leader-following, virtual structure, behavioral). In this chapter, a problem of formation control is discussed in the framework of the leader-following approach. In this approach, each agent is assigned a leader, which may be either another agent or a virtual agent [15], and it is required to track a function of the state of the respective leader (e.g., a constant offset with respect to the leaders' position).

More specifically, we consider the problem of trajectory following by a static formation of N vehicles. For simplicity, only planar operation is discussed; however, the approach can also be extended to operation in the three dimensional Euclidean space. It is assumed that the vehicle motion can be simulated by a system of the form

$$\begin{pmatrix} \dot{x}_i(t) \\ \dot{y}_i(t) \\ \dot{\psi}_i(t) \end{pmatrix} = \begin{pmatrix} u_i(t)\cos(\psi_i(t)) + c_{x,i}(t) \\ u_i(t)\sin(\psi_i(t)) + c_{y,i}(t) \\ \kappa_i(t)u_i(t) \end{pmatrix} \tag{3.1}$$

J.E. da Silva (✉)
School of Engineering, Polytechnic Institute of Porto,
Rua Dr. António Bernardino de Almeida 431, 4200-072 Porto, Portugal
e-mail: jes@isep.ipp.pt

J.B. de Sousa · F. Lobo Pereira
Faculty of Engineering, Porto University, Rua Dr. Roberto Frias, 4200-465 Porto, Portugal
e-mail: jtasso@fe.up.pt

F. Lobo Pereira
e-mail: flp@fe.up.pt

J.H. van Schuppen and T. Villa (eds.), *Coordination Control of Distributed Systems*,
Lecture Notes in Control and Information Sciences 456,
DOI 10.1007/978-3-319-10407-2_3

where $(x_i(t), y_i(t), \psi_i(t))$ is the posture of vehicle $i \in \{1, \ldots, N\}$ at time t, and the bounded input variables $c_{x,i}(t)$ and $c_{y,i}(t)$ model the effect of bounded disturbances (e.g., winds and marine currents) and modeling errors (e.g., in the case of underwater vehicles, the neglected lateral speed). Both control inputs are bounded: $0 \leq u_{i,\min} \leq u_i(t) \leq u_{i,\max}$ and $|\kappa_i(t)| \leq \kappa_{i,\max}$. This model of operation applies not only to surface vehicles and ships but also to submarines and aircrafts operating at constant altitude or heading.

The reference posture $(x_0(t), y_0(t), \psi_0(t))$, where $(x_0(t), y_0(t))$ is the Cartesian position of the virtual leader at time t, and $\psi_0(t)$ is its respective orientation and is generated in real-time by a system of the form

$$\left(\dot{x}_0(t)\ \dot{y}_0(t)\ \dot{\psi}_0(t)\right)' = \left(u_0(t)\cos(\psi_0(t))\ u_0(t)\sin(\psi_0(t))\ \kappa_0(t)u_0(t)\right)' \tag{3.2}$$

where inputs $u_0(t)$ and $\kappa_0(t)$ are bounded ($0 \leq u_{0,\min} \leq u_0(t) \leq u_{0,\max}$ and $|\kappa_0(t)| \leq \kappa_{0,\max}$).

The desired formation shape is specified by the desired relative positions $\mathbf{o}_i = (o_i \cos(\psi_{o,i}), o_i \sin(\psi_{o,i}))$, $i \in \{1, \ldots, N\}$, where o_i and $\psi_{o,i}$ are the respective polar coordinates, of each vehicle with respect to the virtual leader's body fixed frame (x-axis and y-axis oriented with the longitudinal and lateral axis of the vehicle, respectively). The following conditions are also assumed:

- Each vehicle $i \in \{1, \ldots, N\}$ is able to estimate its relative posture with respect to the virtual leader's body fixed frame.
- A set of N disjoint and connected sets $\mathscr{R}_i \subset \mathbb{R}^2$, $i \in \{1, \ldots, N\}$ is given; each subset $\mathscr{R}_i$ corresponds to the a priori acceptable set of locations, relative to the virtual leader's body fixed frame, for the respective vehicle i. These sets are used to prevent collisions between the vehicles. Moreover, these sets can also be chosen in order to provide robustness with respect to estimation errors; this is done by keeping a suitable gap between the sets.

The sets $\mathscr{R}_i$ are time independent and they are defined with no regard to the orientation of the vehicles. The sets $\mathscr{R}_i$ are used to describe constraints on the relative positions of each vehicle in a way that is simple both for system operators and control designers. However, these sets are not required to be invariant with respect to the vehicle dynamics, since, in general, it is hard to define sets with such property; i.e., it is possible for a vehicle i to start inside $\mathscr{R}_i$ and to not be able to stay there afterward. One of the objectives of the control design is to find the maximal invariant subsets of $\mathscr{R}_i \times \mathbb{R}$ with respect to the vehicle dynamics. These invariant subsets will define the truly safe regions of operation for each vehicle. During typical motion, the vehicles may not be able to keep the exact desired formation, but they are required to be inside the respective safe region. Therefore, the research problem can be summarized as follows:

Problem 3.1 given $\mathscr{O}$, where $\mathscr{O} = \{\mathbf{o}_i : i = 1, \ldots, N\}$, find the decentralized feedback control laws $f_i(\mathbf{x}_i)$, where $\mathbf{x}_i$ is the posture of vehicle i relative to the virtual leader, and the safe regions $\mathscr{S}_i \subset \mathscr{R}_i \times \mathbb{R}$ for each vehicle $i \in \{1, \ldots, N\}$, such that, once $\mathbf{x}_i \in \mathscr{S}_i$, the respective vehicle will remain there independently of the leader's trajectory and considered disturbances. Moreover, operation inside $\mathscr{S}_i$ should be optimized with respect to a linear combination of posture error and actuation effort.

The problem of safely driving each vehicle into the respective $\mathscr{S}_i$ is not discussed.

3.2 Motivation

We observe that motion coordination problems can be classified into three main classes: attainability, invariance, and optimization. Some communication constraints, especially those pertaining to range, are easily mapped onto state constraints.

Dynamic optimization provides a uniform way of formulating and solving these classes of problems. Moreover, dynamic optimization may lead to more flexible control schemes. For example, in invariance problems [2], the actual control setting is selected from a set which typically does not consist only of a singleton (except at the boundary of the maximal invariant set). This is related to the notion of control flexibility and robustness discussed in [14]. In fact, the underlying geometric control-space properties have been used to study cooperative dynamic games [19], where each player also performs optimal control selections from a feasible control set. This kind of flexibility allows us, for instance, to consider simultaneous task execution, where one task is executed using the control slack allowed by other tasks. This is why we are interested in developing a uniform DP methodology within which we can find cooperative solutions to these classes of motion coordination problems.

In this chapter, these concepts are illustrated with the help of a static formation control problem. Formation control applications include spacecraft formation flying [13], unmanned aerial vehicles formation [18], military application [10], automated highway systems [11], and ocean sampling [8]. In the last decades, a large body of theoretical work has been developed under the moniker of formation control (see [20] for early results). Several dynamic optimization-based approaches, generally in the framework of receding horizon control, have been proposed (see, e.g., [7, 17, 21] and references therein). However, these approaches are usually more focused on the exact reference tracking. Thorough analysis of flexibility margins and safety envelopes is seldom found. For instance, in [21], the authors consider protected zones consisting of circles around the agents but criteria for the choice of the circle radius is not discussed.

3.3 System Model

Consider the following coordinate transformation:

$$\begin{pmatrix} s_i \\ d_i \\ \psi_{r,i} \end{pmatrix} = \begin{pmatrix} \cos(\psi_0) & \sin(\psi_0) & 0 \\ -\sin(\psi_0) & \cos(\psi_0) & 0 \\ 0 & 0 & 1 \end{pmatrix} \begin{pmatrix} x_i - x_0 \\ y_i - y_0 \\ \psi_i - \psi_0 \end{pmatrix} \tag{3.3}$$

where the $(s_i, d_i, \psi_{r,i})$ defines the posture of vehicle i in the virtual leader's body fixed frame. By considering this transformation, it is possible to formulate the tracking problem for each vehicle in a 3-dimensional space. This is a common transformation, that can be traced as far back as [12], where it is used in the context of pursuit-evasion games. With that in mind, the following model is considered for the motion of each vehicle $i \in \{1, \ldots, N\}$:

$$\begin{pmatrix} \dot{s}_i(t) \\ \dot{d}_i(t) \\ \dot{\psi}_{r,i}(t) \end{pmatrix} = \begin{pmatrix} -u_0(t) + u_i(t)\cos(\psi_{r,i}(t)) + \kappa_0(t)u_0(t)d_i(t) + c_{s,i}(t) \\ u_0(t)\sin(\psi_{r,i}(t)) + \kappa_0(t)u_0(t)s_i(t) + c_{d,i}(t) \\ \kappa_i(t)u_i(t) - \kappa_0(t)u_0(t) \end{pmatrix} \tag{3.4}$$

where the input variables $c_{s,i}(t)$ and $c_{d,i}(t)$ model the effects of disturbances and model uncertainties in the virtual leader's body fixed frame. As in (3.1) and (3.2), all input variables—$u_i(t)$, $\kappa_i(t)$, $u_0(t)$, $\kappa_0(t)$, $c_{s,i}(t)$, and $c_{d,i}(t)$—are bounded. The input sequences $u_0(.)$, $\kappa_0(.)$, $c_{s,i}(.)$, and $c_{d,i}(.)$ are unknown to the follower vehicles. Moreover, all input variables are assumed to be subject to a sample and hold scheme with a period of Δ units of time. Note that, in spite of the continuous-time nature of $c_{s,i}$ and $c_{d,i}$, these are first order disturbances; therefore, if the system is analyzed only at the sampling intervals, the continuous-time input sequence can be simulated by piecewise constant values leading to the same output that would be obtained with the continuous-time sequence.

3.4 Solution Approach

We approach the problem in the framework of zero-sum two-player differential games [12], where the controller for vehicle i is designated as the first player, and the second player is a fictitious entity choosing both the inputs for the virtual leader and the disturbances.

Consider the following cost functional:

$$J_i(\mathbf{x}, \mathbf{u}_i(.), \mathbf{u}_0(.)) = \int_0^\infty L_i(\mathbf{x}_i(\tau), \mathbf{u}_i(\tau))\mathrm{d}\tau \tag{3.5}$$

subject to

$$\dot{\mathbf{x}}_i(t) = f(\mathbf{x}_i(t), \mathbf{u}_i(t), \mathbf{u}_0(t)),\ \mathbf{x}_i(0) = \mathbf{x} \tag{3.6}$$

$$(s_i(t), d_i(t)) \in \mathscr{R}_i,\ \forall t \geq 0 \tag{3.7}$$

where

$$L_i(\mathbf{x}_i, \mathbf{u}_i) = k_{s,i}(s_i - o_i\cos(\psi_{o,i}))^2 + k_{d,i}(d_i - o_i\sin(\psi_{o,i}))^2 + k_{u,i}u_i^2 + k_{\kappa,i}\kappa_i^2 \tag{3.8}$$

$\mathbf{u}_{i\in\{1,\ldots,N\}} = \left(u_i\ \kappa_i\right)'$, $\mathbf{u}_0 = \left(u_0\ \kappa_0\ c_{s,i}\ c_{d,i}\right)'$, and $f(\mathbf{x}_i(t), \mathbf{u}_i(t), \mathbf{u}_0(t))$ is the right hand side of (3.4). Collision avoidance is achieved by considering the state constraints (3.7). Violation of the state constraints corresponds to an infinite cost. By consequence, the cost function is also infinity for states from which it is not possible to meet the state constraints.

It is assumed that, at each instant, $\mathbf{u}_0(t)$ will be chosen in order to maximize the cost. This worst-case approach will lead to a controller that is able to cope with any real-time trajectory generated by (3.2). In line with the worst-case approach, the upper-value solution of the game is considered. This implies the assumption that the choice of $\mathbf{u}_0(t)$ is made with knowledge of $\mathbf{u}_i(t)$. This is a reasonable approach since in practice the control input cannot be changed until the next control instant while the disturbance input may change in continuous-time. This behavior is modeled using the concept of non-anticipative strategies (see [1] or [9] for more details). Therefore, the objective for each vehicle i becomes $\sup_{\beta\in\Gamma} \inf_{\mathbf{u}_i(.)\in\mathscr{U}_i} J_i(\mathbf{x}_i(.), \mathbf{u}_i(.), \beta[\mathbf{u}_i(.)])$ where $\mathscr{U}_i$ is the set of measurable control input signals such that $\mathbf{u}_i(t) \in U_i$, where U_i is a given compact set, and Γ is the set of non-anticipative strategies for the second player.

The control laws are derived applying the dynamic programming techniques presented in [5]. That implies the computation of N value functions of the form

$$V_i(\mathbf{x}) = \sup_{\beta\in\Gamma} \inf_{\mathbf{u}_i(.)\in\mathscr{U}_i} J_i(\mathbf{x}, \mathbf{u}_i(.), \beta[\mathbf{u}_i(.)]) \tag{3.9}$$

and the control law is given by

$$f_i(\mathbf{x}) \in \arg\min_{\mathbf{u}_i\in U_i} \max_{\mathbf{u}_0(.)\in\mathscr{U}_0} \left\{ \int_0^{\Delta} L_i(y_{\Delta,i}(\mathbf{x}, \tau, \mathbf{u}_i, \mathbf{u}_0(.)), \mathbf{u}_i)\mathrm{d}\tau \right.$$
$$\left. + V_i(y_{\Delta,i}(\mathbf{x}, \Delta, \mathbf{u}_i, \mathbf{u}_0(.)))\right\} \tag{3.10}$$

where Δ is the control period and $y_{\Delta,i}(x, t, a, b)$ is the state of vehicle i at time t when driven by the control value a applied during the time interval $[0, \Delta]$, and by the adversarial control signal b from the initial state x.

As implied above, if it is not possible to find a feedback control law ensuring (3.7) for a given initial state $\mathbf{x}$, then the value function will be defined as infinity for $\mathbf{x}$. Therefore, the determination of the safe regions $\mathscr{S}_i$ is made by inspection of the value function, i.e., $\mathscr{S}_i = \{\mathbf{x} : V_i(\mathbf{x}) \neq \infty\}$. In the framework of invariance analysis, $\mathscr{S}_i$ is the maximal (positively robustly controlled) invariant set with respect to system (3.4) with state constraints (3.7).

Numerical schemes for the computation of the value function associated to differential games with nonlinear dynamics are described, for instance, in [16] and [4]. These are grid-based schemes, requiring a trade-off between the grid resolution and the computational requirements (memory and offline computation time). It must also be noted that, in practice, only an approximation of the invariant set can be obtained. The boundary of the invariant set can only be defined accurately up to the grid resolution. In what concerns the implementation of the actual feedback control law, each vehicle's computational system must store a matrix with an off-line computed control for each grid cell (using (3.10)).

3.5 Example

The following example illustrates the behavior of an autonomous underwater vehicle (AUV) with a reference offset of (0, −6) with respect to the virtual leader. The problem data is $u_0 = 1$ m/s, $-1 < u_1 < 2$ (m/s), $|\kappa_1| \leq 0.25$ (m^{-1}). A control rate of 20 Hz is assumed. The AUV is allowed to stay in $\mathscr{R}_1 = \{(s_1, d_1)\colon \sqrt{s_1^2 + (d_1 - 6)^2} = 6\}$.

Two scenarios are considered: straight line tracking with no currents; tracking of trajectories with bounded curvature ($|\kappa_0| \leq 0.05$ (m^{-1})) and subject to currents ($\sqrt{c_s^2 + c_d^2} = 0.5$ (m/s)). The value function $V_1(x)$ was computed using the software described in [5]. The running cost was defined as $L_1(\mathbf{x}_1, \mathbf{u}_1) = s_1^2 + (d_1 - 6)^2$.

The numerical computations were performed using the solver described in [5]. This solver is based on a semi-Lagrangian scheme [4] and value iteration. A regular grid of 61 × 61 × 73 nodes was used to cover the region $[-6, 6] \times [0, -12] \times [0, 2\pi]$ (m × m × rad) of the state space. The corresponding resolution and memory requirements are perfectly acceptable for many AUV. A discrete set of controls with 11 × 11 distinct values was used. The maximal invariant set $\mathscr{S}_1$ is illustrated in Fig. 3.1 for each scenario. As expected, the volume of the maximal invariant set is smaller in the last scenario, showing that the vehicle has to work with tighter safety bounds in order to be able to cope with the unpredictable disturbances and changes of reference direction.

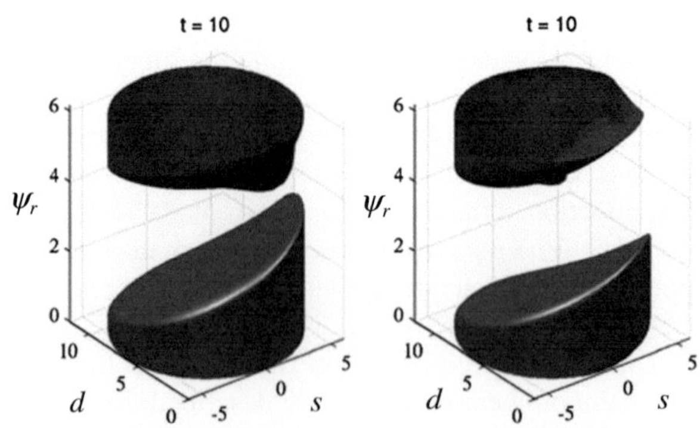

Fig. 3.1 Maximal invariant set $\mathscr{S}_1$ for two scenarios: straight line tracking with no currents; tracking of trajectories with bounded curvature and subject to currents

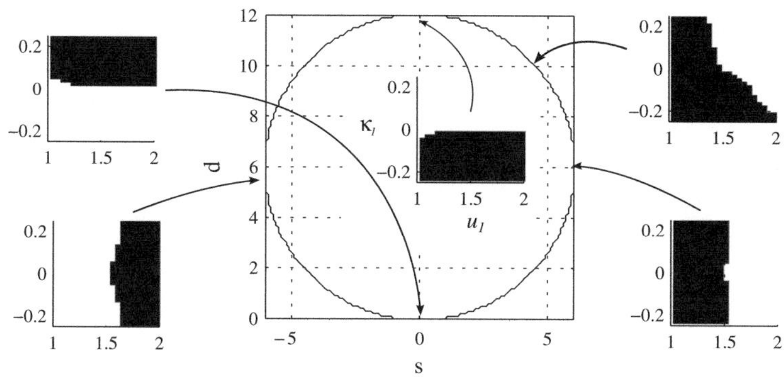

Fig. 3.2 Feasible control actions as a function of the state. Each *sub-plot* shows in *dark* the set of feasible input velocity-curvature pairs (x- and y-axis, respectively) for the indicated relative posture (s, d, ψ_r) (this is a projection of the invariant set for $\psi_{r,1} = 0$)

Simulation results show that, whenever the vehicle initial state is in the maximal invariant set, the vehicle posture converges to the desired reference and, afterward, the vehicle is able to track the real-time generated reference with negligible error.

It must be remarked that the safety regions $\mathcal{S}_i$ do not change with the choice of L_i. Moreover, if the control problem is reduced to the state constraint (3.7), then only controls on the boundary of $\mathcal{S}_i$ have to be considered; in this scenario, the choice of control becomes irrelevant if $\mathbf{x}_i$ is in the interior of $\mathcal{S}_i$. This is illustrated in Fig. 3.2.

3.6 Further Research

Dynamic programming approaches have traditionally been overshadowed by huge computational requirements (the curse of dimensionality and on-line storage requirements for the resulting control law). Current computational systems are able to handle problems of low dimension such as the one described here. However, it must be remarked that, in the proposed approach, a different value function must be computed for each vehicle in the formation. If the desired formation shape changes, then each vehicle will have to receive or compute the new controller. Although the computational burden of the described approach scales linearly with the number of vehicles, a possible avenue for future work would be to find an efficient (with respect to storage and processing requirements) parametric representation of the value function having the desired position of the vehicle in the formation as a parameter.

References

1. Bardi M, Capuzzo-Dolcetta I (1997) Optimal control and viscosity solutions of Hamilton-Jacobi-Bellman equations. Birkhauser, Boston
2. Blanchini F, Miani S (2008) Set-theoretic methods in control. Birkhauser, Boston
3. Chen Y-Q, Wang Z (2005) Formation control: a review and a new consideration. In: 2005 IEEE/RSJ international conference on intelligent robots and systems, 2005 (IROS 2005), pp 3181–3186
4. Cristiani E, Falcone M (2009) Fully-discrete schemes for the value function of pursuit-evasion games with state constraints. Ann Int Soc Dyn Games 10:178–205
5. da Silva JE, de Sousa JB (2011) Dynamic programming techniques for feedback control. In: Proceedings of the IFAC 18th world congress, Milan, August 2011
6. Do KD (2011) Practical formation control of multiple underactuated ships with limited sensing ranges. Robot Auton Syst 59(6):457–471
7. Dunbar WB, Murray RM (2006) Distributed receding horizon control for multi-vehicle formation stabilization. Automatica 42(4):549–558
8. Fiorelli E, Leonard NE, Bhatta P, Paley DA, Bachmayer R, Fratantoni DM (2006) Multi-AUV control and adaptive sampling in Monterey bay. IEEE J Oceanic Eng 31(4):935–948
9. Fleming WH, Soner HM (2006) Controlled Markov processes and viscosity solutions. Springer, New York
10. Healey AJ (2001) Application of formation control for multi-vehicle robotic minesweeping. In: Proceedings of the 40th IEEE conference on decision and control, vol 2, pp 1497–1502
11. Hedrick JK, Tomizuka M, Varaiya P (1994) Control issues in automated highway systems. IEEE Control Syst 14(6):21–32
12. Isaacs R (1965) Differential games; a mathematical theory with applications to warfare and pursuit, control and optimization. Wiley, New York
13. Kristiansen R, Nicklasson PJ (2009) Spacecraft formation flying: a review and new results on state feedback control. Acta Astronaut 65(11–12):1537–1552
14. Lasserre J, Roubellat F (1985) Measuring decision flexibility in production planning. IEEE Trans Autom Control 30(5):447–452
15. Leonard NE, Fiorelli E (2001). Virtual leaders, artificial potentials and coordinated control of groups. In: Proceedings of the 40th IEEE conference on decision and control, vol 3, pp 2968–2973
16. Mitchell IM (2008) The flexible, extensible and efficient toolbox of level set methods. J Sci Comput 35(2–3):300–329
17. Turpin M, Michael N, Kumar V (2012) Decentralized formation control with variable shapes for aerial robots. In: 2012 IEEE international conference on robotics and automation (ICRA), pp 23–30
18. van der Walle D, Fidan B, Sutton A, Yu C, Anderson BDO (2008) Non-hierarchical UAV formation control for surveillance tasks. In: American control conference, 2008, pp 777–782
19. Vincent TL, Leitmann G (1970) Control-space properties of cooperative games. J Optim Theory Appl 6:91–113. doi:10.1007/BF00927045
20. Wang PKC (1989) Navigation strategies for multiple autonomous mobile robots moving in formation. In: Proceedings of the IEEE/RSJ international workshop on intelligent robots and systems ’89. The autonomous mobile robots and its applications (IROS ’89), pp 486–493
21. Zhou C, Lei M, Zhou S, Zhang W (2011) Collision-free UAV formation flight control based on nonlinear MPC. In: 2011 international conference on electronics, communications and control (ICECC), vol 21, pp 1951–1956

Chapter 4
Coordination Challenges in Networked Vehicle Systems: Are We Missing Something?

J. Borges de Sousa and Fernando Lobo Pereira

4.1 Motivation

The rich and exciting research over the past decade concerning the coordination of multiple vehicles has been focused on systems with fixed structure [6]. Structure is typically described in terms of geometric formations, and the properties of formation controllers are studied in the framework of stability and graph theories. Recent developments have also incorporated new results from the theories of network control systems. However, the scope of coordination is still limited to relative motions.

Motion coordination is just one aspect of multi-vehicle coordination. This becomes more evident in networked vehicle systems consisting of heterogeneous ground, air, and ocean vehicles interacting over inter-operated, and possibly intermittent, communication networks [3, 4, 8]. For example, in networked vehicle systems, information and commands are exchanged among multiple vehicles, sensor nodes and operators, and the roles, relative positions, and dependencies of these vehicles and systems change during operations. Moreover, these systems may exhibit properties that are a function of structure, where structure arises from interactions established over physical, sensing, and communication links. Links change over time; the same happens with interactions established over these links. These are *systems with dynamic structure*.

The control of systems with dynamic structure poses new challenges to control engineering and computer science. These challenges entail a shift in the focus of existing methodologies: from prescribing and commanding the behavior of isolated systems, or tightly coupled systems, to prescribing and commanding the behavior of dynamically interacting networked systems—this may be one of the reasons why we are still far from realizing the potential of these systems.

J.B. de Sousa (✉) · F. Lobo Pereira
Departamento de Engenharia Electrotécnica e Computadores, Faculdade de Engenharia da Universidade do Porto, R. Dr. Roberto Frias, 4200-465 Porto, Portugal
e-mail: jtasso@fe.up.pt

J.H. van Schuppen and T. Villa (eds.), *Coordination Control of Distributed Systems*,
Lecture Notes in Control and Information Sciences 456,
DOI 10.1007/978-3-319-10407-2_4

The fact is that, in spite of developments in the control of distributed systems [15], research in control engineering has not yet incorporated fundamental concepts such as link, interaction, and dynamic structure. In contrast, computer scientists were already making strides in this area in the early 1990s, in part because of the pioneering work of Robin Milner. The following quote from Milner highlights these points [10]:

> Dynamic reconfiguration is a common feature of communicating systems. The notion of link, not as a fixed part of the system but as a datum that we can manipulate, is essential for understanding such systems. What is the mathematics of linkage? The theories of computation are evolving from notions like value, evaluation, and function to those of link, interaction, and process.

Milner's questions were partially addressed in the Pi-calculus [11] (a continuation of Milner's work on the process calculus CCS (Calculus of Communicating Systems) [9]), a calculus of communicating systems in which the component agents of a system may be arbitrarily linked and the communication over linked neighbors may carry information which changes that linkage.

Meanwhile, the advent of ubiquitous mobile computing introduced a new modeling challenge. While the Pi-calculus deals well with mobile connectivity, it does not handle mobile locality. Ubiquitous systems need both. This was the motivation behind the development of the theory of bi-graphical reactive systems (BRS's) by Milner and co-workers [12]. The theory is based on a graphical model of mobile computation that emphasizes both locality and connectivity. The theory evolved from process calculi, especially the calculus of mobile ambients (invented by Cardelli and Gordon [2] deals with spatial reconfiguration) and the Pi-calculus. A bi-graph comprises a place graph, representing locations of computational nodes, and a link graph, representing interconnection of these nodes. Mobile connectivity and locality are expressed with BRS's by defining a set of reaction rules. A reaction rule is a pair of bi-graphs, *redex* and *reactum*, where the *redex* defines a pattern to be matched with a bi-graph modeling the current state of a system. A reaction is simply the substitution of a *redex* with a *reactum*. In this model, systems of autonomous agents interact and move among each other, or within each other.

A careful examination of these developments in computer science may prove invaluable to control engineering, especially in what concerns the coordination of networked vehicle systems. First, because they draw our attention to models of mobile connectivity and mobile locality, which are intrinsic to dynamic structure and coupled dynamics, this is the true essence of cyber-physical systems [1]. Second, because they do not seem to tackle the control of mobile connectivity and mobile locality, namely how to "guide" systems of autonomous agents to interact and move among each other, or within each other, according to some specification.

In the theory of BRS's, the structure of locations of a system is modeled with a place graph, which is restricted to have a tree structure. The assumption is that the topography of a system can be modeled as a set of domains or objects contained within each other. Connections between objects or domains are modeled by links. A place graph may fail to capture several types of geometric relations occurring in networked vehicle systems. First, locations may move, change geometry, and intersect. Second, locations may be permanently associated with mobile computational nodes (e.g., communications range of a physical device). The link graph may fail to capture the intrinsic hierarchical structure of links and interactions among networked vehicles. This is because communication links, which are location-dependent, enable interactions among computational nodes; the failure of a communication link may entail the failure of a complex structure of interactions (which are basically another type of links). The BRS's models of mobility allow the migration of computational nodes, a capability that may open a completely new research direction for control engineering, but fail to capture the fine-grained space-time dynamics of interacting vehicle systems. In addition, the mobility of vehicles may entail the mobility of locations, but this relation does not hold for computational processes.

Control engineers have approached the design of complex systems in the framework of control architectures [13], in which a complex design problem is partitioned into a number of more manageable subproblems. There are several partitioning techniques, being layering the most used one in real applications [14]. However, the language of control architectures, with the exception of developments in the framework of dynamic networks of hybrid automata,[1] has been missing the semantically rich concepts evoked by mobile connectivity and locality. On the other hand, research on BRS's models is missing the principled design approaches associated with control architectures. The interesting question is then: *What is the architectural organization required to support mobile connectivity and locality in a networked vehicle system tasked to satisfy some high-level control specification?*

The architectural organization will have to include mechanisms for context awareness (to adapt the behavior depending on the "context" at hand), for robustness (to sustain computational interactions in the presence of failures of communication links), for estimation of external behavior (to estimate the evolution of components out of communications range), for state and data propagation (to ensure delivery of data and state updates in the presence of intermittent communications), and for setting up controller structures (to evolve the architecture).

This essay is about the computation and control challenges posed by systems exhibiting both (coupled) physical and computational dynamics and dynamic structure. Section 4.2 presents an example to illustrate these two aspects of the behavior of networked vehicle systems. The example draws from experience in designing, building, and deploying networked vehicle systems and was motivated by the developments from the *Control for Coordination* FP7 project. Section 4.3 discusses elements of a control model for these systems. First, we introduce a clear distinction between physical and computation entities and briefly describe the underlying state and control spaces. Second, we explain the couplings between the physical and computation

[1] Informally, dynamic networks of hybrid automata [5] allows for interacting automata to create and destroy links among themselves, and for the creation and destruction of automata. Formally, for each hybrid automaton, there are two types of interactions (mediated by means of communications): (1) the differential inclusions, guards, jump, and reset relations are also functions of variables from other automata and (2) exchange of events among automata. At the level of software implementation, the mechanisms by which software modules interact are called models of computation. The choice of the model of computation (or mix of models) is quite application dependent [7]. This is particularly difficult for dynamic networks of hybrid automata.

dynamics and show how these affect the selection controls. Section 4.4 shows how behavior specifications can be phased in terms of concepts used in traditional control specifications when we consider the modeling concepts introduced previously. Finally, in Sect. 4.5, we discuss the computation and control challenges in this modeling framework. Directions for a research agenda are discussed with special emphasis on the aspects of coupled dynamics and dynamic structure that seem to be missing in the literature.

4.2 Example

Consider the problem of managing a team of autonomous vehicles operating 24/7 in a remote region. The operation consists of monitoring a geographically distributed phenomena (e.g., the levels of radiation at sea). The vehicles operate from a base, which is used for refueling and mission planning. There are no direct communication links between the base and the remote region. Vehicles are used as data mules to transport data between the base and the remote region. Short-range communications are used for team coordination in the remote region. Team coordination is done by a team controller, a computational entity which runs on a designated vehicle, the team leader. The team controller migrates to a new vehicle when the vehicle where it resides returns to the base for refueling—this is an instance of the coupling between physical and computational dynamics. There is another controller at the base to control the overall operation. Data arriving from the remote region is used to update estimates of the status of the remote operations. Based on these updates, the base controller may generate a new controller for the remote team leader. The new controller is sent to the team leader by vehicles departing to the region.

4.3 System Models

This is an example of coordination problems for systems consisting of entities that evolve, interact, and communicate in a common environment that can be modified through the actions of these entities. There are two types of entities in these systems: physical and computational entities. The former are governed by the laws of physics, the later by the laws of computation. Physical entities may interact among themselves and with the environment and can be "composed" to form other physical entities. Computational entities interact through communications. Physical entities may affect computational entities through sensing links. Examples of physical entities include vehicles, sensors, communication devices, computers, and human operators. Physical entities have attributes, which may change with time, e.g., consumable resources, and lodge computational entities. Computational entities may create other computational entities and may be deleted as well. Some computational entities may be the capability to migrate between physical entities over communication channels.

Communications may be geographically constrained. Physical entities can be used to transport computational entities and information across regions where communications are not available. This is also used for maintaining knowledge representation and consistency across the system. The "composition" of computational entities is either local, with respect to the one physical entity where they reside, or distributed, over communicating physical entities.

In what follows, we consider systems of the form *System* = (*PhysicalEntities*, *Environment*, *ComputationalEntities*, *SystemConstraints*).

PhysicalEntity = (*StaticAttributes*, *Dynamics*, *Outputs*, *Constraints*). *StaticAttributes* is the vector of the time-invariant attributes such as type, computational, and communication capabilities; *Dynamics* are the continuous and discrete dynamics which may affect, and be affected, by the environment (this may lead to non-intended consequences or side effects through causal pathways); *Outputs* are the vector of outputs; and *Constraints* represent the state and control constraints. The discrete dynamics has set-valued state variables to model (dynamic) physical and computational interactions with other entities.

Environment models the environment where the elements of *PhysicalEntities* evolve. It has a controlled component, to model environmental aspects that depend on the actions of physical entities (e.g., electromagnetic radiation generated by a set of radars), and an uncontrolled component, to model the aspects of the environment which do not depend on these actions (e.g., terrain and wind fields).

ComputationalEntity is a generic term for software components that encode controllers and other computations. There are two types of computational entities: atomic and composed. An atomic computational entity resides on a physical entity; a composed entity may be distributed over a network in strict accordance to composition rules to ensure that these are well formed. Composition is dynamic in that it can evolve over time, for example, in a dynamic communication network. Atomic computational entities may be allowed to migrate between physical entities over a communication channel. Computational entities can be created and deleted on the fly. Each *PhysicalEntity* is abstracted by one atomic computational entity to bridge the physical and computational worlds. Abstractions of physical entities are not allowed to migrate.

SystemConstraints model constraints in the complex state space of *System*. In the previous example, the team controller runs on a team leader. The team leader exists only in the given region and it changes over time.

4.4 Specifications

The modeling concepts from the previous section allow the specification of behaviors for a networked vehicle system in terms of traditional specification patterns from control engineering. We discuss specifications for a few representative problems to deepen our understanding of the underlying computational and control challenges.

Invariance. The generic problem of invariance involves a pair (*System*, S), where S is a set in the state space of *System*. In this problem, the state of the *System* is required remain in S if the initial state is in S. This generic formulation allows us to express constraints on the controlled component of the environment, as well as on physical and computational mobility, physical interactions, communication links, etc. For example, we may require a controller to "stay" in a given region independently of the physical entity where it resides; or we may want to have at least one vehicle in a given region.

Attainability. In the problem of attainability, we require the state of the *System* to "attain" a set Γ within a given time interval τ. As with invariance, this specification allows us to consider complex physical and computational target sets.

Optimization. The specification of optimization problems involves departure and target sets, state constraints, information structures, control spaces, cost functions, and the "mood" of the problem (cooperative, adversarial, etc.).

As before, departure and target sets and state constraints are defined in the complex state space of the *System*. This enables us to encode non-standard specifications (e.g., permissions for the migration of computational entities or for network access).

Full-state information may not be accessible in the *System*. Information structures, which concern who knows what and what is sent to whom, are affected by the mobility of physical entities in communications challenged environments. This leads to dynamic information structures, i.e., those depending on the state of the system.

Control spaces and control constraints can be very complex. Each physical entity may affect other physical entities and the environment. This may lead to some level of indirectness when it comes to finding optimal controls. For example, the cost function may depend on the environment, which may be affected by the motions of physical entity A which, in turn, may be disabled by the actions of physical entity B. We need to identify causal control pathways, which link actions to their effects, with causal constraints not only based on commitments in the past, but potentially in the future. The challenge is that causal control pathways may be dynamic since future commitments might change with time. In addition, the effects of control actions may be significantly delayed (e.g., dropping a bomb or migration of a computational entity). Finally, it is up to the designer to specify the control space for the controllers in a system (e.g., change control authority and add/remove state constraints or permissions for establishing links of communication).

The global performance (or cost) of a set of interacting computational entities and supporting physical entities depends on the initial, terminal, integral, and switching costs (incurred when switching between discrete controls). Each of these costs may have terms associated to physical and computational interactions (e.g., cost may depend on the structure), in addition to terms associated to physical and computational entities (e.g., the cost of computations). Cost functions may also depend on predicates on the state of the world (e.g., in military operations, we may want to switch from minimum risk to optimal time formulations when the level of threat drops below some threshold), thus introducing non-Lipchitz dependencies. The performance evaluation of persistent 24/7 operations also presents new challenges to optimization. This is partially related to the fact that physical entities enter and leave the system.

A high-level interpretation of the behaviors exhibited by the systems under consideration is in order. Generally speaking, these systems evolve through phases. In each phase, a subset of the constituent vehicles may operate on their own, while the remaining vehicles may form clusters where several types of interactions may take place. Switching between consecutive phases is triggered by events such as the achievement of partial or global goals, failures, or environmental changes—vehicles can modify and sense the environment, which can be used for signaling. The switching logic triggers the formation of new clusters, the generation of the corresponding goals, and distributed goal allocation. The new goals should enable communications at the end of the phase, so that the process can start again—this is what keeps the system alive. Basically the system alternates between the computation of goals and the control of itself to reach these goals.

An abstract control interpretation of these behaviors is as follows. In each phase, there is a set of concurrent, and possibly coupled, invariance, attainability, and optimal controllers; phase switching entails changing controllers and associated interactions. The hypothesis is that control-inspired specifications suffice to specify the behaviors for a large class of systems, if not for all networked vehicle systems.

4.5 Challenges

The problem of designing controllers for physical entities, operating either in isolation or in a system with fixed structure, is generally well understood. This is not the case with a networked vehicle system, where loosely coupled physical and computational entities interact in communications challenged environments.

Given a generic specification for the behavior of a system, the design problem consists of deriving a structure of computational entities which, when "composed" with the system, will satisfy the specification in some sense to be defined.

This design problem presents new challenges to computation and control: (1) these systems have complex state and control spaces and coupled physical and computational dynamics; (2) physical and computational dynamics may depend both on physical and computational interactions through complex pathways of causality; (3) physical and computational interactions are dynamic and have constraints on location and linking; (4) networks of physical and computational entities have properties which depend on the structure of these networks; (5) physical entities may enter and leave the system, while computational entities may be created/destroyed on the fly; (6) physical and computational entities are distributed over the underlying physical and computational spaces; (7) physical entities may have limited autonomy, thus requiring periodic refueling; (8) control actions available to computational entities may include the generation of new controllers (this requires controllers to know how to generate other controllers); and (9) state may not be directly accessible by all computational entities. These challenges are not unique to networked vehicle systems. This discussion may lead to new insights into other fields, such as biology or ecology.

References

1. Baheti R, Gill H (2011) The impact of control technology, chapter cyberphysical systems. IEEE Control Systems Society, pp 161–166
2. Cardelli L, Gordon AD (1988) Mobile ambients. In: Nivat M (ed) Proceedings of the first international conference on foundations of software science and computation structure, Lecture notes in computer science: 1378. Springer, Berlin, pp 140–155
3. de Sousa JB, Johansson KH, da Silva JE, Speranzon A (2005) A verified hierarchical control architecture for coordinated multi-vehicle operations. Int J Adapt Control Sig Process 21(2–3):159–188 (Special Issue: Autonomous and adaptive control of vehicles in formation)
4. de Sousa JB, Maciel B, Pereira FL (2009) Sensor systems on networked vehicles. Netw Heterogen Media 4(2):223–247 (American Institute of Mathematical Sciences)
5. Deshpande A, Gollu A, Semenzato L (1997) The shift programming language and run-time system for dynamic networks of hybrid automata. Technical Report UCB-ITS-PRR-97-7, California PATH
6. Girard AR, Borges de Sousa J, Hedrick JK (2005) A selection of recent advances in networked multi-vehicle systems. Proc I MECH E Part I J Syst Control Eng 219:1–14
7. Lee E, Sangiovanni-Vincentelli A (1996) Comparing models of computation
8. McGuillivary P, de Sousa JB, Martins R (2012) Connecting the dots. networking maritime fleets of autonomous systems for science and surveillance. Marine Technology Reporter (October)
9. Milner R (1980) A Calculus of Communicating Systems. Springer, Berlin (1980)
10. Milner R (1996) Semantic ideas in computing. In: Wand I, Milner R (eds) Computing tomorrow: future research directions in computer science. Cambridge University Press, Cambridge, pp 246–283
11. Milner R (1999) Communicating and mobile systems: the Π-calculus. Cambridge University Press, Cambridge
12. Milner R (2009) The space and motion of communicating agents. Cambridge University Press, Cambridge
13. Varaiya P (1972) Theory of hierarchical, multilevel systems. IEEE Trans Autom Control 17(2):280–281
14. Varaiya P (1997) Towards a layered view of control. In: Proceedings of the 36th IEEE conference on decision and control, pp 1187–90
15. Varaiya P, Simsek T, de Sousa JB (2001) Communication and control of distributed hybrid systems—tutorial session. In: Proceedings of the 2001 American control conference, pp 4968–83. IEEE

Chapter 5
Leader–Follower Coordination Control for Urban Traffic

René Boel and Nicolae Marinică

5.1 Motivation

Open-loop control of traffic lights in an urban environment is typically tuned so as to generate *green waves*, which ideally should ensure that vehicles never have to stop, avoiding waste of capacity of the intersections. A green wave minimizes average delay under ideal, noiseless assumptions, by selecting the phase shift between the switching times of successive traffic lights, so that this phase shift corresponds to the travel time between intersections. Achieving green waves is easier if the vehicles travel together in large platoons along major axes of traffic flow. In a large network with dispersed two-way traffic and with different time delays along different roads connecting successive intersections, it becomes difficult to find a perfect green wave switching schedule. Moreover vehicles entering from uncontrolled side streets or parking lots, unexpected delays along connecting link roads, and all other sources of noise also lead to a deterioration of the performance of the open-loop strategy.

Local feedback control tries to avoid this deterioration by adjusting the switching times of the traffic lights to currently available data on the expected arrival times of vehicles. This feedback control tends to destroy the green wave synchronization. Combining a green wave with feedback control thus requires active coordination among the neighboring feedback control agents. This problem is not critical when the traffic load is very light, since then all specifications are very easy to meet. It is also not relevant under very heavy load since then the effect of the random perturbations is small compared to the average load, and no green wave solution can exist anyway. However, under intermediate load, when there are a few critical intersections that

R. Boel (✉) · N. Marinică
SYSTeMS Research Group, Ghent University, Technologiepark-Zwijnaarde 914,
9052 Gent, Belgium
e-mail: rene.boel@ugent.be

N. Marinică
e-mail: nicolaeemanuel.marinica@ugent.be

J.H. van Schuppen and T. Villa (eds.), *Coordination Control of Distributed Systems*,
Lecture Notes in Control and Information Sciences 456,
DOI 10.1007/978-3-319-10407-2_5

have difficulty achieving their local specifications, the leader–follower coordination as proposed here can significantly improve performance.

It is then a good solution to designate the critical intersections as leader intersections and to allow these leader intersections to generate additional specifications on the time intervals in which their upstream neighboring follower intersections allow platoons of vehicles to start traveling toward their downstream leader intersection. This reduces the waste of capacity at the leader intersection, improving the overall performance of the network, taking into account that the closer an intersection is to saturation, the higher the risk that random perturbations will lead to overflowing queues. Remember that the average queue size under stationary behavior for a simple queueing model behaves like the inverse of $1/(1-\rho)$, where ρ is the ratio of the average arrival rate over the average departure rate. To make matters worse, the variance of the stationary queue size grows like the square of $1/(1-\rho)^2$, and the growth rate increases even more for realistic models with more irregular arrival and departure rates. Clearly, the risk that queues grow so long as to cause blocking of upstream intersections and gridlock is extremely sensitive to how close this load ρ is to 1. The control agents for the critical intersections ($1-\rho$ small) are therefore assigned as leaders, while the control agents for their less heavily loaded neighbors are given the role of followers.

This essay only deals with the coordination control at the layer of the local controllers, not with the hierarchically higher supervisory layer. We assume that the roles of leaders and followers are assigned by a higher-level supervisor. We only discuss the coordination strategies that enable the different components to help each other to perform satisfactorily, whether these local goals are formulated via local specifications or as online optimization of a local cost criterion or a combination of both. As long as the leader–follower assignment corresponds to the current overall network conditions this coordination strategy will lead to good overall performance, since by definition, followers are lightly loaded and thus can easily satisfy their goals. If the average traffic flow changes, making a follower heavily loaded, or in case a follower receives contradictory requests from two neighboring leaders, then this assumption may become invalid. Hence, the supervisor needs to constantly monitor the average traffic flow and change the leader/follower assignments when the relative load of neighboring intersections changes or when followers inform the supervisor that they have problems satisfying their specifications. This essay assumes stationary conditions of average traffic flow, so that leader and followers are fixed forever.

5.2 Examples

Such a leader–follower approach can be applied to many networks of interacting components under intermediate, unevenly distributed load, where a load increase might lead to inefficient use of the capacity of the network. Some examples areas follows:

- traffic lights in an urban traffic network;
- on-ramp metering in control of freeway traffic, taking overflow into neighboring roads into account;
- coupled tanks with fluids in process control problems;
- flood control, or dually irrigation channel control, where controllable gates can regulate the flow of water;
- air traffic management;
- communication networks.

To simplify the discussion, we use the urban traffic terminology and leave it to the reader to consider the other cases. Each system corresponds to a graph of interconnected hybrid dynamical systems HA_m, each with a local control agent CA_m. In the traffic example, HA_m models an intersection, with controlled traffic lights, and the evolution of the traffic in its surrounding region with its access roads. Two neighboring components interact when traffic flows from the upstream component HA_m to the downstream component HA_ℓ. At the boundary, further on called a *port*, traffic variables like platoon arrival time and platoon size (or for more aggregated models: traffic density and average speed), remain unchanged. Sensors located at these boundaries (or ports) measure the characteristics of the traffic flowing from HA_m to HA_ℓ. In the intermediate load cases that we consider for the leader/follower approach, it is usually obvious which variable is an input to a node. In the traffic case, the flow rate is always determined by the upstream intersection, unless the downstream intersection is blocked. If the leader manages to keep queues from overflowing even in the intermediate traffic load, and that is the goal of our controllers, then there is a clear concept of upstream intersections generating flow into a downstream intersection. In general, our leader/follower approach is applicable when it is possible for upstream nodes to control variables influencing, i.e., acting as input variables to, their downstream neighbor.

The interconnection between neighboring nodes HA_n in any network of physical systems can be described by ports, expressing the flow of material, power, messages,... from one node to the next node. Port variables take the same value in both nodes connected by the port. A port variable is an input variable for one of these nodes, and an output variables for the other node (otherwise, the model would be incomplete or contradictory). Usually, the port variables in networked systems theory come in dual pairs, like current and voltage or velocity and force, the product having the dimension of power, but this is not really necessary and is not applicable to traffic problems.

5.3 Urban Traffic: Case Study for Coordination Control

The case study considered in this essay proposes a design methodology for coordinating feedback control of the red/green switching times of traffic lights in an urban traffic network under intermediate load. In order to explain the leader/follower

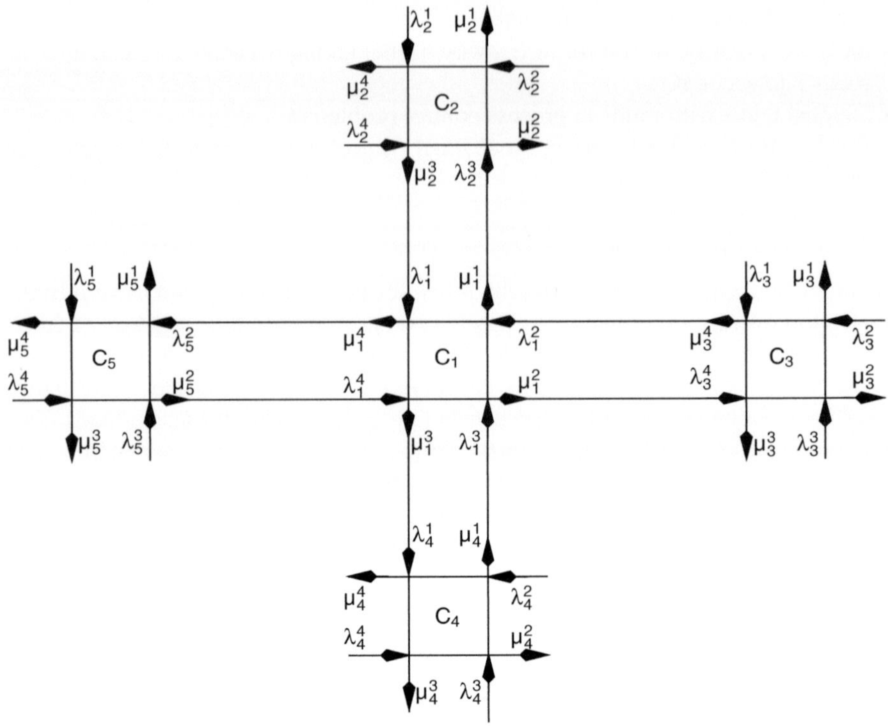

Fig. 5.1 Case study area with one leader (intersection c_1) and 4 follower intersections

paradigm as simply as possible, we consider the example of one heavily loaded leader intersection, with 2-way roads connecting it to 4 neighboring less heavily loaded follower intersections, as shown in Fig. 5.1. Control agent CA_i selects the next switching time of the traffic lights at intersection i (not just the cycle time, or the red/green split). CA_i receives measurements from local sensors which we assume to be sufficient to provide the approximate information on the arrival times of vehicles at intersection i in the near future (see [1] for a state estimator that can help to achieve this).

A model that describes traffic at the granularity of a few seconds, the average time between two successive vehicles crossing the intersection, is needed. Microscopic models, describing the movement of each vehicle individually, are computationally too expensive for practical applications. We proposed and validated in [2] a model that describes urban traffic using platoons of vehicles traveling close together at approximately the same speed. Assume that at time t, there are $K(t)$ platoons, $Platoon_k, k = 1, \dots, K(t)$, present in the network under study. The state X_t of the system at time t is obtained by specifying for each of these $K(t)$ platoons the location $Loc_k(t)$ of the first vehicle of $Platoon_k$ and the size $Size_k(t)$ of vehicles in this k-th platoon. Details on the platoon-based model can be found in [2]. While the leader/follower paradigm can in principle be applied for other traffic models as well, we use from now on the concept of platoons of vehicles in describing the behavior of our controlled urban traffic system.

The feedback control law, as implemented by the leader control agent at a critical intersection, selects the switching times of the local traffic light so as to minimize the waste of capacity at this leader intersection, averaged over all flow directions for that intersection. This is intuitively sufficient in order to minimize the average waiting time for all vehicles using this leader intersection. Other costs, like the environmental impact (noise, pollution) and the fraction of vehicles that have to stop and then accelerate at the intersection, are closely related to this waste of capacity. This wasted capacity can actually be calculated over a certain prediction horizon, analytically or via simulation, for each choice of switching times, as a function of the predicted arrival times of platoons and of the current switching state (taking into account minimal and maximal green periods). Because there may be discrete choices to be made (one needs to compare cases allowing a platoon to pass in the ℓ-th cycle of the traffic light or to make this platoon wait for service until the $\ell+1$-th cycle), this optimization problem is a complicated, generally non-convex, constraint satisfaction problem. One possible implementation of the leader–follower paradigm is a model-predictive control-style solution: generate at time t_k the costs for different future switching patterns and select the one with the smallest wasted capacity, if the best switching pattern requires switching at time t_{k+1} implement the switch; repeat the comparison between different switching patterns every Δ seconds or as soon as another switching of the traffic lights is allowed (remember that there is a minimal green time for each direction for safety reasons and for performance reasons).

The optimization carried out at the leader agent will also indicate, using knowledge of the travel time between upstream follower and downstream leader intersection, which (intervals of) switching times at the upstream follower intersections lead to platoons of vehicles arriving at the leader at a time when the leader is giving green to that direction anyway in its optimal (or in a near optimal) pattern of switching times. Based on this solution, the leader intersection $Inter_{\text{leader}}$ should therefore send information to each of its upstream follower intersections $Inter_i$ (where i belongs to the set of follower neighbors of $Inter_{\text{leader}}$) a message that sets lower and upper bounds on the switching times at $Inter_i$. These bounds should be such that they allow the leader intersection to find a good solution, with very little wasted capacity. In practice, this amounts to selecting follower switching times so that the platoons released by the follower arrive at the leader at a time when it is feasible for the leader intersection to give this platoon a green light, without giving red to non-empty queues, and at a time when platoons arriving from other upstream followers of the leader intersection can also use this green period. Thus, the approach not only will synchronize the operations of leaders and followers, but it will also indirectly synchronize operations by different followers of the same leader. In [3], it is shown how a cost can be defined that should be minimized by the follower control agent, using messages from the leader about its planned switching times, so that the optimal switching time decisions by the follower control agent do indeed achieve good synchronization. Note that in this approach synchronization is achieved—as will be shown by the results in section 5—even though the leader and follower control agents only perform local optimizations.

5.4 Theory and Concepts

The leader–follower feedback coordination paradigm described above is applicable to a broad class of problems. In order to prove properties of a networked system with the leader/follower control paradigm, it is necessary to represent this system as a network $\mathcal{N} = \{node_i, i = 1, \ldots, I; link_j = \{i_j, \ell_j\}, j = 1, \ldots, J\}$ consisting of I nodes and J links. Each node is described as a timed or hybrid controllable input–output automaton further denoted as $node_i = HA_i$. The different nodes interact with each other via their ports that send output variables to and receive input variables transmitted from their neighbors. Formally, there are J channels $link_j = \{i_j, \ell_j\}, j = 1, \ldots, J\}$ that transmit variables from the output port of a node connected to input node $^*link_j$, to their output port $link_j^*$. A port corresponds in the urban traffic case with the location, where a sensor measures the flow of traffic leaving the region upstream from the sensor, and enters the downstream region.

Each node HA_i can be controlled by a local control agent CA_i (in the traffic example CA_i selects the switching times of the traffic lights). The actions of CA_i influence the future evolution of the internal state variables of HA_i. The local cost at HA_i (e.g., the average waiting time for all traffic using $Inter_i$) depends on the control actions of CA_i, on the evolution of the port variables connecting HA_i to its upstream neighbors (the traffic entering region j), and on the local noise acting on HA_i. The control actions at HA_i determine the port variables at the output links $link_j$ of HA_i (all traffic sent to the downstream links such that $node_i =^* link_j$).

In order to guarantee that global performance specifications are met by the controlled system the leader–follower coordinating control approach assigns some critical nodes to act as leaders. Leader nodes are those nodes where it is most difficult to satisfy all the specifications and where failure to meet these specifications leads to a high risk of global failure of the networked system. The CA_i at leader i solves a local cost criterion and at the same time generates specifications that must be included as additional constraints on the control decision algorithms of the neighboring follower nodes of HA_i. Note that the leader–follower approach does not provide any global optimization. Each leader CA_i optimizes its local cost (e.g., minimizing wasted capacity), without taking other leaders (let alone followers) into account. It is assumed that the assignment of leader agents is such that no neighbors can simultaneously be leaders. The interaction between the different leader nodes is via the specifications they impose on their (possibly common) follower nodes. Exact performance proofs are very difficult in general requiring tricky assume–commit arguments.

5.5 Overview of Research Contributions

The proposed leader/follower coordinations strategy has been applied to a simulated urban traffic network with 5 nodes, including one leader node, as in Fig. 5.1 (using the SUMO microsimulator [4]). In [3], a simulator using the platoon-based model is implemented and used for a particle filter that transforms local raw noisy sensor output data into an online recursive particle filter with as output reliable estimates of the state of the local node (arrival time and size of platoons moving to intersection). In order to clearly separate the estimation and the control issues, the results discussed below were obtained assuming exact knowledge of arriving platoons over a given prediction horizon, corresponding to the travel time of vehicles from sensor to intersection. At the leader intersection c_1, the control agent in our simulation experiments selected switching times over a control horizon of 120 s, using as information the exact platoon arrival times as far as available, extending with average arrival rates for the remaining part of the control horizon. The cost criterion that is minimized is the minimal expected integral of the queue sizes at the different approach directions (this is proportional to average waiting times). This optimization is repeated every Δ time units ($\Delta = 1$ s in our experiments, but slower updating may ne sufficient for many applications). The traffic light is switched if the optimization indicates this is optimal at the next second, otherwise no local action is taken. The leader sends a message each time a new optimal schedule is found, telling its 4 followers during which intervals they should switch their traffic light so as ro minimize waste of capacity at the leader. The followers then also select their local switching strategy, minimizing over the same prediction horizon a cost that is a weighted sum of the average local waiting times and a cost for not satisfying the request sent b y their leader. The relative weight between local cost at the follower and incremental cost at the leader is denoted by $\omega(=100$ in the case reported in Fig. 5.2). The cost of not following the requests of the leader is calculated as the average increase in waiting time at the leader, assuming that the leader follows a simple fixed-cycle strategy for its switching times in the distant future. We also assume that the, hopefully small, incremental cost at the leader is additive over the 4 followers.

The graph on the right of Fig. 5.2 illustrates the improvement achieved by the leader–follower controller. This figure shows the histogram of the average waiting times observed over 30 simulation runs, for the leader and for the followers, for different values of ω. Especially for values $\omega = 50, 100, 500$, a significant reduction in the waiting is observed in almost all sample runs, not only at the leader, but also to a lesser extent at the followers. The explanation for this improvement is the synchronization which is obtained not by imposing it in advance, but through the local optimization and the information about desirable switching times that is sent from leader to follower. The graph on the left of Fig. 5.2 illustrates how after an initial transient the switching times at the 5 intersections take on approximately equal red/green cycles. Notice that the time origins should be shifted in order to take travel time from follower to leader into account in order to properly see that also the followers get synchronized. Further results are reported in [3].

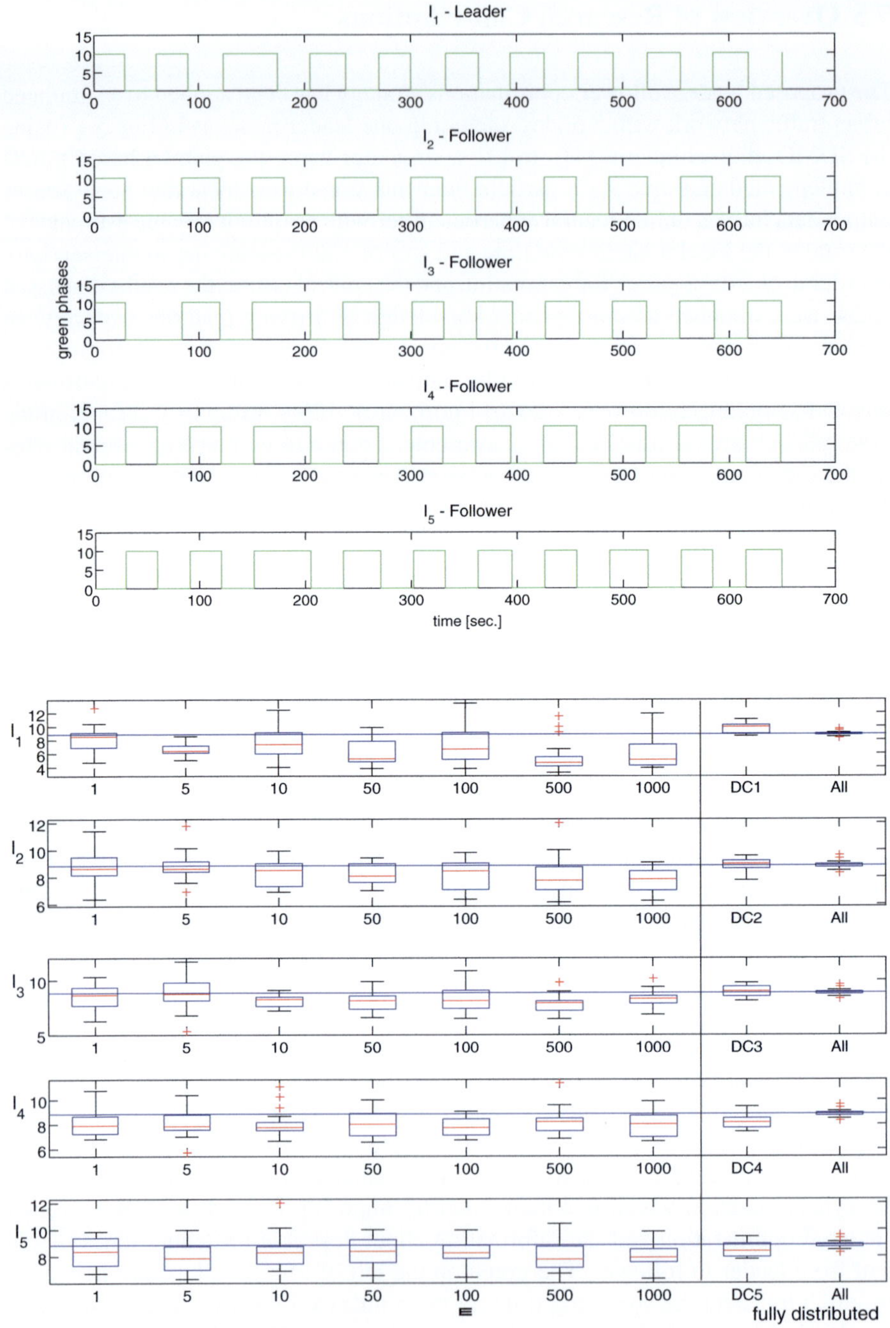

Fig. 5.2 Coordinated traffic light switching times, and histogram of average delays, from simulation runs using the leader/follower controller

So far only cases with one single leader and a number of followers have been investigated. Future work will have to confirm the usefulness of this leader/follower approach to networks with many leaders. No problems should occur if each follower has only one leader, i.e., if the distance between any two leader nodes is at least 2. However, as soon as some followers have to satisfy requests by two different leaders, problems may occur. In those cases, the supervisory layer in the control hierarchy will become important. A control structure with different layers of leaders, or a treelike hierarchy of influence among nodes, needs to be investigated further.

5.6 Conclusions and References to Current Methodology

Most traffic light coordination studies deal with adaptive open-loop approaches, where the cycle time, red/green split and phase shift of the intersections in an urban area are adjusted, according to predefined rules or scenarios. A good survey of classical approaches can be found in chapter 5 of the book [5]. Many commercial systems [6, 7] for urban traffic control are available, incorporating some form of coordination. The Australian SCATS system [7], that has evolved through a period of more than 20 years of practical experience, in fact uses the notion of a leader for a region, with regions being selected by a (human or automatic) supervisor. A disadvantage of many of these commercial tools, from the point of view of the transport authority using them, is that the algorithm becomes so complicated that it is impossible for the users to understand intuitively what the effects will be of any changes made to the system (new infrastructure, or changes to the traffic rules). The purpose of this paper is to overcome this problem by developing model-based strategies for traffic lights. Only a few papers develop feedback control strategies based on a dynamical model, mainly for one single intersection (see, e.g., as a good reference [8]), with [9] as only model-based paper including coordination. This paper and [10] also emphasizes the need for determining what is the minimal communication necessary for achieving good coordination in urban traffic networks. This can help to improve intuitive understanding of the effects of feedback control.

References

1. Marinică NE, Sarlette A, Boel RK (2013) Distributed particle filter for urban traffic networks using a platoon based model. IEEE Trans Intell Transp Syst 14(4):1918–1929
2. Marinică N, Boel R (2012) Platoon based model for urban traffic control. In: Proceedings of American control conference, pp 6563–6568
3. Marinică NE (2014) Distributed estimation and control of interacting hybrid systems for traffic applications. PhD thesis, Ghent University
4. Behrisch M, Bieker L, Erdmann J, Krajzewicz D (2011) Sumo-simulation of urban mobility: an overview. In: Proceedings of 3rd international conference on advances in system simulation, pp 55–60

5. Daganzo C (1997) Fundamentals of transportation and traffic operations. Elsevier, Pergamon
6. Hansen BG, Martin PT, Perrin HJ (2000) SCOOT real-time adaptive control in a CORSIM simulation environment. Transpo Res Rec J Transp Res Board 1727:27–30
7. Roads and Traffic Authority. SCATS. NSW Government, www.scats.com.au/index.html
8. De Schutter B (1999) Optimal traffic light control for a single intersection. In: Proceedings of 1999 American control conference, San Diego
9. Porsche I, Lafortune S (1998) Coordination of local adaptive traffic signal controllers. In: Proceedings of 1998 American control conference
10. Marinică N, Boel R (2012) A leader/follower approach for distributed coordination of interacting components. In: Proceedings of the 20th international symposium on mathematical theory of networks and systems, 2012

Chapter 6
Prediction of Traffic Flow in a Road Network

Yubin Wang, Jan H. van Schuppen and Jos L.M. Vrancken

6.1 Introduction

The chapter presents two algorithms which together produce predictions of traffic flows in a large-scale road network for a horizon of 30 min.

The motivation to predict traffic flow in a road network is to detect troublesome traffic situations, for example traffic queues, before they start to appear, and to evaluate one or more control scenarios for the traffic flow in a road network. The latter approach is often described as a decision support unit for traffic control. The prediction algorithms could be implemented at the traffic control centers currently in operation in The Netherlands.

The approach to prediction of traffic flow in a large-scale network consists of two steps. The first step is to predict the traffic flow at all boundary points of the network including motorway inflows and on-ramp inflows. The second step is to predict the traffic flow in the network. Due to the need to predict in a relatively short time, less than a second and the need to have robust algorithms, the organization of the computations is best formulated in a distributed way. Hence, the authors have chosen for a distributed prediction algorithm. However, coordination between the various distributed predictions is necessary; hence, the approach will be described as a coordinated–distributed prediction algorithm. It will be argued that the complexity of this coordinated–distributed approach is very good for large networks.

Y. Wang (✉) · J.L.M. Vrancken
Delft University of Technology, 2600 GA Delft, The Netherlands
e-mail: yubin.wang@tudelft.nl

J.L.M. Vrancken
e-mail: j.l.m.vrancken@tudelft.nl

J.H. van Schuppen
Van Schuppen Control Research, Gouden Leeuw 143, 1103 KB Amsterdam, The Netherlands
e-mail: jan.h.van.schuppen@xs4all.nl

J.H. van Schuppen and T. Villa (eds.), *Coordination Control of Distributed Systems*,
Lecture Notes in Control and Information Sciences 456,
DOI 10.1007/978-3-319-10407-2_6

6.2 Problem Statement

In The Netherlands, there are five traffic control centers for online monitoring and control of motorway networks. To assist the operators with their control tasks, predictions of traffic flow have to be generated for a prediction horizon and one or more control scenarios have to be evaluated for their effect on the traffic flow. For example, the traffic control center of the Northwest Netherlands in the town of Velsen monitors and stores the traffic flow data on the motorways and on several provincial roads on an hourly, daily, weekly, and yearly basis more or less for the road network in the province of North Holland. Control measures such as on-ramp metering, dynamic speed control, routing advice, and other measures are taken partly automatically and partly by the road operators.

Currently, there is apparently no traffic control center in the world which has an online prediction algorithm as described above. The authors are not aware of any publication of an algorithm for this type of prediction.

Prediction of traffic flow in a road network requires attention for two subproblems. The first subproblem is to predict the traffic inflows (1) of motorways at the boundary of the road network and (2) of all on-ramps of the network. The second subproblem is to produce the predictions of the traffic flow in all links of the road network including the motorway outflows of the network and all outflows to off-ramps. Both subproblems are discussed in this chapter.

To clarify the need for the solution of these subproblems, consider first a road network with a boundary where motorways enter or exit. The traffic control center does not have detailed information on the state of traffic upstream from those boundary points. Therefore, the predictions have to be determined based on the values of the observed traffic flows at the boundary of the road network. Prediction of traffic flow in a full road network implemented in a centralized way is expected to require rather long computation times. In addition, the centralized computation makes the system sensitive to failures. The advantage of a distributed algorithm for computation of predictions is that the computation time is much less due to the parallel computations and that the system is more robust if parts of the system fail or if the communication network partly fails. For the province of North Holland, it is likely that there will be more than 100 subnetworks and for all of The Netherlands more than 500 subnetworks.

Problem 6.1 The problem considered in this chapter is to predict every 10 min the traffic flow in a large-scale motorway network, for example, in a 80 by 50 km network, for a prediction horizon of 30 min as quick as possible (within a fraction of a second). The numbers were selected by the authors, and the algorithms will also operate for slightly different values.

The prediction algorithm should produce predictions of (1) the inflows of motorways at boundaries of the road network and of all inflows of on-ramps in the network; (2) the densities and speeds of all sections of the networks; and (3) the outflows to all motorways at boundary points and of all off-ramp traffic flows in the full network. The approach taken is to formulate a coordinated–distributed prediction algorithm.

Assume available at any prediction time of a multiple of 10 min, observations in the recent past of the traffic inflows at boundary points of the road network and of on-ramps. In addition, assume available an estimate of the state of the full road network considered. The latter estimate is produced by a filter which is already implemented in traffic control centers.

6.3 Literature Review

The problem of prediction of a signal has been considered in several domains. The authors have chosen to apply an algorithm based on modeling, the Kalman filter, and a recursive parameter estimation because it is reported to work well for models of several domains and because it requires almost no tuning. An alternative is artificial neural nets, but such an algorithm requires much tuning such as for every on-ramp. References to alternative prediction algorithms are exponential smoothing [5], support vector regression (SVR) [3], and support vector machines (SVMs) [2]. More references are provided in [11].

Coordinated–distributed prediction problems have been developed for several domains. A frequently used approach is to let all subsystems coordinate their one-step results after every time step. But the complexity of such an approach is too high for the traffic application. References include the partitioning of networks [4], a conservative algorithm [6], and a glueing algorithm [7, 8].

6.4 Prediction of Traffic Flow at Boundaries of a Road Network

In this section, an adaptive prediction algorithm is described for the traffic flow at the boundaries of a road network. The approach of adaptive prediction is well known in control theory but has not been applied to traffic flow prediction. The known approach must be adjusted for this application of prediction of traffic flow. The results of this section have been published in the journal paper [11] and in the conference paper [9].

Consider then a road network and at the boundary of the network a motorway inflow to the network. Assume available the traffic flow data at that location for each step of about 5 s for the past hour.

The problem of this section is to predict the traffic flow of this motorway inflow for a prediction horizon of 30 min. The same prediction problem is to be solved for any on-ramp in the network based on observations of the traffic flow. It is to be noted that the prediction of the traffic flow in a large-scale network is critically dependent on having accurate predictions of the boundary inflows. The algorithm used for the predictions has to be fast and easy to implement at various locations so as to make the traffic control center efficient.

The approach of prediction of traffic flow is based on an adaptive Kalman filter, where the adaptation comes from a recursive least-square algorithm. The underlying model and the algorithm are nonlinear though based on the Kalman filter.

The overall algorithm for the prediction of traffic flow at boundary points of a road network consists of the following steps: (1) a week profile and its predictor; (2) computation of the differences between the measurements of a period within a week and the prediction of the same period computed a week earlier, the relative period errors: a model for the relative period errors in the form of a Gaussian system and an adaptive prediction algorithm for the relative period errors; and (3) the prediction algorithm based on the predictions of the week profile and the predictions of the relative period errors. The algorithm is summarized below.

The definitions of the symbols follow: $\hat{w}(s+1|s,t)$ denotes the prediction of the traffic flow for period t in week $s+1$ based on data available till week s for the same period. The recursion for the prediction is driven by the observations of traffic flow $y(s,t)$ for the same week and period as described before. The symbols q, q_{v0}, q_w, q_{ya} denote various variances of the variables in the subindex, and the symbol $k_w(s)$ denotes the time-varying Kalman gain. The mod operator is to be able to refer to a particular period within the week profile.

(1) The predictor for week $s \in \mathbb{Z}$ and period $t \in T = \{1, 2, \ldots, n_w\}$ is then

$$k_w(s) = q_w(s|s-1)[q_w(s|s-1) + q_{v_2}]^{-1}, \tag{6.1}$$

$$\begin{aligned} \hat{w}(s+1|s, t \bmod n_w) &= \hat{w}(s|s-1, t \bmod n_w) \\ &\quad + k_w(s)[y(s,t) - \hat{w}(s|s-1, t \bmod n_w)], \end{aligned} \tag{6.2}$$

$$\hat{w}(0|-1) = m_{w0},$$

$$q_w(s+1|s) = q_w(s|s-1) + q_{v_1} - q_w(s|s-1)^2[q_w(s|s-1) + q_{v_2}]^{-1},$$

$$q_w(0|-1) = q_{v0}, \tag{6.3}$$

$$q_{ya}(s+1|s) = q_w(s+1|s) + q_{v_2}. \tag{6.4}$$

A verbal description of the algorithm follows. The algorithm is initialized by a week profile obtained by averaging the week profiles of several preceding weeks. After receipt of an observation, the algorithm produces a prediction of the week profile for that particular period next week according to the Kalman predictor.

(2) Define the *relative error* of week s and of period t as

$$e(s,t) = [y(s,t) - \hat{w}(s|s-1,t)]/\hat{w}(s|s-1,t). \tag{6.5}$$

The model for the relative period errors is that it is a stationary Gaussian process, and, moreover, the output of a Gaussian system of dimension $n \in \mathbb{Z}$. Based on such a model, there exists an adaptive prediction algorithm which produces predictions of the relative error process for as many steps as needed (see [1, 11]). The prediction horizon is 30 min; hence, the program computes three periods of 10 min each. The predictions are denoted as follows (in week s and at time t):

$$\hat{e}(s, t+1|t), \hat{e}(s, t+2|t), \hat{e}(s, t+3|t), \ \forall\, t \in \{1, \ldots, n_w\}, \ s \in \mathbb{Z}_+.$$

(3) Finally, the predictions of the traffic flow data are then described by the following expressions:

$$\hat{y}(s, t+i) = \hat{w}(s|s-1, t+i) + \hat{w}(s|s-1, t+i)\hat{e}(s, t+i|t), \ i = 1, 2, 3. \quad (6.6)$$

As an example, the adaptive prediction algorithm is illustrated for the ring network of Amsterdam. There are four motorways (A1, A2, A4, and A8) which end on the ring road of Amsterdam (A10). There are two kinds of inflows at the boundary of the network: (1) from four motorways and (2) from on-ramps. Only, the data of motorways are shown in this paper. In a related journal paper, [11], the predictions of on-ramp traffic flows are shown.

In order to test the algorithm for traffic flow from motorways, the authors collected traffic flow data from four sites on the motorways ending on the ring road of Amsterdam at a short distance before the merge point. The traffic data of the four sites (A1, A2, A4, and A8) were collected from May 20, 2010, until June 24, 2010. The 1-min average traffic data over 5 weeks were collected by the sensors of the MONICA system (velocity flow measurement points). The adaptive predictions of Tuesday's traffic flow is shown in Fig. 6.1.

Table 6.1 shows the performance of the adaptive predictor. Note that the variance of the traffic flow increases when the prediction period lies further away in the future. We can conclude that the adaptive filter performs very well and can considerably better predict the traffic flow than the week profile.

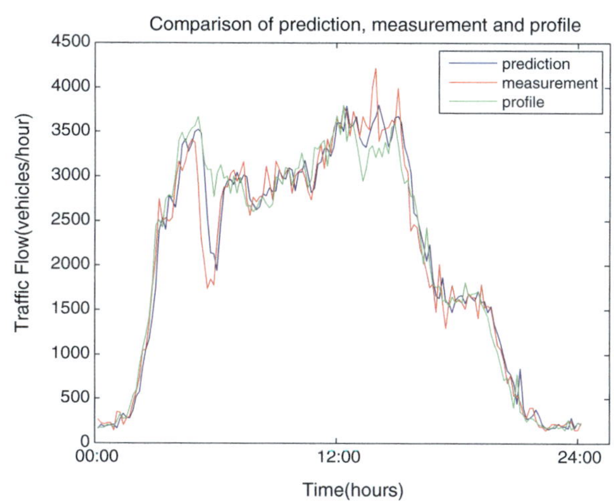

Fig. 6.1 Adaptive predictions of Tuesday's traffic flow

Table 6.1 Comparison of the performance of the adaptive predictor and of the week profile

	First step prediction (SP)	Second SP	Third SP	Week profile
RMSE	247.6	315.4	386.4	390.0
RMPE	12.4	14.2	15.3	15.1
VAF	95.3	92.3	88.1	88.5

The performance criteria are as follows: root mean-square error (RMSE), the root mean-square prediction error (RMPE), and the variance accounted for (VAF)

6.5 Coordinated–Distributed Prediction

The problem is to predict the traffic flow in a large-scale network over a horizon of about 30 min. As argued in Sect. 6.2, the approach consists of a coordinated–distributed algorithm to meet the practical constraints of traffic control centers.

The approach for the prediction problem consists of the following steps: (1) Partition the full network into two or more subnetworks, each of which has only a one-directional traffic flow. (2) Predict for all subnetworks the traffic flows on all road sections based on control systems for traffic flow and on predictions of the traffic inflows into those subnetworks from motorways and from on-ramps. (3) Take care of the consistency of the predictions produced by the subnetworks; thus, if the outflow of one subnetwork is to be equal to the inflow of another network, then take care that these traffic flows are approximately equal.

The choices of the first two steps described above are not discussed in detail. The prediction of traffic flow in a subnetwork is computed by simulation of the control system because there are available predictions of traffic inflows of motorways at the boundary points of the network and of on-ramps (see the previous section).

The main problem of coordinated–distributed prediction is as follows: How to achieve consistency of the traffic flow predictions computed in a distributed way by the subnetworks? Note that the outflow of one subnetwork may be the inflow of the next downstream network. The consistency of the equality of these traffic flows can be formulated as solving a fixed point equation for traffic flows. A method to solve this fixed point equation is the successive approximation method.

Definition 6.1 (*The fixed point method to compute traffic flow predictions in a road network*). Consider at time $t \in T$ the states $\hat{x}_j(t)$ of all subsystems $j \in J$, the traffic inflows from motorways $\{\hat{u}^{(0)}_{m,j}(t+1 : t+t_p),\ j \in J\}$, and the traffic inflows from on-ramps $\{\hat{u}^{(0)}_{on,j}(t+1 : t+t_p),\ j \in J\}$.

The definitions of the variables and of the formulas follow: $\hat{u}^{(0)}_{m,j}$ is the available motorway inflow of subnetwork $j \in J$ in the first iteration (iteration 0). $\hat{u}_{on,j}$ is the available motorway inflow from on-ramps of the same subnetwork. $\hat{x}_j(t)$ is the available estimated state of the same network. $\hat{z}^{(0)}_{m,j}$ is the computed prediction of the traffic flow of the motorway outflows of subnetwork $j \in J$ for the indicated prediction horizon, $t+1 : t+t_p$. $\hat{z}^{(0)}_m$ is the computed prediction of the traffic flows

Table 6.2 Comparison of computation time and communication time

Method	Computation (s)	Communication (s)
Traditional parallel simulation	0.002	0.0260
Centralized simulation	0.004	0
Proposed parallel simulation	Best 0.002	0.0001
	Worst 0.004	0.0001

of the motorway outflows of all subnetworks combined. The function $f_{h,net}$ is based only on the topography of the network, and it relates the outflows of any subnetwork to the inflow of another subnetwork if any; if the outflow is not connected to any inflow of a subnetwork, then the function equals the traffic inflow of the previous iteration. $\hat{u}_m^{(1)}$ is the computed motorway inflow of all subnetworks in the second iteration.

The computations are summarized by the following equations:

$$\hat{z}_{m,j}^{(0)}(t+1:t+t_p|t) = f_{h,j}(\hat{x}_j(t), \hat{u}_{m,j}^{(0)}(t+1:t+t_p|t), \hat{u}_{on}(t+1:t+p|t)); \tag{6.7}$$

$$\text{check if} \tag{6.8}$$

$$\|\hat{u}_m^{(0)}(t+1:t+t+t_p|t) - f_{h,net}(\hat{z}_m^{(0)}(t+1:t+t_p|t))\| < \varepsilon;$$

$$\text{if the condition is not met, then iterate } \hat{u}_m^{(0)} \mapsto \hat{u}_m^{(1)} \text{ etc.}$$

$$\hat{u}_m^{(1)}(t+1:t+t_p|t) = f_r(\hat{u}_m^{(0)}(t+1:t+t_p|t), f_{h,net}(\hat{z}_m^{(0)}(t+1:t+t_p|t))). \tag{6.9}$$

$$\begin{aligned}\hat{u}_m(t+1:t+t_p|t) = f_r(&\hat{u}_m(t+1:t+t_p|t),\\ &f_{h,\text{net}}(f_h(\hat{x}(t), \hat{u}_m(t+1:t+t_p|t), \hat{u}_{on}(t+1:t+t_p|t)))),\\ &\text{is the fixed point equation}\\ &\text{for the traffic inflows } \hat{u}_m(t+1:t+t_p|t).\end{aligned}$$

Call the system with the state transition function f_h with as components $f_{h,j}$, the *prediction system of motorway outflows*.

Due to an enforcement of equality of traffic demand and of supply of adjacent subnetworks not described above, the spillback of one subnetwork to an upstream subnetwork is correctly incorporated in the prediction algorithm.

A preliminary result is summarized in Table 6.2.

6.6 Further Research and Further Reading

Further research. (1) *Time and approximation complexity for coordinated prediction*. A comparison will be made of the computation time and the simulation accuracy for parallel prediction and for known alternatives. (2) *Partitioning of a road network into subnetworks*. (3) *Control*. Multilevel control of a large-scale road network is to be investigated. Further Reading. The adaptive prediction algorithm for traffic flow at the boundary of a network is described in the publications [11, 12]. The coordinated–distributed prediction algorithm is described in the conference papers [9, 10], while a journal publication is in preparation.

References

1. Bohlin T (1976) Four cases of identification of changing systems. In Mehra RK, Lainiotis DG (eds) System identification advances and case studies, vol 126 of Mathematics in Science and Engineering. Elsevier, Amsterdam, pp 441–518
2. Davarynejad M, Wang Y, Vrancken J, van den Berg J (2011) Multi-phase time series models for motorway flow forecasting. In: 14th international IEEE annual conference on intelligent transportation systems, October 2011
3. Hong WC, Pai PF, Yang SL, Theng R (2006) Highway traffic forecasting by support vector regression model with tabu search algorithms. In: IEEE International joint conference on neural networks, 2006 (IJCNN'06), pp 1617–1621
4. Klefstad R, Zhang Y, Lai M, Jayakrishnan R, Lavanya R (2005) A distributed, scalable, and synchronized framework for large-scale microscopic traffic simulation. In: IEEE Proceedings of intelligent transportation systems, 2005, pp 813–818, September 2005
5. Messer CJ (1993) Advanced freeway system ramp metering strategies for Texas. Research report, Texas Transportation Institute
6. Perumalla KS (2006) Parallel and distributed simulation: traditional techniques and recent advances. In: Proceedings of the 38th conference on Winter simulation (WSC'06), pp 84–95, December 2006
7. Robbins BA, Zavala VM (2011) Convergence analysis of a parallel newton scheme for dynamic power grid simulations. In: Proceedings of the first international workshop on High performance computing, networking and analytics for the power grid, HiPCNA-PG'11, pp 3–10, New York, NY, USA, 2011. ACM
8. Wang J, Ma Z-D, Hulbert GM (2003) A gluing algorithm for distributed simulation of multibody systems. In: Nonlinear dynamics, pp 159–188
9. Wang Y, van Schuppen JH, Vrancken J (2012) On-line distributed prediction of traffic flow in a large-scale traffic network. In: Proceedings of ITS World 2012
10. Wang Y, van Schuppen JH, Vrancken J (2013) Parallel traffic prediction with a nonlinear model. In: Proceedings ITS World Congress 2013, Tokyo
11. Yubin W, van Schuppen JH, Vrancken J (2014) Prediction of traffic flow at the boundary of a motorway network. IEEE Trans Intell Transp Syst 15:214–227
12. Wang Y, van Schuppen JH, Vrancken JLM (2011) Adaptive prediction of traffic flow into a motorway network. In: Proceedings of ITS World Congress 2011

Chapter 7
Zone-Control-Based Traffic Control of Automated Guided Vehicles

Qin Li, Jan Tijmen Udding and Alexander Pogromsky

7.1 Introduction and Motivation

Automated guided vehicles (AGV) normally mean mobile robots (or unmanned vehicles) used for transportation purposes. They were traditionally employed in manufacturing systems and have recently extended their popularity to many other industrial applications, such as goods transportation in warehouses and container transshipment at container terminals. See [5] for a comprehensive survey of the research on the design of AGV systems. Here is an example application of an AGV system to quayside (waterside) container transshipment at a container terminal. In this practice, when discharging a vessel, a couple of quay cranes (QCs) take the containers off the vessel and put them in the associated container buffers in the quay area (QA). Each container will be picked up by some AGV later on and transported across the transportation area (TA) to a container buffer of some yard stacker (YS) in the yard area (YA). There it will be put in a container stack by that YS. The other way around, in loading a vessel, each container is collected from a certain yard stack by a YS and transported to a QC buffer by some AGV (see Fig. 7.1).

Usually, traffic control is needed to resolve possible motion conflicts among the vehicles in an AGV system. The most popular and widely discussed traffic control strategy for AGV systems is called zone-control strategy, in which the guide-path network (we call it road network in this chapter) is composed of a number of zones. This strategy eases the avoidance of inter-vehicle collisions by demanding that each

Q. Li (✉)
Statoil Research Centre, Hydrovegen 67, 3936 Porsgrunn, Norway
e-mail: qinli01@gmail.com

J.T. Udding
Eindhoven University of Technology, Eindhoven, The Netherlands
e-mail: j.t.udding@home.nl

A. Pogromsky
Eindhoven University of Technology, P.O. Box 513, 5600 MB Eindhoven, The Netherlands
e-mail: A.Pogromsky@tue.nl

J.H. van Schuppen and T. Villa (eds.), *Coordination Control of Distributed Systems*,
Lecture Notes in Control and Information Sciences 456,
DOI 10.1007/978-3-319-10407-2_7

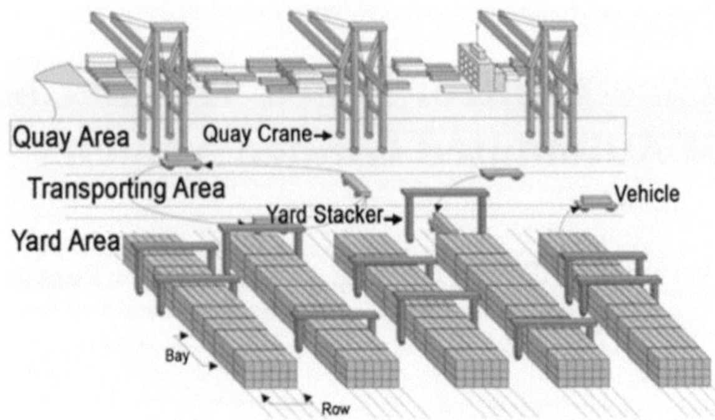

Fig. 7.1 Quayside container transportation at an automated container terminal

zone can be occupied by at most one vehicle. The key traffic control issue is then to keep the vehicles away from deadlocks caused by competing for the traffic resource: the zones. As pointed out by [5] and [3], the relatively large numbers of vehicles and zones in certain applications such as container terminals defy most existing deadlock avoidance approaches which rely on either algebraic operations or complex searches. See [1, 3, 4, 6] for some related works. In this chapter, we propose a time-efficient traffic control scheme based on a novel discrete-event zone-control model.

7.2 Discrete-Event Zone-Control Model

7.2.1 Building Blocks of the Road Network

Lane and zone: A lane is a finite sequence of zones. Vehicles moving on a lane must visit the zones according to their order in the sequence (so that the lane is unidirectional). We use c_k^i to denote the kth zone of the lane i. Particularly, we call the first zone and the last zone of a lane the *starting zone* (SZ) and *ending zone* (EZ) of the lane, respectively. In practice, physically, a zone is large enough to accommodate one vehicle.

A *depot* is a zone that can accommodate all vehicles. It needs to be emphasized that a depot is not a zone in any lane. Each depot is affiliated with at least one *entry lane* and one *exit lane*. Physically, an entry lane (resp. exit lane) of a depot is a lane that allows a vehicle to move into (resp. out of) the depot. We denote the set of all zones by $\mathscr{C}$.

Crossing: A crossing i is affiliated with a set of *in-lanes* and a set of *out-lanes*. Each in-lane of a crossing has at least one out-lane of the crossing as its *neighboring lane(s)*. Vehicles can move from an in-lane of a crossing on to any of its neighboring lanes by passing the crossing. A zone pair (c_1, c_2) is called a *crossing–passing zone pair of crossing* i if c_1 is the EZ of some in-lane, say lane j, of crossing i and c_2 is the SZ of a neighboring lane of lane j. We denote all the crossing–passing zone pairs of crossing i by $\mathscr{R}_i$. For each zone pair $(c_1, c_2) \in \mathscr{R}_i$, there is a subset (maybe empty) of $\mathscr{R}_i$, denoted by $\mathscr{X}_i(c_1, c_2)$, in which the zone pairs are called the *conflicting crossing–passing zone pairs of* (c_1, c_2) *at the crossing* i. Physically, if $(c_3, c_4) \in \mathscr{X}_i(c_1, c_2)$, then a vehicle passing the crossing i by moving from c_1 to c_2 can collide with another vehicle passing the same crossing by moving from c_3 to c_4. In view of this, for any crossing i, it is considered that $(c_3, c_4) \in \mathscr{X}_i(c_1, c_2)$ if and only if $(c_1, c_2) \in \mathscr{X}_i(c_3, c_4)$. For each crossing i, we call the set of EZs of all its in-lanes *at-crossing zones of the crossing* i. A zone c is called an *off-crossing zone* if it is not an at-crossing zone of any crossing.

Neighboring zone: The set of the neighboring zones of a depot consists of the SZs of all its exit lanes. The neighboring zone of a non-EZ zone c_k^i is c_{k+1}^i. The neighboring zone of the EZ of any entry lane of a depot is the depot. The neighboring zones of the EZ of an in-lane of a crossing are the SZs of all its neighboring lanes. We use Υ_c to denote the set of neighboring zones of zone c.

Regarding the layout of the road network, we make the following assumptions: (a) $c_{k_1}^{i_1} \neq c_{k_2}^{i_2}$ if either $i_1 \neq i_2$ or $k_1 \neq k_2$. (b) Each lane is either an in-lane of a unique crossing or an entry lane of a unique depot, but cannot be both. (c) Each lane can be either an exit lane of at most one depot or an out-lane of at most one crossing, but cannot be both the exit lane of a depot and the out-lane of a crossing. (d) Each lane has at least two zones if it is not an entry lane or exit lane of a depot. In these assumptions, in fact, (a) says that each lane neither self-intersects nor intersects with any other lane; (b) implies that each lane connects to a depot or to another lane via a crossing; (d) is key to the deadlock avoidance by the traffic control presented later. Also note that (c) does not exclude the case that some lane is neither an exit lane of any depot nor an out-lane of any crossing. The assumptions guarantee that each zone has at least one neighboring zone.

7.2.2 Vehicle States and Events

Each vehicle can have two types of states: "in c_1 (with the next zone c_2)" and "moving from c_1 to c_2 (with the next zone c_2)," where c_1 can be any zone in $\mathscr{C}$ and c_2 is one of the neighboring zones of c_1 (i.e., $c_2 \in \Upsilon_{c_1}$). In the former case, we write that the state of a vehicle is (c_1, c_1, c_2), while in the latter, it is (c_1, c_2, c_2). (It can be shown that $c_2 \neq c_1$ if $c_2 \in \Upsilon_{c_1}$, so that no confusions arise when using these notations.) We say that a vehicle *occupies* the zone c_1 (resp. both c_1 and c_2) if its state is (c_1, c_1, c_2) [resp. (c_1, c_2, c_2)]. A zone is said to be *available* if it is not occupied by any vehicle. The set of all possible vehicle states is the union of the two sets: $\mathscr{S}_1 := \{(c_1, c_1, c_2) \in \mathscr{C}^3 : c_2 \in \Upsilon_{c_1}\}$ and $\mathscr{S}_2 := \{(c_1, c_2, c_2) \in \mathscr{C}^3 : c_2 \in \Upsilon_{c_1}\}$.

Each vehicle is given an initial state. As the system runs, the state of each vehicle changes only at the occurrences of *vehicle events*. There are two vehicle events: "leave" and "arrival," denoted in short by LEA and ARR, respectively. The following simple rules specify the *feasible events* for any vehicle state s and the state transitions they lead to

(a) if $s = (c_1, c_1, c_2) \in \mathscr{S}_1$, then the only feasible event for s is the LEA, which changes s to be $(c_1, c_2, c_2) \in \mathscr{S}_2$;
(b) if $s = (c_1, c_2, c_2) \in \mathscr{S}_2$, then the only feasible event for s is the ARR, which changes s to be $(c_2, c_2, c_3) \in \mathscr{S}_1$, where c_3 can be any zone in Υ_{c_2}.

We say that, in (a), the vehicle *leaves* c_1 by the LEA; in (b), the vehicle *arrives at* c_2 by the ARR. The full freedom of selecting the zone c_3 in (b) indicates that there are no constraints on routing vehicles for zone-to-zone traveling. Furthermore, a route of a vehicle can be either prefixed or established online. For example, a fixed route $c_1, c_2, \ldots, c_n$ may be preassigned to a vehicle that is initially in c_1; then, to follow the route, the vehicle should trigger LEA and ARR events in an alternate fashion such that it leaves c_1, arrives at c_2, leaves c_2, arrives at c_3, and so forth. The vehicle, on the other hand, may only know the next zone c_2 on its way when it is in c_1; before arriving at c_2, the vehicle can pick any zone c_3 out of Υ_{c_2} to be the next zone after c_2 on its route. In both cases, there is exactly one feasible event for each vehicle at any time.

It must be emphasized that an event is feasible for some vehicle state does not mean that the event is always allowed. Indeed, the traffic rules that we will present in the following are used to judge if a feasible event can be permitted without causing traffic problems, i.e., inter-vehicle collisions and deadlocks. Roughly speaking, the traffic rules say that each vehicle is free to trigger an ARR yet must avoid colliding with another vehicle as well as ending up within a cycle of vehicles when it intends to trigger a LEA.

7.3 Collision and Deadlock Avoidance

To specify the traffic rules, we need to address different types of LEA events. A LEA is called an *at-crossing LEA* (at the crossing k) if it causes a vehicle to leave some at-crossing zone (of the crossing k). In this case, we also say that the vehicle triggers a (an at-crossing) LEA to *pass* the crossing k. A LEA is called an *at-depot LEA* if it causes a vehicle to leave a depot.

7.3.1 Inter-vehicle Collision Avoidance

A primary goal of the traffic control is to prevent vehicles from colliding with each other. First, of course, we need to define no-collision cases using our zone modeling language. An AGV system is said to be collision free if the following two conditions are satisfied:

(C1) Each non-depot zone is occupied by at most one vehicle.
(C2) If two vehicles are with the states (c_1, c_2, c_2) and (c_3, c_4, c_4), respectively, then $(c_1, c_2) \notin \mathscr{X}_i(c_3, c_4)$ for any crossing i.

The condition C2 reflects our motivation for defining the notion of "conflicting crossing–passing zone pairs" (see Sect. 7.2).

The following traffic rules are sufficient to guarantee that an AGV system is collision free if the two conditions C1 and C2 are satisfied initially:

Rule 1: For each depot, at most one vehicle is allowed to trigger an at-depot LEA at the depot at any time.

Rule 2: For any crossing, there is at most one vehicle that can trigger a LEA to pass the crossing at any time.

Rule 3: If the vehicle has the state (c_1, c_1, c_2), with c_1 an off-crossing zone, then the LEA, which changes the state of the vehicle to (c_1, c_2, c_2), is allowed to be triggered if c_2 is available.

Rule 4: If the vehicle has the state (c_1, c_1, c_2), with c_1 an at-crossing zone of the crossing i, then the LEA, which changes the state of the vehicle to (c_1, c_2, c_2), is allowed to be triggered if (1) c_2 is available and (2) there is no vehicle with the state (c_3, c_4, c_4), where $(c_3, c_4) \in \mathscr{X}_i(c_1, c_2)$.

Rules 1 and 2 are imposed to exclude two vehicles occupying the same SZ of some exit lane of a depot or some out-lane of a crossing.

These two rules can be realized by requiring any vehicle to hold exclusively a unique local crossing (resp. depot) token whenever triggering an at-crossing (resp. at-depot) LEA at the crossing (resp. depot) and to release that token after the triggering was successful or was prohibited by the traffic rules. Rule 3 and the first part of Rule 4 say that a vehicle cannot move to its next zone if the zone is occupied. The second part of Rule 4 prevents the triggering of an at-crossing LEA that leads to conflicting passings of the crossing.

7.3.2 Deadlock Avoidance

Consider now a generic operation scenario of the AGV system: each vehicle is designated to visit a series of zones and eventually parks in some depot and does not move on. Thus, each operational vehicle can terminate its operation after triggering a finite number of events.

We say that a vehicle is *operational* if it still has the intention to trigger a new vehicle event (because, for example, it has not finished its given route) in finite time. An operational vehicle is said to be *blocked* if the intended event of the vehicle is not allowed to be triggered in finite time. It is easy to show that if, at any time, at least one operational vehicle is not blocked, then all vehicles can terminate their operations (as they all involve a finite number of events). This observation motivates us to give the definition of deadlock of the AGV system as follows: The AGV system has a *deadlock* if all operational vehicles are blocked. Now, it should be clear that all vehicles can terminate their operations if the AGV system is free of deadlocks.

It turns out that to prevent deadlocks, it is key to ensure the absence of black cycles defined below (together with the operation graph) which bear the potential of inter-blocking of a chain of vehicles. An *operation graph* is a time-dependent directed graph, where the vertex set is composed of all zones in the road network and the arc set consists of such zone pairs (c_1, c_2) that either $c_2 \in \Upsilon_{c_1}$ if c_1 is an off-crossing zone or c_1 and c_2 are, respectively, the current zone and next zone of some operational vehicle if c_1 is an at-crossing zone. A *cycle* in the operation graph is a sequence of vertices $c_1, c_2 \cdots c_n, c_1$, $n \geq 2$, such that $c_1, c_2, \cdots, c_n$ are all distinct and (c_1, c_2), (c_2, c_3), $\cdots$, (c_{n-1}, c_n), (c_n, c_1) all belong to the arc set. A *black cycle* in the operation graph is a cycle with the property that for each vertex (zone) contained in the cycle, there is a vehicle either in it or moving from some zone to it.

Here comes our main claim: An AGV system is guaranteed free of collisions and deadlocks if initially the conditions C1 and C2 hold and the operation graph has no black cycle, and the occurrence of the LEAs is governed by Rules 1, 2′, 3, and 4′, where the Rules 2′ and 4′ are as follows:

Rule 2′: There is at most one vehicle that can trigger an at-crossing LEA at any time.

Rule 4′: If the vehicle has the state (c_1, c_1, c_2), with c_1 an at-crossing zone of crossing i, then the LEA, which changes the vehicle state to (c_1, c_2, c_2), is allowed to be triggered if (1) c_2 is available; (2) there is no vehicle with the state (c_3, c_4, c_4), where $(c_3, c_4) \in \mathscr{X}_i(c_1, c_2)$; and (3) this will not lead to a black cycle containing c_2.

Compared with Rule 2, Rule 2′ is more strict and can be realized by imposing the use of a *global crossing token*; i.e., each vehicle that intends to trigger an at-crossing LEA must request the crossing token, check with Rule 4′ while holding the token, and release the token after the checking (no matter whether the LEA is allowed to be triggered or not).

7.4 Characteristics of the Traffic Control

The most appealing characteristic of the proposed traffic control, as aforementioned, is that it imposes no restrictions on routing the vehicles. The route of each vehicle can be either pregiven or constructed online. This property not only benefits the throughput optimization of the system but also helps deal with unexpected disturbances; e.g., if some obstacle or vehicle breakdown appears, the (other) vehicles can be re-routed to bypass it.

Regarding implementation, the traffic control has the following two merits:

1. It has a small time complexity and thus is suitable for real-time executions. In fact, we have proposed a time-efficient algorithm for the black cycle prediction

required in Rule 4′. The time complexity of the algorithm depends quadratically on the number of vehicles.
2. It can be implemented in an almost distributed mode, demanding moderate data communication and slight intervention from a central controller. The central controller is needed here mainly because of the crossing token assignment.

7.5 Application of the Traffic Control to the Example Application

Let us consider the container transshipment scenario described at the beginning of this note; suppose that a road network with zones and crossings defined as in Sect. 7.2 covers the workspace of the AGVs (i.e., the combined area of the QA, TA, and YA in Fig. 7.1), with the buffer of each QC or YS modeled as one or more zone(s). In this setting, each AGV can be seen as assigned a sequence of tasks, each of which is a transportation of a container from one zone to another. Note that this operation scenario of the AGVs becomes a special case of what we described at the beginning of Sect. 7.3.2 if the last task of each AGV ends up with parking in some depot. Then, by applying the traffic control presented above, it is guaranteed that each vehicle will eventually finish all the given tasks without colliding with others (The tasks can be prefixed off-line or online scheduled).

The application is discussed in detail in [2], where we also present a routing algorithm and a layout design of the road network based on the practical setting of a container terminal. Computer simulations have demonstrated the efficiency of the resulted AGV system in terms of a set of practical performance measures.

7.6 Further Topics

There are several interesting topics for further research on the proposed event-based zone-control strategy. First, in our traffic control strategy, it is required that each lane of the road network must contain at least two zones if it is neither an entry lane nor exit lane of a depot. To somehow weaken this constraint may be important for some applications where the workspace of the AGVs is very limited. Secondly, to improve the efficiency of the AGV system, effort is deserved to relax the token-holding requirement in the traffic rules so that multiple vehicles can leave different at-crossing zones simultaneously. As far as we could foresee, it may lead to a successful trial to use local crossing tokens (instead of a global one required in this work) together with some inter-crossing communication and coordination mechanisms. Another interesting work would be to equip the AGV system with some fault-tolerant mechanisms so that it can still run safely and smoothly in the presence of unexpected events. Some results of this type would appear in our later papers.

7.7 Further Reading

The interested reader is advised to read [5] for a good survey on the design and control of AGV systems and [1, 3, 6] for the most related works to this chapter.

References

1. Fanti MP, Maione B, Mascolo S, Turchiano B (1997) Event-based feedback control for deadlock avoidance in flexible production systems. IEEE Trans Robot Autom 13:347–363
2. Li Q, Adriaansen AC, Udding JT, Pogromsky AY (2011) Design and control of automated guided vehicle systems: a case study. In: Proceedings of the 18th world congress of the IFAC, pp 13852–13857
3. Rajeeva LM, Wee HG, Ng WC, Teo CP, Yang NS (2003) Cyclic deadlock prediction and avoidance for zone-controlled agv system. Int J Prod Econ 83:309–324
4. Reveliotis SA (2002) Conflict resolution in agv systems. IIE Trans 32:647–659
5. Vis IFA (2006) Survey of research in the design and control of automated guided vehicle systems. Eur J Oper Res 170:677–709
6. Yeh MS, Yeh WC (1998) Deadlock prediction and avoidance for zone-control agvs. Int J Prod Res 36:2879–2889

Chapter 8
Coordination Control of Complex Machines

Jos C.M. Baeten, Bert van Beek, Jasen Markovski and Lou J.A.M. Somers

8.1 Background and Motivation

In the last few decades, control software development has taken a more central role due to the ever-increasing complexity of the machines, demands for higher quality and performance, and improved safety and ease of use [1]. Traditionally, the control software requirements are formulated informally in some sort of specification documents by domain engineers, to be translated into control software by software engineers. This is a time-consuming and an error-prone process, since control requirements are often ambiguous and change frequently, so the produced software needs to be validated against the machine. If validation fails, the code must be rewritten, leading to a *define–validate–redefine loop*.

This issue in control software design gave rise to supervisory control theory [2–5], where models of high-level supervisory controllers, referred to as *supervisors*, are synthesized automatically based upon a formal model of the uncontrolled system, known as *plant*, and a model of *the control requirements*. Supervisory controllers observe the discrete-event behavior of the system and make control decisions based on the observed information. The supervisory controller synthesis problem is to achieve the greatest possible allowed behavior, which is specified by the control requirements, based on a given observable behavior.

J.C.M. Baeten · B. van Beek (✉) · J. Markovski · L.J.A.M. Somers
Eindhoven University of Technology, P.O. Box 513, 5600 MB Eindhoven, The Netherlands
e-mail: d.a.v.beek@tue.nl

J.C.M. Baeten
e-mail: j.c.m.baeten@tue.nl

J. Markovski
e-mail: j.markovski@tue.nl

L.J.A.M. Somers
e-mail: l.j.a.m.somers@tue.nl

J.H. van Schuppen and T. Villa (eds.), *Coordination Control of Distributed Systems*,
Lecture Notes in Control and Information Sciences 456,
DOI 10.1007/978-3-319-10407-2_8

By applying automated control software generation based upon formal behavioral models, we aim to:

1. Shift the focus of software engineers from writing and debugging code to modeling the plant and the control requirements.
2. Increase the use of models. Software engineers need not think in terms of code, since it is generated automatically. Instead, they make the formal models in terms of behavior, thus improving their communication with the domain engineers and other involved parties.
3. Shorten the design loop. Adjustments in the control specifications result in model changes, followed by automated synthesis of control software, which is likely to be less time-consuming than directly adapting control code.
4. Use the formal behavioral models for simulation. We can validate the supervisor before expensive prototypes are built increasing confidence in the design. It also provides opportunity for verification, performance analysis, and reliability analysis.
5. Reuse of models and improved evolvability of the process. When developing a new product, models can be reused more easily than specific code since small adjustments are incorporated more easily. Furthermore, the changes in the control requirements are more easily managed as we directly synthesize probably correct models.

Reflecting on the points above, we aim to increase product quality by dealing with increased machine complexity more easily; reduce costs by validating controllers before expensive prototypes are in place; and reduce time to market by improving communication between the engineers and shortening the design loop.

8.2 Model-Based Systems Engineering Framework

To structure the process of supervisory control synthesis, we employ the framework depicted in Fig. 8.1 [6, 7], which is a refinement and extension of the framework defined in [8]. Domain engineers define the desired overall system specification (Specification *Controlled System* in Fig. 8.1), that is later elaborated into a design document (Design *Controlled System*) by domain and software engineers together. The design defines the architecture of the system and its composition into subsystems. In this case, we consider as subsystems the controller and uncontrolled systems, the latter also referred to as ‘plant.’

The subsystems are likewise (informally) specified, resulting in the documents Specification *Control Requirements* and Specification *Plant*, in Fig. 8.1. The supervisory control synthesis framework then facilitates formal specification of control requirements (Model *Control Requirements*), instead of specification of a controller. Supervisory control synthesis also requires a discrete-event model of the uncontrolled system (plant).

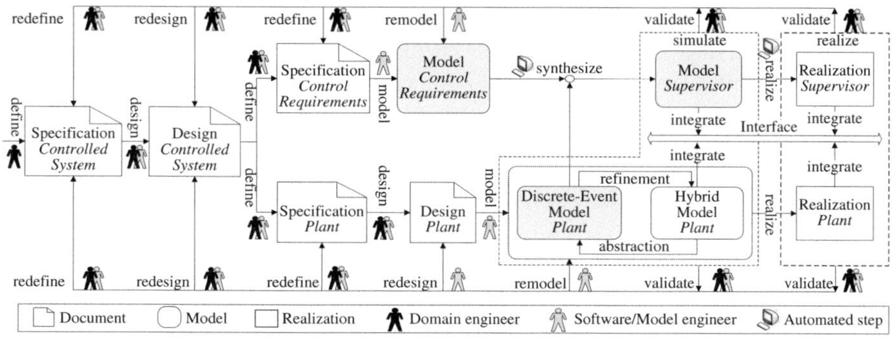

Fig. 8.1 Model-based systems engineering framework for supervisory control synthesis

Plants typically contain hybrid (continuous and discrete-event) behavior suitable for simulation. For synthesis purposes [2, 5], continuous behavior is abstracted from, i.e., a pure high-level discrete-event model is derived. Alternatively, a discrete-event model can be made and subsequently refined to a hybrid model.

The synthesis algorithm generates a minimally restrictive supervisor, based on the models of the control requirements and plant. Such a supervisor, by construction, satisfies the control requirements and is nonblocking, which means that from any reachable state, always a marked state can be reached. Many different kinds of supervisory control synthesis algorithms exist (see for example [2, 5, 9, 10]).

The model of the generated supervisor can then be combined with the (hybrid) model of the plant, synchronizing on events and location names, for early, model-based validation of the controlled system. Such a coupling is possible as the supervisor only considers the discrete-event behavior of the system, which is orthogonal to the continuous dynamics. This validation should ensure that the observed controlled system behavior satisfies the system behavior and control requirements as specified in the informal documents of Fig. 8.1. If validation fails, remodeling the control requirements or even complete revision proves necessary.

As a last step, the control software is generated automatically from the validated models. Our framework also supports other options for early integration, such as hardware-in-the-loop simulation, by coupling of the realization of the supervisor with the hybrid real-time simulation model of the plant.

The tooling used for supervisory control synthesis was based on the first version of the Compositional Interchange Format (CIF 1) [11]. For simulation-based validation, CIF 2 [12] was used. CIF is an expressive formalism, with a formal semantics, based on hybrid automata. CIF interfaces with many different tools by means of model transformations. Recently, the formal semantics and modeling concepts have been considerably simplified, resulting in CIF 3 [13]. The CIF toolset is now implemented in Eclipse [14], using Java as implementation language, and scalable vector graphics (SVG) [15] for interactive, simulation-based visualization. Recent improvements to the CIF 3 development environment are an Eclipse editor with integrated real-time background parsing and type checking. Furthermore, the CIF 3 simulator has been redesigned for fast simulation by employing run-time code generation.

8.3 Case Study: High-Tech Printers

We illustrate the modeling process on a case study involving coordination of maintenance procedures of a printing process of a high-end Océ printer of [16]. Due to confidentiality concerns, we can only present an obfuscated part of the case study.

We abstractly depict a printing process function in Fig. 8.2, where the control architecture of the printer is given to the left. We coordinate the function responsible for the maintenance of the printing process. The printer executes print jobs in run mode of operation, whereas several maintenance operations to preserve print quality have to be carried out in standby mode. Maintenance operations are scheduled based on the amount of pages printed since the last maintenance. Soft deadlines denote that a maintenance *can* be scheduled, and hard deadlines denote that the maintenance *must* be scheduled. Maintenance procedures with expired soft deadlines can be postponed if there is an ongoing print job, but hard deadlines must be respected.

A printing process function comprising one maintenance operation is depicted in Fig. 8.2. The supervisory control problem is to synthesize a model of the Status Procedure, which is responsible for coordinating the other procedures given input from the controllers. The plant that models the printing process function is given in Fig. 8.3. For modeling, we employ state-labeled finite automata. The state labels are employed to keep track of the state of the plant and can be referenced in the control requirements. We employ two types of state-based control requirements [6]: ϕ and $\mathrm{e} \implies \phi$, for some event e, and some Boolean formula ϕ over the state labels, using the logical operators $\wedge$, $\vee$, $\neg$, and $\implies$. The first requirement type specifies an invariant ϕ that must hold for every state of the supervised system, whereas the second type specifies necessary conditions, given by ϕ, for enabling event e.

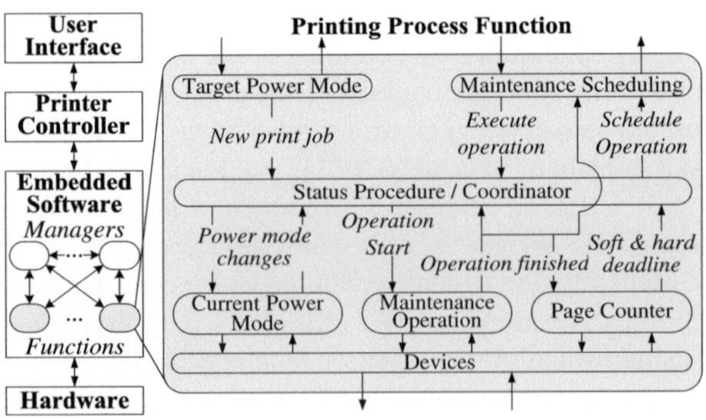

Fig. 8.2 Printing process function

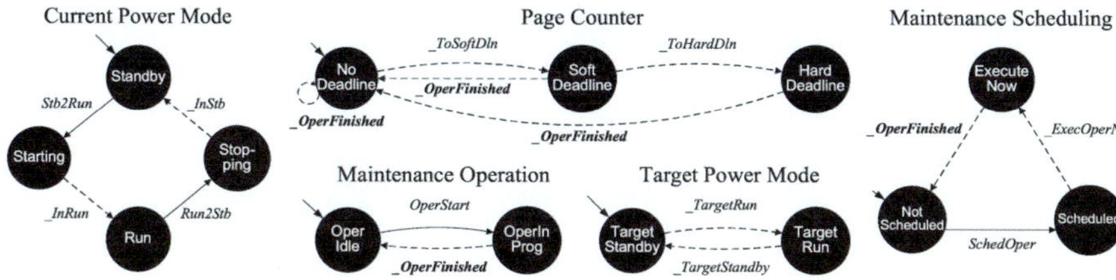

Fig. 8.3 Printing process plant

Uncontrollable events are underscored. Initial states have incoming arrows, whereas marked states, which specify states in which the system is considered to have successfully completed some task [2], coincide with the initial states, as we are dealing with a reactive system. The plant is formed by the synchronization of the automata in Fig. 8.3. Current power mode sets the power mode to run or standby, using *Stb2Run* or *Run2Stb*, respectively, and sends back feedback by employing *_InRun* and *_InStb*, respectively. Maintenance operation either carries out a maintenance operation, started by *OperStart*, or is idle. The confirmation is sent back by the event *_OperFinished*, which synchronizes with maintenance scheduling and page counter. Page counter announces when soft or hard deadlines are reached using *_ToSoftDln* and *_ToHardDln*, respectively. The page counter is reset, triggered by the synchronization on *_OperFinished*, each time maintenance is finished. The controller target power mode sends signals regarding incoming print jobs to Status Procedure. The event *_TargetRun* should set the printing process to run mode for printing. When the print job is finished, the event *_TargetStandby* is activated. Maintenance scheduling receives a request for maintenance with respect to expiration of page counter from Status Procedure, by the event *SchedOper*, and forwards it to the manager. The manager confirms the scheduling with the other functions and sends a response back to the Status Procedure, using *_ExecOperNow*. It also receives feedback from Maintenance Operation that the maintenance is finished in order to reset the scheduling, again triggered by *_OperFinished*.

The coordination is performed according to the following requirements: (1) maintenance operations can be performed only when printing process function is in standby; (2) maintenance operations can be scheduled only if a soft deadline has been reached and there are no print jobs in progress, or a hard deadline has passed; (3) only scheduled maintenance operations can be started; and (4) the power mode of the printing process must follow the power mode dictated by the managers, unless overridden by a pending maintenance operation. For a detailed account of the model-based systems engineering process and specification and formalization of the control requirements, we refer to [16].

(1) To model this requirement, we consider the states from current power mode and Maintenance Operation, and we require that it must always hold (R1) $OperInProg \Rightarrow Standby$.
(2) The states labeled by *SoftDeadline* and *HardDeadline* indicate when soft and hard deadlines are reached, respectively. State *TargetRun* of target power mode

states that there is a print job in progress. The event *SchedOper* is responsible for scheduling maintenance procedures. We specify the requirement as (R2) $SchedOper \Rightarrow (SoftDeadline \wedge \neg TargetRun) \vee HardDeadline$.

(3) The maintenance operation can be started when the maintenance scheduling is completed, which is modeled as (R3) $OperStart \Rightarrow ExecuteNow$.

(4) The last condition is modeled by two separate requirements for switching from run to standby mode and vice versa. We can change from run to standby mode if this is required by the manager, i.e., identified by state *TargetRun*, and there is no need to start a maintenance operation, identified by $\neg ExecuteNow$. The transitions labeled by *Stb2Run* are enabled by (R4) $Stb2Run \Rightarrow TargetRun \wedge \neg ExecuteNow$. In the other direction, we have (R5) $Run2Stb \Rightarrow TargetStandby \vee ExecuteNow$.

Finally, we synthesize a supervisor by employing the plant of Fig. 8.2 and the control requirements (R1)–(R5).

8.4 Research Contributions

The study of the informal specification documents revealed that engineers use a state-based approach, i.e., they give relations when a certain activity may be performed with respect to the state of the machine. Unfortunately, the corresponding synthesis tool [4] requires as input the exact opposite, i.e., the modeler has to specify which behavior is undesired, in the form of a negation of a conjunction of automaton locations. The latter leads to less intuitive specifications and results in a large number of control requirements.

To improve the modeling process and support greater modeling convenience, we generalize the control requirements to enable unrestricted use of propositional logic that we found in the specification documents. Thereafter, we automatically translate the generalized control requirements to an input suitable for the synthesis tool, which is a structurally restricted conjunctive normal form of the control requirements. This enabled us to specify the complete set of coordination rules for the printing process. For a detailed discussion of this transformation, including a formal definition, proof of correctness, and details on the implementation, we refer the interested reader to [6]. As an illustration of the effectiveness of our method, a part of the case study modeled by 23 generalized control requirements resulted in 500+ requirements in the original form. Admittedly, if one were to model the system in the original setting, a more optimal approach might have been possible, but then the modeling process would be to cleverly specify coordination rules, in an unintuitive form, which is both a time-consuming and an error-prone modeling process.

The issues of Fig. 8.2 were resolved by the supervisory controller, validated by means of simulation, and tested via a prototype implementation of the control software. Additionally, we could specify forms of state-based expressions which do not fit the supervisory control format: Instead of state-event *exclusion* expressions of the form e $\Longrightarrow \phi$, we also used state-event *inclusion* expressions of the form $\phi \Longrightarrow$ e. Since e $\Longrightarrow \phi$ is equivalent to $\neg\phi \Longrightarrow \overset{e}{\nrightarrow}$, the state-event exclusion predicate defines a state $\neg\phi$ where the event *e* is disabled, whereas the state-event inclusion defines a state where the event is enabled. The state-event inclusion predicates can be interpreted as verification properties of plant functionalities, which enables us to combine synthesis and verification. Ongoing research investigates the use of so-called reactive supervisor synthesis that aims to verify that desired behavior will be present in the supervised system during the synthesis procedure. Namely, supervisor synthesis caters only for safety properties, whereas we aim to guarantee progress properties of the supervised system, which ensure that the desired functionality is preserved.

We find that employing formal models is a key element for successful application of a synthesis-centric systems engineering process. Model-based specifications are consistent and less ambiguous than informal specification documents, forcing the engineers to clarify all aspects of the system. The proposed framework most importantly affects the control software development process, switching the focus from interpreting requirements, coding, and testing, to analyzing requirements, modeling, and validating the behavior of the system.

8.5 Further Research

Despite being able to show a proof of concept and to generate control software for a prototype of a future high-tech printer, the proposed approach needs further improvement, and we foresee the need for advancement in several important aspects. Techniques are needed that can directly synthesize supervisors for plants incorporating data, somewhat mitigating the state explosion problem. The control requirements should be reinforced with specific efficiently computable liveness or progress properties, related to the aforementioned reactive supervisor synthesis. Finally, an investigation is needed into suitable software and hardware architectures for automatic control software synthesis (as existing implementations vary per case), fitting the synthesized software in existing environments tailored for manual control software development. By working on and answering the above challenges, we hope to provide valuable model-based development techniques and tools for the software engineers of the future.

References

1. Leveson NG (1990) The challenge of building process-control software. IEEE Softw 7(6): 55–62
2. Cassandras C, Lafortune S (2008) Introduction to discrete event systems, 2nd edn. Springer, Berlin

3. Wonham WM (2012) Supervisory control of discrete-event systems. Department of Electrical and Computer Engineering, University of Toronto, http://www.control.toronto.edu/DES/
4. Ma C, Wonham WM (2005) Nonblocking supervisory control of state tree structures. Lecture notes in control and information sciences, vol 317. Springer, Berlin
5. Ramadge PJ, Wonham WM (1987) Supervisory control of a class of discrete event processes. SIAM J Control Optim 25(1):206–230
6. Markovski J, van Beek DA, Theunissen RJM, Jacobs KGM, Rooda JE (2010) A state-based framework for supervisory control synthesis and verification. In: Proceedings of CDC 2010. IEEE, pp 3481–3486
7. Schiffelers RRH, Theunissen RJM, van Beek DA, Rooda JE (2009) Model-based engineering of supervisory controllers using CIF. In: Proceedings of the 3rd international workshop on multi-paradigm modeling, electronic communications of the EASST, vol 21. Denver, pp 1–10
8. Braspenning NCWM, van de Mortel-Fronczak JM, Rooda JE (2006) A model-based integration and testing method to reduce system development effort. Electron Notes Theor Comput Sci 164:13–28
9. Chuan M, Wonham WM (2006) Nonblocking supervisory control of state tree structures. IEEE Trans Autom Control 51(5):782–793
10. Su R, van Schuppen JH, Rooda JE (2010) Aggregative synthesis of distributed supervisors based on automaton abstraction. IEEE Trans Autom Control 55(7):1627–1640
11. van Beek DA, Reniers MA, Schiffelers RRH, Rooda JE (2007) Foundations of an interchange format for hybrid systems. In: Bemporad A, Bicchi A, Butazzo G (eds) Hybrid systems: computation and control, 10th international workshop. Lecture notes in computer science, vol 4416. Springer, Pisa, pp 587–600
12. Nadales Agut DE, van Beek DA, Rooda JE (2013) Syntax and semantics of the compositional interchange format for hybrid systems. J Logic Algebraic Program 82(1):1–52
13. Systems Engineering Group TU/e (2013) CIF toolset, http://cif.se.wtb.tue.nl
14. The Eclipse Foundation (2013) Eclipse, http://www.eclipse.org/
15. W3C (2013) W3C SVG working group, http://www.w3.org/Graphics/SVG/
16. Markovski J, Jacobs KGM, van Beek DA, Somers LJAM, Rooda JE (2010) Coordination of resources using generalized state-based requirements. In: Proceedings of WODES 2010. IFAC, pp 300–305

Chapter 9
Coordinated Model-Predictive Control for Avoiding Voltage Collapse in an Electric Power Transmission Net

René Boel, Mohammad Moradzadeh and Lieven Vandevelde

9.1 Motivation

In this essay, we introduce the coordinating model-predictive control (CMPC) paradigm for coordination of interacting feedback controllers. The goal is to achieve global stability, or meeting global specifications, of the network of interacting dynamical systems with very limited requirements on the communication between local control agents (CAs), and without requiring detailed global model knowledge. If the local CAs act independently, compensating for perturbations observed in their local area, these actions might cause further perturbations in neighboring areas which the neighboring CA reacts to, and so forth, eventually causing a destabilizing positive feedback loop, leading to collapse of the whole system. In order to avoid such a global collapse, the local CAs must have some way of anticipating not only how the variables in their local area will react to the "locally observed" perturbation and to the planned local control actions, but also how the input port variables that depend on actions at the neighboring components will evolve in the near future. This is what can be achieved if the CAs exchange information about their future control actions.

Consider a dynamical system represented by a graph $\mathscr{G} = (\mathscr{V}, \mathscr{E})$ of interacting hybrid dynamical systems. Each of the $\sharp\mathscr{V} = M$ nodes of $\mathscr{G}$ behaves as an input–output hybrid automaton HA_m, $m = 1, \ldots, M$, controlled by the local control agent

R. Boel (✉)
SYSTeMS Research Group, Ghent University, Technologiepark-Zwijnaarde 914, 9052 Gent, Belgium
e-mail: rene.boel@ugent.be

M. Moradzadeh · L. Vandevelde
Electrical Energy Laboratory, Ghent University, Sint-Pietersnieuwstraat 41, 9000 Gent, Belgium
e-mail: mohammad.moradzadeh@ugent.be

L. Vandevelde
e-mail: lieven.vandevelde@ugent.be

J.H. van Schuppen and T. Villa (eds.), *Coordination Control of Distributed Systems*, Lecture Notes in Control and Information Sciences 456,
DOI 10.1007/978-3-319-10407-2_9

CA_m, $m = 1, \ldots, M$. The $\sharp\mathcal{E} = L$ edges $\text{Link}_\ell = \{i_\ell, j_\ell\}$, $\ell = 1, \ldots, L$ connect the neighboring nodes HA_{i_ℓ} and HA_{j_ℓ} (we abuse notation by using the same notation HA_m for a node in $\mathcal{G}$ and for its input-output model). Link_n defines port variables $P_\ell(t)$ which at each instant in time t take the same value in HA_{i_ℓ} and HA_{j_ℓ}. Typically, $P_\ell(t)$ is a vector of pairs of complimentary variables (such as voltage and current, or force and speed), where one edge of the link determines one element of the pair, and the other edge determines the complimentary variable. These edges Link_n thus define the interactions between the dynamical systems at the nodes of $\mathcal{G}$.

Control agent CA_m (acting as a discrete time or as a discrete event controller) selects at time t_k the next value u_k for the controllable input. A simple anticipating controller CA_m would implement a model-predictive control (MPC) anticipating the effects of its control actions by taking into account at time t_k all currently available local observations (the history of all data collected by sensors in node HA_m up to time t_k) and using the detailed model HA_m. Without communication among neighboring CAs, information on the future evolution of input port variables $P_\ell(t)$ of HA_m is completely ignored, leading to predictions that may be completely wrong. The future behavior of these port variables depends directly on the planned control actions of neighboring CAs, and indirectly also on the planned future actions of the CAs of the neighbors of these neighbors, and so on.

Clearly, the local predictions of CA_m will be improved if CA_m has some information on the future behavior of the port variables $P_\ell(t), \exists j \ni \text{Link}_\ell = \{j, m\}$. Bandwidth limitations and unreliability of communication links as well the inevitable frequent changes in the model of the network (the dynamics HA_j of the neighboring vertices, and even the interconnectivity may change from time to time) imply that a robust control design must avoid dependence on detailed models of distant parts of the network and that messages should be kept as simple as possible. The CMPC paradigm tries to use the idea of neighboring CAs exchanging information only on their planned control actions in such a way as to be robust against modeling and communication errors.

In order to develop a coordinating and anticipating control paradigm that is robust against model uncertainty, we therefore assume that detailed knowledge is available only for the model of the local node, while less and less detailed information is used for more and more distant nodes. Formally, consider the sets of indices $\mathcal{N}_i = \{j \mid (v_i, v_j) \in \mathcal{E},\ j \neq i\}$ corresponding to the immediate neighbors, and $\mathcal{R}_i = \mathcal{V} \backslash \{i \bigcup \mathcal{N}_i\}$ representing the remote/distant areas. Each CA_i, $i \in \{1, \ldots, M\}$ is implemented in this essay as a local MPC controller MPC_i, $i \in \{1, \ldots, M\}$, with control horizon N and prediction horizon $H \geq N$, updating the planned control sequence at discrete points in time $t_k = k \cdot \Delta$. The predicted sequence of local state variables and the planned sequence of control values at times $k + h$, $h \in \{1, \ldots, H\}$ for MPC_m, $m = 1, \ldots, M$, based on the information available at time instant k is denoted, respectively, by $\mathbf{x_m}$ and $\mathbf{u_{i,m}}$, where

$$\mathbf{x}_m(k) = \{x_m(k+1|k), \ldots, x_m(k+h|k)\}$$

$$\mathbf{u}_m(k) = \{\underbrace{u_m(k|k), \ldots, u_m(k+h-1|k)}_{N}, \underbrace{0, \ldots, 0}_{H-N+1}\}$$

The planned control actions for $h \in \{N+1, \ldots, H\}$ are denoted as 0 indicating that some other, simpler approach is used for selecting $u_m(k)$ at these times (e.g., do nothing).

The anticipation that is implicit in the predicted sequences $x_m(k+\ell|k)$ uses the detailed local model HA_m, the planned control sequence $\mathbf{u}_i(k-1)$, $j \in \mathcal{N}_m$ which the close neighbors transmitted at the previous control update step, as well as an approximate model $\widetilde{\text{HA}}_j$, $j \in \mathcal{N}_m$. Every Δ time units MPC_m, $m = 1, \ldots, M$ evaluates this local prediction for a finite set of feasible sequences of control actions $\mathbf{u}_m(k)$ over the horizon $[t_{k+1}, \ldots, t_{k+H}]$, and evaluates a local quantitative performance measure for each of these control sequences. Each MPC_m assumes that neighboring CAs will use their most recently communicated control sequence over the prediction horizon. Each CA_m then selects, immediately after this evaluation, which control sequence $\mathbf{u}_m^*(k)$ achieves the best local performance. CA_m at time t_{k+1} locally implements the first step $u_m^*(k)$ of this control sequence for the next Δ time units, and communicates this planned control sequence $\mathbf{u}_m^*(k)$ to its neighbors in $\mathcal{N}_m$. This distributed MPC approach is inspired by the work of Camponegara et al. [1].

Since CA_m knows an approximate model of its neighbors HA_j, $j \in \mathcal{N}_m$ it can predict approximately how its port variables evolve over its prediction horizon $[t_{k+1}, \ldots, t_{k+H}]$ (note that the predictions of these port variables will depend on $\mathbf{u}_m(k)$). This exchange of information on planned control sequences hopefully leads to coordination of the actions of neighboring CAs, since each agent whether it should solve a problem itself, or whether it can expect that the planned actions of its neighbors will be sufficient to solve the problem it observes. Intuitively, this ensures that each time a perturbation is detected in the system, at least one CA will try to solve it, and there will be no overreaction by several CAs all trying to solve the problem at the same time without knowing of each others actions. This overreaction is what might lead to a destabilizing loop. Of course, there is a risk that each local CA will see that at least one of its neighbors plans to solve the problem, and that therefore none of the CAs take proper action. The inherent delay in the communication exchange hopefully solves this problem.

The proposed CMPC paradigm is intuitively attractive for coordination and can be applied to a large collection of networked systems as listed in the next section. However, the components in these applications are typically highly nonlinear, often hybrid, systems. Proving stability under reasonable conditions is hence very difficult. Therefore, we studied this CMPC approach for a specific case study of electric power systems (see [2, 3]), and we describe in Sect. 9.3 how the results of these papers show that CMPC does indeed contribute to improving the coordination control for intelligent power grids.

9.2 Examples

This coordination strategy can be applied in principle to any network of interacting components where a local perturbation can lead to global performance degradation. Some examples are:

- voltage control in multi-area power systems (as described in Sect. 9.3);
- smart grids, consisting of interacting distribution networks, with some uncontrollable generators [such as photovoltaic (PV) or wind generators, and some controllable loads];
- traffic lights in an urban traffic network;
- on-ramp metering in control of freeway traffic, taking overflow into neighboring roads into account;
- large logistics networks with local task schedulers;
- flood/irrigation control, where controllable gates can regulate the flow of water.

9.3 Case Study

Consider an electric power system consisting of a set $\mathscr{V}$ of nodes, where each node represents a subnetwork consisting of at least one bus (or several electrically close buses) to which some generators and some loads are connected. Active and reactive power is transmitted between nodes via a set $\mathscr{E}$ of links. Each node m in $\mathscr{V}$ is connected via transformers to the transmission lines that connect it to its neighbors in $\mathscr{N}_m$. The transmitted power at time t depends on the voltage phasors $V_{m,\ell}(t)$ and $V_{\ell,m}(t)$ at the points where the transmission line is connected via a transformer to the bus of component m, respectively, the bus of component ℓ. Due to the impedance of the transmission lines active and reactive power is consumed in the connection of two neighboring nodes. In order to make the case study fit into the general framework described in Sect. 9.1, the port connecting two neighboring nodes is taken at any selected point along the transmission line (e.g., the middle, or either end). The part of the transmission line included with node m, as well as the dynamics of all generators, loads and internal lines of node m, are included in the model HA_m.

In order to understand the practical issues involved in this case study, it is important to remember that voltage is a local property of each node, where we make the assumption that buses that belong to the same HA_m are electrically so close to each other that they share the same voltage. Frequency is a global property of the network $\mathscr{V}$. It is a well-known problem in electrical power systems that local perturbations (such as disconnecting a generator or a line short-circuit) cause local voltage drops, that in some cases propagate to global voltage drop, and eventually to complete system failure, a blackout. This can sometimes be prevented by controlling the flow of reactive power between different nodes, as shown in [4–6]. While the frequency depends mainly on the global balance between generated and consumed active power, the local voltages are determined largely by the locally generated reactive power, and by the exchange of reactive power with neighboring nodes. How much reactive power can be generated locally depends on the state of the local synchronous machines, including some discrete states like over-excitation limiters. The flow of reactive power between different nodes depends on the voltages at the end of the transmission lines linking these nodes. These voltages can be influenced by adjusting the winding ratio of the transformers connecting a node to a transmission line. Load tap changing transformers (LTCs) are slowly acting discrete devices operating at time scale of several seconds that allow to adjust the winding ratio. Typical LTCs adjust their winding ratio in small steps, over a narrow range of values (e.g., 10 steps up and down, with a maximal change of 5 % from nominal winding ratio). Adjustment typically occur by taking one step up or down, and technical limitations of LTCs impose a delay of approximately 10 s between a request to move up or down, and the actual change of the winding ratio.

The case study discussed in this essay uses adjustments to the winding ratio of the LTC as the only actuators for each controllable input–output HA_m that are controlled by the local control agent CA_m. We simplify the analysis by assuming that each node m consists of one single generator, connected via an LTC to a single bus, and that this bus is connected to the bus of each neighbor $j \in \mathcal{N}_m$ via transmission line $\mathrm{Link}_{m,j}$. Under traditional deadband control of LTC, the winding ratio is adjusted, as a function of the local voltage of the connected bus, in order to locally maintain its associated bus voltage between a lower and upper bound. If the voltage drops below a threshold then the tap position is increased (and vice versa if the voltage becomes too high), with a preset delay, so as to increase (respectively, decrease) the bus voltage. This may, however, destabilize if the higher voltage leads to an increase in the voltage-dependent load or more transfer of reactive power toward a neighboring node. If the local load increases beyond the maximal power generation capabilities, then further voltage reductions will lead to further adjustments of the tap position, leading to even lower voltage. The system may eventually becomes unstable. One way to solve this could be to use a local model HA_m to anticipate the long-term effects of control actions. This means implementing a control agent CA_m that applies an MPC_m controller, using local model HA_m to anticipate the local effects of the control actions including the possible future events, such as reaching the maximal tap position or the generator reaching its maximal power limitations. Standard excitation controllers, including over-excitation limitations, for influencing the reactive power generated by the local synchronous generator are included in the model HA_m. Classical, uncoordinated methods for avoiding voltage collapse can be found in e.g., [1, 4, 5].

In Sect. 9.5, we show that in some cases, this local anticipatory control is not sufficient to stabilize the power system. A local perturbation in node m may be corrected by local control actions that cause reactive power flows from or to neighboring nodes $\mathcal{N}_m$, causing perturbations to the voltage at these neighbors, which in turn cause undesirable adjustments to LTCs located in neighboring areas. This interaction between local control loops may eventually lead to instability because some neighboring LTCs may reach their maximum physical limits, or neighboring generators may reach over-excitation limits, withdrawing voltage support from that

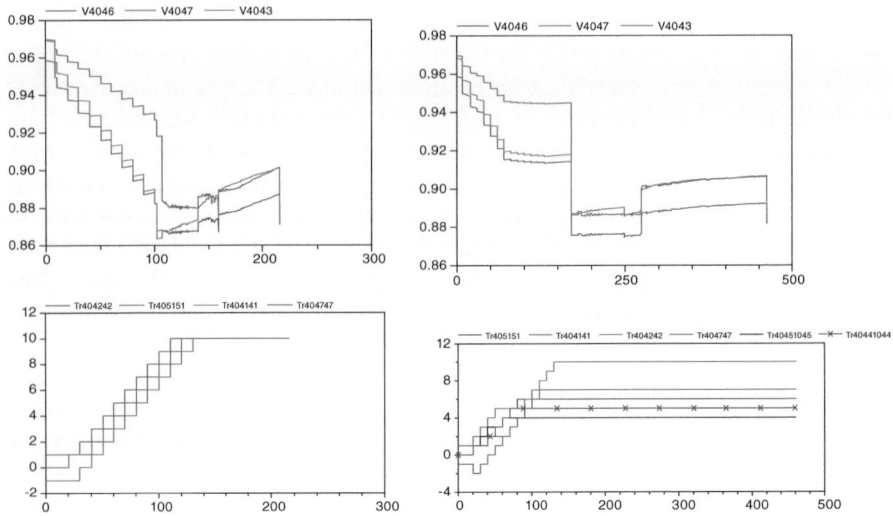

Fig. 9.1 Nordic32 response under uncoordinated decentralized deadband control of LTCs (*left*) and under uncoordinated decentralized MPC (*right*)

point onwards. This uncoordinated interaction of local control actions is one of the most likely driving mechanisms for voltage collapse in the long term. In [3, 7], we analyzed a fairly realistic case and observed that this anticipatory local MPC control cannot avoid voltage collapse. The results of a careful simulation show this, see Fig. 9.1. In the next section, we explain how CMPC as introduced in Sect. 9.3 may provide the coordination between neighboring CAs that is necessary in order to avoid this voltage collapse. In Sect. 9.5, we show that CMPC control can indeed avoid the voltage collapse shown in Fig. 9.1.

9.4 Theory and Concepts

In order to avoid these destabilizing loop interactions, it is necessary to coordinate the actions of neighboring CAs. By so doing, the anticipating/coordinating local voltage control not only anticipates, within the prediction horizon window, e.g., the activation of local over-excitation limiters or locally reaching the maximum physical tap limits for LTCs, but also anticipates the evolution of the reactive power exchange with neighboring nodes. The controller will then efficiently use this anticipation not only of its own planned actions but also of the actions of its neighbors, and thus coordinate its actions with those planned by its neighbors. This coordination must reduce the set of perturbations that cause voltage collapse.

In order to apply CMPC for voltage control, using LTCs as actuators, it is necessary that each control agent CA_m for node m simulates for each possible sequence $\mathbf{u}_m$ the evolution of all local system variables (internal voltages and currents, tap positions, over-excitation variables). Given that at each time $t_{k+h}, h = 1, \ldots, N$ in the control horizon, the actuator LTC can only plan 3 possible actions $\{up, no\ change, down\}$ it is obvious that $\mathbf{u}_m$ has at most 3^N possible cases to consider. The predictions used by the local MPC$_m$ use a hybrid systems version HA$_m$ of the classical quasi steady-state (QSS) model of power systems [8]. The information $\mathbf{u}_j^*(k)$, $j \in \mathcal{N}_m$ received from the neighbors is translated into approximate predictions of the port variables over the horizon $[t_{k+1}, t_{k+H}]$ by using an approximate model $\tilde{\text{HA}}_j$ for the neighbors $j \in \mathcal{N}_m$. This reduces the sensitivity of the CMPC control to uncertainty about models of distant parts of the network. For each local control sequence $\mathbf{u}_m$, control agent CA$_m$ calculates a local performance measure (e.g., a quadratic cost for the deviation of the predicted voltage from the nominal voltage at some local reference buses, or the maximal voltage deviation, or the number of tap changes needed to keep the system within specifications).

According to the CMPC approach, the first element $u^*(t_m)$ of the selected best sequence $\mathbf{u}_m^*$ calculated at time t_k is then implemented at t_k by CA$_m$. Moreover, the selected control sequence $\mathbf{u}_m^*$ is sent to each neighboring agents CA$_j$, $j \in \mathcal{N}_m$ (this requires at most $N \log(3)$ bits). All these calculations are repeated iteratively, using new observations and messages that become available in the interval (t_k, t_{k+1}), at the next time instant t_{k+1}, this time predicting performance over a shifted window with the size of $H \cdot \Delta$. In this way, each MPC$_i$, $i \in \{1, \ldots, M\}$, knowing approximately the model of its neighbors, can predict how the planned control actions of its neighbor will approximately influence the evolution of its own state variables. At the same time, each CA can anticipate to some extent the risk that neighboring nodes will not be able to further support the voltage control due to reaching an over-excitation limit or a limit on their LTCs. This is a weak form of coordination that can be implemented with very limited communications requirements. Only local measurements, and a few simple ternary messages from neighbors, are needed in order for MPC$_m$ to select its control actions. No obvious approach is available for calculating the set of perturbations for which the CMPC approach can actually avoid voltage collapse. It is obvious that the method cannot work for all perturbations, since some perturbations simply correspond to a load that is higher than the maximal amount of power that can be generated. Therefore, in the next section, we give some results on a simulations study that shows an example where at least the set of allowed perturbations is enlarged.

9.5 Review of the Research Contribution

Reference [3, 7] illustrates by simulation results on the realistic-size well-known Nordic32 test system. This system contains 10 nodes, most nodes being connected to 2 or 3 neighbors. A perturbation, corresponding to the disappearance of one transmission line, i.e., one link, that carries a significant amount of active and reactive power, was introduced at time 10 s after the start of the simulations. Figure 9.1 shows the evolution of the voltages, and the tap positions, for 3 buses that are close to the perturbation. Both deadband control and uncoordinated local MPC control leads eventually to a collapse of the voltage, respectively, after 210 s, and after 460 s. The collapse in both cases is due to some distant generators reaching their over-excitation limit and to tap position of LTCs reaching their upper limit. As a result, the system eventually fails to produce enough reactive power to maintain the voltage.

Under exactly the same experimental conditions, the simulations were repeated using the proposed CMCP controller in order to identify the distinct contribution of local feedback coordination to improved performance of the CMPC. Figure 9.2 shows that CMPC succeeds in maintaining the voltage within the specified bounds (10 % up and down around the nominal value). In fact, the coordination of the LTCs leads to a significant reduction in the number of adjustments in the tap positions: often local MPCs use information, indicating that their neighbors will increase the voltage and that therefore, it is best if they do not take any action. This coordination ensures that the risk becomes small that generators or transformers reach limits, thus reducing the risk of voltage collapse.

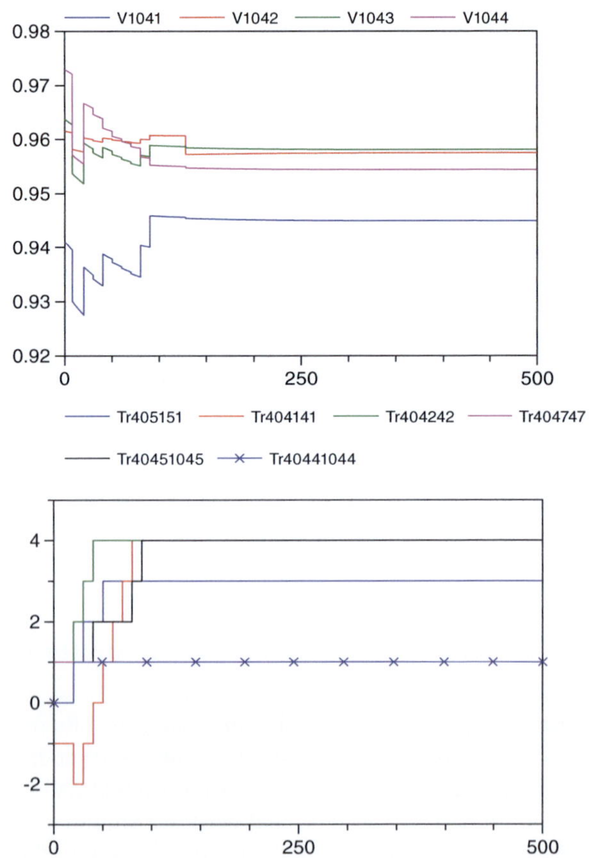

Fig. 9.2 Nordic32 response under CMPC scheme

9.6 Conclusions

This essay introduces a CMPC scheme, which combines the anticipation feature of MPC with additional active coordinating signal among CAs, for control of large networks of interacting systems. It illustrates the good performance of CMPC for voltage control in large-scale multi-area power systems. Simulation results illustrate that anticipating/coordinating voltage control may effectively stabilize the system voltages in circumstances where uncoordinated solutions leads to a final collapse. The proposed approach is generally applicable to many large networks of interacting dynamic components. It is useful whenever both anticipation (implemented via the analysis of the effects of a local control action over a sufficiently long window of time) and coordination (avoiding that the overall system is destabilized by unintended interactions between local control actions) are needed.

References

1. Camponogara E, Jia D, Krogh BH et al (2002) Distributed model predictive control. IEEE Control Syst Mag 22(1):44–52
2. Moradzadeh M, Bhojwani L, Boel R (2011) Coordinated voltage control via distributed model predictive control. In: Proceedings of CCDC2011, pp 1612–1618
3. Moradzadeh M, Boel R, Vandevelde L (2013) Voltage coordination in multi-area power systems via distributed model predictive control. IEEE Trans Power Syst 28(1):513–521
4. Phulpin Y, Begovic M, Petit M et al (2009) Evaluation of network equivalents for voltage optimization in multi-area power systems. IEEE Trans Power Syst 24(2):729–743
5. Glavic M, Hajian M, Rosehart W, Van Cutsem T (2011) Receding-horizon multi-step optimization to correct non viable or unstable transmission voltages. IEEE Trans Power Syst 26(3):1641–1650
6. Beccuti AG, Demiray TH, Andersson G et al (2010) A Lagrangian decomposition algorithm for optimal emergency voltage control. IEEE Trans Power Syst 25(4):1769–1779
7. Moradzadeh M, Boel R, Vandevelde L (2014) Anticipating and coordinating voltage control for interconnected power systems. Energies 7(2):1027–1047
8. Moradzadeh M, Boel R (2010) A hybrid framework for coordinated voltage control of power systems. In: Proceedings of IPEC2010, pp 304–309

Part II
Architectures of Distributed Systems and Their Control

Both the chapters of this part are of introductory character. The reader will learn in these chapters the classification of system architectures of distributed and of multilevel systems, and their related control architectures used in the rest of the book.

Chapter 10
System Architectures of Distributed and of Multilevel Systems

Jan H. van Schuppen

10.1 Motivation

The purpose of this chapter is (1) to describe system architectures of distributed systems and of multilevel systems, and (2) to classify such architectures in regard to their structure and their form of interaction. In this chapter, the term *multilevel system* is used. The term of a hierarchical system will not be used, it could be regarded as a special case of a multilevel system with a graph structure in the form of a tree.

Examples of distributed systems include sets of road vehicles, of underwater vehicles, of sensors in a network, of manufacturing machines, etc. Examples of multilevel systems include telephone networks, computer networks, communication networks, road networks, power systems, etc. The five C4C Case Studies each have their own architecture.

Control theory will benefit from distinguishing distributed and multilevel systems in several classes and from developing control theory for each class separately. There is such an enormous variety of such systems and so many examples that one becomes quickly convinced of the usefulness of control theory per subclass of systems.

The contents of this chapter consist of sections with examples, problem issues, concepts, suggestions for further research, and suggestions for further reading.

10.2 Examples

A description follows of several of the system architectures used for the case studies of the C4C Project.

J.H. van Schuppen (✉)
Van Schuppen Control Research, Gouden Leeuw 143, 1103 KB Amsterdam, Netherlands
e-mail: jan.h.van.schuppen@xs4all.nl

J.H. van Schuppen and T. Villa (eds.), *Coordination Control of Distributed Systems*,
Lecture Notes in Control and Information Sciences 456,
DOI 10.1007/978-3-319-10407-2_10

Autonomous Underwater, Aerial, and Guided Vehicles. In the C4C Case Studies of underwater vehicles and aerial vehicles, the system architecture is such that the network of subsystems is regularly changing. The changes are due to vehicles joining the network arriving from their base or departing from the network to take care of other activities or to return to their base.

For modeling, distinguish the domains of the real vehicles, the domain of the engineering vehicle models, and the domain of the vehicle systems. Each vehicle is modeled by an engineering vehicle model and subsequently by a vehicle subsystem. Any vehicle subsystem can interact with other vehicles subsystems via a communication channel. In certain situations, the vehicles could also be physically connected which would then have the corresponding relation between the vehicle subsystems.

For the communication of underwater vehicles, sonar communication is used and the range with limited energy is typically 500–800 m at a particular energy level used. For the communication of aerial vehicles, the range of communication is much larger. The configuration of these networks will in general change quickly over time because new vehicles appear or present vehicles disconnect from the network.

Road Networks. The system architecture of traffic control centers for road networks is rather static compared with that of vehicle systems, and it is in general fixed and changes only in case new investments in roads or in equipment are made.

The system architecture used in the C4C Case Study for the provincial road network in the province of North-Holland is multilevel. Distinguish the levels, from the bottom up, a section level, a link level consisting of several sections, a ring level, the level of a focus area, the level of a provincial network, etc. Each subsystem is connected to two or more child subsystems at the next-lower level. In addition, it is connected to one parent subsystem at the next-higher level. A subsystem, say a road section, may be directly linked to an adjacent road section either upstream or downstream depending on the structure of the network. See the Chaps. 6, and 28, 29.

In another case study, a semi-urban road network is considered in which the subsystem of each intersection is related to the nearest neighbor subsections corresponding to the road connection and the communication network connection. See Chap. 5.

In regard to the operations, a multilevel system directs the operation from the top to the bottom, while the information flow is usually from the bottom to the top.

Automated Guided Vehicles. The system structure of the automated guided vehicles (AGVs) on a container terminal of the C4C Case Study is similar to that of the underwater vehicles and the aerial vehicles discussed above. Each vehicle is regarded as a subsystem with interaction with other such subsystems. The vehicles are likely to encounter frequently other vehicles at intersections and at the stacks on the yard. The system structure changes dynamically with the movement of the vehicles. The system architecture of automated guided vehicles may be found in Chap. 7.

Complex Machines. The system architecture of the complex machines such as high-speed printers is rather static compared to a vehicle system, it is fixed with the construction of the machine. Of course, subsystems can be added or removed from the machine over time. For any operation, particular subsystems are needed and other subsystems are not required, but they remain available in the system architecture. The different sensors, actuators, and controllers are connected by a fixed communication or computer network. The system architecture is best described as a multilevel one, but the form of the interaction between the subsystems will depend on the purpose of the machine and has to be formulated for each individual machine. See Chap. 8.

10.3 Problem

Problem 10.1 The problem of formulating system architectures is to define subclasses of systems and to relate these subclasses.

10.4 Concepts of System Architectures

In the literature of control and system theory, see for example the books [19, 20, 22] and the journals *IEEE Transactions Automatic Control* and *Automatica*, there is neither a standard for classification of system architectures nor one for control architectures. Therefore, a classification of system architectures is introduced below. It should be clear that the classification proposed in this chapter is preliminary, it may have to be modified later on.

In this chapter, a system is defined as a control system with states, inputs, and outputs, as understood in control theory, see [23]. A *networked control system* is defined as a set of control systems and a network relation. The network relation describes for each subsystem how its inputs depend on outputs of which other subsystems and how its outputs become the inputs of which other subsystems.

For the interaction of subsystems, one often uses a graph. Each node of the graph represents a subsystem and a directed edge exists between two nodes in a graph if the corresponding subsystems interact in the corresponding direction. The interaction of real vehicles then corresponds in the model to an interaction relation between the corresponding subsystems of the same level. In a multilevel system, when regarding two subsystems at adjacent levels, one refers to the subsystem at the highest level as the *parent subsystem* of the subsystem at the lowest level, while one refers to the subsystem at the lowest level as the *child subsystem* of the subsystem at the next-highest level. The interaction with between two adjacent levels in a multilevel system then corresponds mostly to either communication from a parent subsystem to a child subsystem at the next-lower level or communication from a child subsystem to a parent subsystem at the next-higher level. See for system architectures of underwater vehicles the Chaps. 2 and 3.

Definition 10.1 Define the following *classes of system architectures*.

Define a *multilevel system* as a networked control system in which are distinguished two or more levels, at each level there are one or several subsystems, and in which each subsystem interacts with one subsystem at the next-higher level and

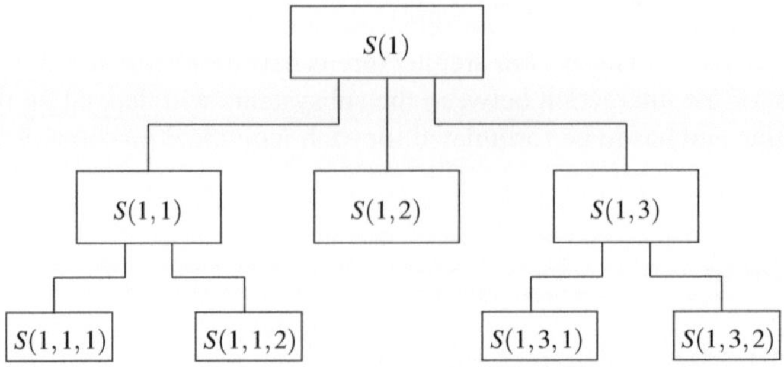

Fig. 10.1 Diagram of a multilevel system with a parent–child network architecture

Fig. 10.2 Diagram of a multilevel system with a peer-to-peer network structure

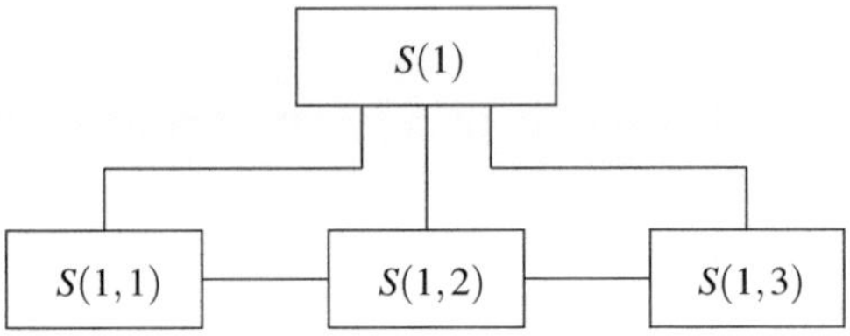

with one or more subsystems at the next-lower level. In addition, each subsystem at a particular level may interact with other subsystems at the same level.

Define a *distributed system* as a networked control system consisting of two or more subsystems. A *decentralized system* is a monolithic system (meaning not distinguishable into subsystems) with two or more pairs of inputs and outputs. A distributed system may occur at a level of a multilevel system.

An example of a multilevel system is a telephone network in which the levels correspond with the neighborhood, a local part of the town, the region, the province, the country, and the international network.

The diagrams of two multilevel systems are displayed in the Figs. 10.1 and 10.2. The diagrams of a decentralized and of a distributed system are displayed in Fig. 10.3. The distinction between a decentralized system and a distributed system is only in the structure of the system. An example of a distributed system is the interconnection of a large power system in which a local subsystem represents the power system of a small region. A decentralized system is obtained when abstraction is used from a distributed systems, the individual subsystems are no longer distinguished.

Definition 10.2 Within the system architectures of multilevel and of distributed systems, define the *network system architecture* as the relation between the various subsystems of the system. In terms of a graph, the relation is the set of directed edges which relate a tuple of nodes and where the nodes represent subsystems.

Distinguish the following *network system architectures of a distributed system* or of a level of a multilevel system:

Fig. 10.3 Diagram of a decentralized system, *left*, and of a distributed system, *right*

- A *ring network*, in which each subsystem is connected to two neighboring subsystems.
- A *line network*, a line-piece network is a more accurate label, in which each subsystem is connected to two neighbors except for the two subsystems at the end of the line which are only connected to one subsystem.
- A *grid network*, either in dimension two or in higher dimensions, which is defined as a network of parallel lines in every available dimension.

Distinguish the following *network system architectures of a multilevel system* concerning two adjacent levels:

- A *multilevel network with a child–parent architecture*. Each subsystem, the parent, is related to one or more subsystems at the next-lower level called the *children* (except for the subsystems at the lowest level). Correspondingly, each subsystem is connected to only one subsystem at the next-higher level called then the *parent* of the subsystem considered (except for the subsystem at the next-highest level). This definition excludes the case of a subsystem having two parents and relations between parents and their grandchildren.
- A *multilevel network with one child per parent*. This is a special case of a multilevel network with a child–parent architecture in which each subsystem, the parent, is related to only one subsystem at the next-lower level (except for the subsystems at the lowest level).
- A *multilevel network with peer relations*. This is a special case of a multilevel network with a child–parent architecture in which each subsystem at a particular level is also related to one or more subsystems at the same level called the *peers* or the *peer subsystems* of the subsystem considered. See Fig. 10.2.
- A *coordinated system*. This is a special case of a multilevel network with a child–parent architecture in which there are only two levels with at the highest level one subsystem which is related to each of the two or more subsystems at the next-lower level. The relation from parent to child is one of command and control, the parent restricts the behavior of the child, while there is no restriction from child to parent. The subsystem at the highest level is then called the *coordinator* of the coordinated system. See Chap. 14 for a particular definition of a coordinated linear system.
- A *mammillary system*. This is a special case of a multilevel network with a child–parent architecture in which there are only two levels with at the highest level one subsystem whose output feeds into all subsystems at the next-lower level and whose input comes from all subsystems at the next-lower level. In such a system, both parents and children restrict each other's behavior. The subsystem at the highest level is called the *coordinator* of the coordinated system. In the domain

of compartmental systems, the coordinator is also called the *mother compartment* based on the analogy with the species of mammals, see [9].

The concepts of system architectures of distributed and multilevel systems are intuitively clear from the engineering interpretations. The usefulness of a multilevel system with one child per parent is in abstraction of a subsystem by one another subsystem at the next-higher level. Coordinated systems are used in Part 3 of this book. They represent an elementary form of a multilevel system.

The procedure of abstraction of a subsystem to an abstracted subsystem is known for a long time. The basis is in the concept of an equivalence relation of a set. There is then the concept of a quotient set based on the equivalence relation. If this is put in the context of a dynamic system, then one obtains an abstracted dynamic system. This approach has been used for linear systems, see [28], for discrete-event systems in the form of automata, see [29], and undoubtedly for other class of dynamic systems. In each of these cases, one obtains a multilevel system with one child per parent. Another approach of abstraction is by dynamic approximation. This is also termed time-scale reduction. See for references on this approach for stochastic systems [11].

Definition 10.3 The *interactions and relations* of the subsystems of a distributed system can be of the following types:

- *Physical interaction.* The interaction is physically as an electric wire connection of two regional power systems or a flow connection between two vessels in a chemical plant. In this case, the delay in the communication or in the interaction between the subsystems is often neglected. The modeling choice of whether or not to neglect the delay depends on the example considered and has to be made with care.
- *Communication interaction.* The interconnection is by communication exchange as between underwater vehicles. In this case, there could be delays between the communications.

Distinguish a multilevel or a distributed system in regard to the *interaction relation of subsystems*: (1) The *command relation*: A communication of a subsystem sent to another subsystem is regarded as restricting the behavior of the receiving subsystem. (2) The *information relation*: A communication of a subsystem sent to another subsystem is information only and need not restrict the behavior of the receiving subsystem.

Distinguish distributed or multilevel systems based on their relation with respect to the time axis:

- *Synchronous subsystems.* The clocks of all subsystems are identical.
- *Asynchronous subsystems.* The clocks of the subsystems can be different and these clocks may drift with respect to each other, see [2–4].

10.5 Further Reading

The reader may want to check the chapters of the C4C Case Studies for concrete examples of system architectures, see Part I of the book.

The books by D.D. Šiljak provide examples and concepts of system architectures, see [19–22].

The research area of power systems provides many examples of distributed systems, see the books [5, 8, 10, 15]. Multilevel systems were developed in telephone networks, computer networks, and communication networks, see [6, 17, 24, 25]. Multilevel systems in control engineering are described in the books [13] and in papers, see [18]. Multilevel systems in control theory are partly based on the concept of abstraction. Multilevel discrete-event systems are described in [16, 27]. Multilevel systems of nonlinear or hybrid type are described in [12, 14].

Relevant for a description of multilevel systems is also the research topic of multi-level optimization problems which started with a paper by K. Arrow and L. Hurwicz, [1, 7]. In that approach, the relation between adjacent levels in a multilevel structure derives from the cost function rather than from the structure of the underlying system.

References

1. Arrow KJ, Hurwicz L (1963) Decentralization and computation in resource allocation. In: Pfouts RW (ed) Essays in economics and econometrics. University of North Carolina Press, Chapel Hill, pp 34–104
2. Benveniste A, Fabre E, Haar S, Jard C (2003) Diagnosis of asynchronous discrete-event systems: a net unfolding approach. IEEE Trans Autom Control 48:714–727
3. Brzozwoski JA, Negulescu R (2000) Automata of asynchronous behaviors. Theor Comput Sci 231:113–128
4. Cristian F, Fetzer C (1999) The timed asynchronous distributed system model. IEEE Trans Parallel Distrib Sys 10:642–657
5. van Cutsem T, Vournas C (2001) Voltage stability of electric power systems, second printing edition edn. Kluwer Academic Publishers, Dordrecht
6. Dandamudi SP, Eager DL (1990) Hierarchical interconnection networks for multicomputer systems. IEEE Trans Comput 39:786–797
7. Findeisen W, Bailey FN, Brdys M, Malinowski K, Tatjewski P, Wozniak A (1980) Control and coordination in hierarchical systems. Wiley, Chichester
8. Ilic MD, Liu S (1996) Hierarchical power systems control. Springer, London
9. Jacquez JA (1985) Compartmental analysis in biology and medicine, 2nd edn. The University of Michigan Press, Ann Arbor
10. Kundur P (1994) Power system stability and control. McGraw Hill Inc., New York
11. Kushner HJ, Dupuis PG (2001) Numerical methods for stochastic control problems in continuous time (2nd edn). Number 24 in applications of mathematics. Springer, New York
12. Lygeros J (1996) Hierarchical, hybrid control of large scale systems. PhD thesis, University of California, Berkeley
13. Mesarovič MD, Macko D, Takahara Y (1970) Theory of hierarchical, multi-level systems. Academic Press, New York
14. Pappas G, Laferriere G, Sastry S (2000) Hierarchically consistent control systems. IEEE Trans Autom Control 45:1144–1160

15. Schavemaker P, van der Sluis L (2008) Electrical power systems essentials. Wiley, Chichester. bookjhvs, gift of Lou van der Sluis.
16. Schmidt K, Moor T, Perk S (2008) Nonblocking hierarchical control of decentralized discrete event systems. IEEE Trans Autom Control 53:2252–2265
17. Schwartz M (1987) Telecommunication networks. Addison-Wesley Publishing Co., Reading
18. Shankar AU (1991) Modular design principles for protocols with an application to the transport layer. Proc IEEE 79:1687–1707
19. Šiljak DD (1978) Large-scale dynamic systems: stability and structure. North-Holland Books, New York
20. Šiljak DD (1990) Decentralized control of complex systems. Academic Press, New York
21. Šiljak DD (1996) Decentralized control and computations: status and prospects. Annu Rev Contr 20:131–141
22. Šiljak DD (2007) Large-scale dynamic systems: stability and structure. Dover Publications Inc, Mineola
23. Sontag ED (1998) Mathematical control theory: deterministic finite dimensional systems (2nd edn). Number 6 in graduate text in applied mathematics. Springer, New York
24. Tanenbaum AS (1981) Computer networks. Prentice-Hall International Inc., London
25. Walrand J, Varaiya PP (2000) High-performance communication networks, 2nd edn. Morgan Kaufmann, San Francisco
26. Ren W, Cao Y (2011) Distributed coordination of multi-agent networks. Communications and control engineering. Springer, London
27. Wong KC, Wonham WM (1996) Hierarchical control of discrete-event systems. Discrete Event Dyn Sys 6:241–273
28. Wonham WM (1979) Linear multivariable control: a geometric approach. Springer, Berlin
29. Wonham WM (2008) Supervisory control of discrete-event systems. Publisher W.M. Wonham, Toronto

Chapter 11
Control Architectures

Jan H. van Schuppen

11.1 Introduction

Distributed control systems can be of various forms as described in Chap. 10. For control of such systems, one also needs a variety of control structures. Engineering experiences illustrate that a wide variety of control architectures are useful. A classification of control structures is provided below.

A control architecture is the description of the various controllers of a distributed or multilevel control system and the ways in which these controllers are functioning. An example is the control architecture of a communication network where there is a controller at every node and, for example, a controller for the full network. Another example is that controllers of a set of underwater vehicles with one controller per vehicle and a coordinating controller at the surface vessel. To distinguish a distributed system as a system without inputs from a control system with inputs, the term distributed control system will be used. A multilevel control system is then defined correspondingly. The various controllers can be operating synchronously or asynchronously, there are examples of both cases in the literature.

For control design and control synthesis, the control engineer has to make a choice of the control architecture. Often the choice is based on the system architecture of the system under consideration, see Chap. 10 for a description of system architectures. The main choices of the control engineer are whether or not there is only one central controller or a set of controllers. If the choice is for a set of controllers, then the second choice is whether the various controllers should communicate directly by message passing. In a multilevel control architecture, the specification includes how the controllers of various levels communicate and when. The closed-loop system then is a composition of all controllers with all subsystems of the control system.

J.H. van Schuppen (✉)
Van Schuppen Control Research, Gouden Leeuw 143, 1103 KB Amsterdam, The Netherlands
e-mail: jan.h.van.schuppen@xs4all.nl

J.H. van Schuppen and T. Villa (eds.), *Coordination Control of Distributed Systems*,
Lecture Notes in Control and Information Sciences 456,
DOI 10.1007/978-3-319-10407-2_11

11.2 Classification

In the literature of control theory, there is no standard for classification of control architectures. Therefore, one is introduced below. It should be clear that the classification proposed in this chapter is tentative, it may have to be modified later on. Yet, for the purpose of this book, it is better to have a preliminary classification than not to have one. The distinctions between the classification categories are not yet as deep as one would like. This classification was first published in [7].

The main guideline used by the author for the classification of control architectures is the distinction for distributed and for multilevel control systems in terms of the degree of coordination of the subsystems and of the complexity of the levels of a multilevel system. The coordination can be regarded as a restriction on the behavior of the other subsystems. A complexity concept of multilevel control systems is not yet formulated. In Chap. 44, this is stated as an important open research issue.

The classification consists of the following subclasses:

- Centralized control.
- Distributed control.
- Distributed control with direct communication between controllers.
- Coordinated control.
- Multilevel control.

Below, these subclasses are described. Afterward, a comparison is made of the control architectures and principles are formulated for choosing one of these. The control objectives need not depend on the control architecture chosen. However, if a control architecture has been chosen, then the overall control objective has to subdivided over the various control subsystems. In addition, then the control synthesis has to be carried out either centrally or in a decentralized or distributed way. The control synthesis of each of the control architectures is described in the subsequent parts of the book.

11.3 Central Control Architecture

In this case, there is one controller which controls the complete distributed system. An engineering example is the controller of a printer, see Chap. 8. The advantages of the central control architecture is the simplicity of control synthesis. The main disadvantage is the complexity of control synthesis, the communication efforts required, and the lack of robustness in case of failures. The centralized control architecture is not described further because it is treated in every textbook of control engineering.

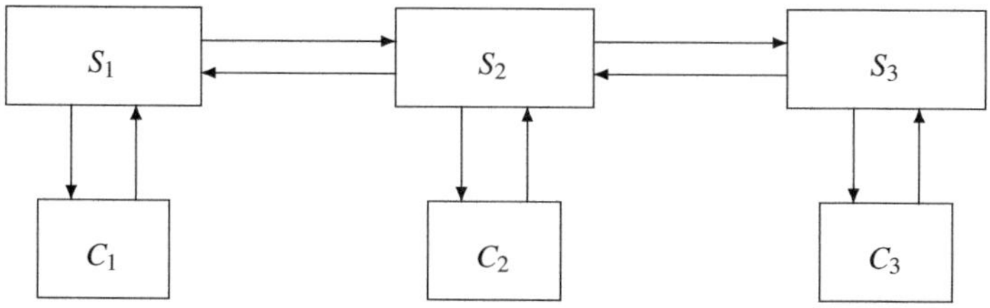

Fig. 11.1 Diagram of the distributed control architecture of an example of a distributed system. The subsystems are indicated by S_i and the corresponding controllers by C_i

11.4 Distributed Control

In the subclass of distributed control architectures, there are two or more control laws or controllers. Each controller receives an observation stream directly from the associated subsystem and produces an input to the distributed subsystem, see in Fig. 11.1, the arrows from one of the S_i boxes to the corresponding C_i boxes and back. There is no direct communication whatsoever between different controllers. See Fig. 11.1 for a diagram of the control architecture of distributed control. An example of this control architecture is the control of a large-scale power system if there is no communication between the controllers at various subnetworks.

Control synthesis of distributed control is difficult. Consider a cost function of the tuple of control laws. The problem is to determine a tuple of control laws which achieves the lowest cost. Information about distributed control may be found in the chapters on team theory, see Chaps. 18 and 19.

11.5 Distributed Control with Direct Communication Between Controllers

In the subclass of control architectures of distributed control with direct communication between controllers, there are two or more control laws or controllers. Each controller receives an observation stream directly from the distributed system and also receives one or more observation streams directly from other controllers. The observation streams from other controllers need not arrive at every time step, meaning the information is not sent after every observation or it is a strict subset of the state information. The diagram of this control architecture is displayed in Fig. 11.2.

There always exists indirect communication between controllers via the control system. With direct communication is meant the direct communication link between controllers, see the arrows between C_1 and C_2 in Fig. 11.2.

Distributed control with communication is a control architecture used for many control engineering problems. The alternating bit protocol of communication networks is a particular example of this control architecture which has been analyzed

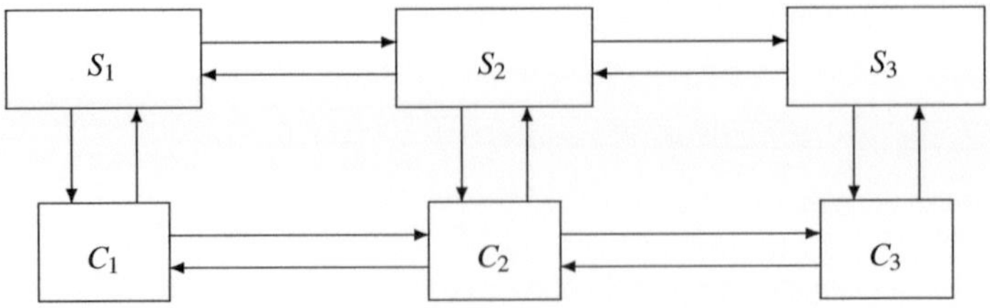

Fig. 11.2 Diagram of the control architecture of distributed control with communication of an example of a distributed system in which additional direct links between the subsystems S_1 and S_3, and C_1 and C_3 are not indicated in the figure. The solid lines between the C_i boxes indicate that the communication takes place every time step

as a problem of a discrete-event system. In this case, there is communication from the receiver to the sender. The backpressure algorithm for the routing of messages in a communication network is another example, see Chap. 30. Another example is a platoon of cars on a motorway in which each car only communicates with its nearest neighbors, the car directly in front and the car directly behind. It has been proven that this control architecture cannot stabilize the platoon, see Chap. 22. The communication requires financial resources and energy. But those costs seem to be less than the loss in performance if no communication is used. That communication between controllers is also an economic issue was already remarked by J. Marschak and R. Radner.

Control synthesis of distributed control with communication is even more difficult than distributed control without communication. First, a communication law has to be determined when extra observations are to be requested or are to be sent. Next the additional observations have to be integrated into a state estimator. Finally, a tuple of control laws has to be determined. Because of these difficulties, there is no substantial theory for this case, except of the case of distributed control with communication of discrete-event systems. Yet, in practice, researchers formulate control laws of which the performance is satisfactory as in the alternating bit protocol.

11.6 Coordinated Control

In the subclass of coordinated control architectures, one considers a coordinated system and a corresponding coordinated control architecture. It applies only to a coordinated system as defined in Chap. 10. Recall from Chap. 10 that a coordinated system consists of a coordinator and two or more subsystems such that conditioned on the coordinator the subsystems are independent. In a coordinated control architecture, there is a controller for the coordinator and a controller for each of the subsystems. The controller influences the subsystems while the subsystems do not influence the coordinator. In the Chapter System Architectures, it is described that the task of a

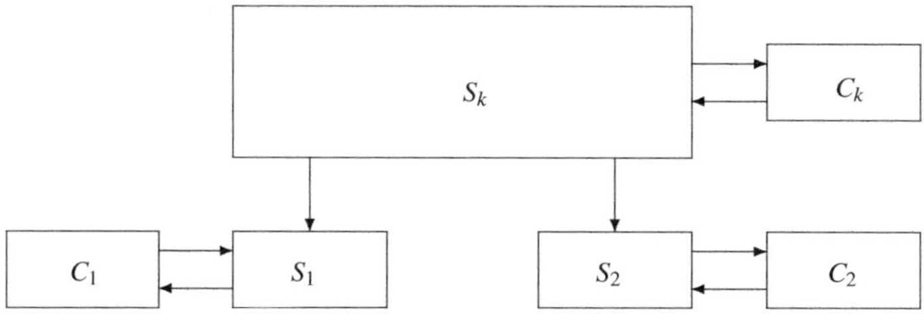

Fig. 11.3 Diagram of the control architecture of coordinated control of an example of a coordinated system in which S_k denotes the coordinator, S_1 and S_2 denote the subsystems, and the various controllers are indicated by the symbol C_i

coordinator is to restrict the behavior of the two subsystems often inspired by control objectives or other properties.

See Fig. 11.3 for the diagram of this control architecture.

An example of coordination control is where a surface vessel ship coordinates the actions of two underwater vehicles.

Coordination control becomes of interest when the control objectives cannot be met by distributed control with or without communication. It is then necessary to impose a degree of centralized control here called coordination control. An example is coordination control of coordinated discrete-event system where, without coordinator, blocking occurs meaning the distributed system does not function as needed. In a coordinated system, there is communication from the coordinator to the subsystems.

There is also the class of M systems, see Chap. 10, in which the control architecture is almost the same as that of a coordination control architecture except that the coordinator and each of the subsystems communicate in both directions.

Control synthesis of a distributed control system with the coordination control architecture is a little involved. Please have a look at the corresponding chapters in this collection, the Chaps. 10, 28, and 29.

11.7 Multilevel Control

In this subclass of control architectures, one considers a multilevel control system and a corresponding multilevel control architecture. Recall from Chap. 10 that a multilevel control system, see Fig. 11.4 consists of two or more levels in which each subsystem of a level is related to one or more subsystems at the next-lower level and related to one subsystem at the next-higher level. In a multilevel control architecture, there is one control law or controller per subsystem. Thus, a controller of one subsystem is connected to those controllers of subsystems at the next-lower level to which the subsystem is related and it is connected to the controller of the subsystem at the next-higher level to which the subsystem is related. A multilevel control architecture

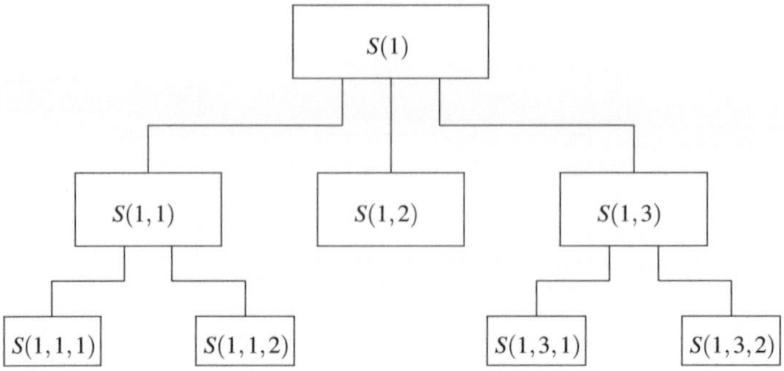

Fig. 11.4 Diagram of the control architecture of multilevel control of an example multilevel control system. A *box* represents a closed-loop system consisting of a subsystem and its controller so as not to make the figure too complicated. A link between two boxes is always a two-directional physical or communication channel. Controllers of subsystems which are linked to the same parent subsystem could have a communication link between them not indicated in the figure

is a generalization of the coordination control architecture defined above or of the M-system control architecture.

A technological example of a multilevel control architecture is a telephone network. There is a controller at every level and at the highest level a controller for the computation of the routing tables. The multilevel structure proposed for computer networks in the book by A. Tannenbaum is another example. The protocols for locating a subscriber in mobile communication networks is another example. Multilevel systems have been used in the society of the Aztecs and for the organization of the Roman army.

Multilevel control for particular systems has been investigated. For discrete-event systems, multilevel control has been investigated but it seems to concern mostly the case of one subsystem at each level. More theory is needed for multilevel control.

11.8 Comparison and Choice Principles

The control architectures defined above are listed in the order of increasing degree of dependence between the controllers. It is possible to consider other orders on the set of subsystems. Below a multilevel system is regarded as a special case of a distributed system which can be obtained by ignoring the structure of the multilevel system.

Problem 11.1 Formulate guidelines for the selection of a control architecture for any particular distributed control system.

Preliminary guidelines follow.

1. Use the distributed control architecture as much as possible at the subsystem level. This approach requires the least restriction on the behaviors of the subsystems. See Part IV of this book for the relevant chapters.

2. If the closed-loop system with the distributed control architecture cannot meet the control objectives, then consider the architecture of distributed control with direct communication between controllers. The nearest-neighbor control architecture may be explored first. See Part V of this book for the relevant chapters.
3. If the performance with respect to the control objectives is still unsatisfactory, then the guideline is to use the coordination control architecture. The particular level of satisfaction of the control objectives has to be set in each individual case. See Part III if this book on coordination control for the corresponding control synthesis method.
4. If the complexity of the distributed control system is very large, then a multilevel system with the multilevel control architecture seems best. See Part VI of this book for chapters on this topic.

Needed is a complexity theory for multilevel control systems. No such theory is currently available. Chapter 44 formulates the formulation of a complexity of multilevel systems as an important research issue. The last step of the guideline depends on such a complexity theory.

11.9 Further Reading

A reader novel to the subject is advised to read the book [5] or the papers [7, 8].

Books on control of decentralized systems include [2, 3, 6, 9]. Books on multilevel systems include [1, 4].

References

1. Findeisen W, Bailey FN, Brdys M, Malinowski K, Tatjewski P, Wozniak A (1980) Control and coordination in hierarchical systems. Wiley, Chichester
2. Hamilton SC, Broucke ME (2012) Geometric control of patterned linear systems. Number 428 in LNCIS. Springer, Berlin
3. Lunze J (1992) Feedback control of large scale systems. Prentice-Hall, New Jersey
4. Mesarovič MD, Macko D, Takahara Y (1970) Theory of hierarchical, multi-level systems. Academic Press, New York
5. Šiljak DD (1978) Large-scale dynamic systems—stability and structure. North-Holland Books, New York
6. Šiljak DD (1990) Decentralized control of complex systems. Academic Press, New York
7. van Schuppen JH, Boutin O, Kempker PL, Komenda J, Masopust T, Pambakian N, Ran A (2011) Control of distributed systems: tutorial and overview. Europ J Control 17:579–602
8. van Schuppen JH (2011) Control of distributed stochastic systems—introduction, problems, and approaches. Proceedings of IFAC World Congress 2011, pp 6029–6035
9. Yüksel S, Basar T (2013) Stochastic networked control systems—stabilization and optimization under information constraints. Birkhäuser, Boston

Part III
Coordination Control

Coordination control is introduced in Chap. 12. It is followed by several research chapters on coordination control of linear systems and of discrete-event systems. Chapter. 16 provides an introduction to control of discrete-event systems.

Chapter 12
What is Coordination Control?

Jan H. van Schuppen

12.1 Introduction

The purpose of this chapter is to introduce the reader to coordination control which is then further described in this part of the book.

Coordination control is relevant for control of distributed systems. Distinguish in a distributed system a coordinator subsystem and the other subsystems. The task of the coordinator subsystem is to coordinate the operation or the activities of the other subsystems. It does so by restricting their behavior or activities so that the overall control objectives of the distributed system are met.

In control theory of distributed discrete-event systems, it is well known that without coordination control, two or more subsystems may block each other, meaning that the overall system at a particular state cannot make any transition. Even for linear distributed systems, there are examples where safety cannot be met without coordination being imposed on the subsystems. Hence, coordination control is motivated.

Coordination control then addresses how to achieve the coordination control objectives. In addition, it addresses how to construct a coordinator subsystem of minimal complexity. Though coordination control has been discussed in various application domains, the fundamental control theoretic problems remain to be investigated further. Coordination control is a form of control of multilevel systems.

The next section contains three examples of coordination control. Section 12.3 presents coordinated linear systems with equations in a rather simple setting. Coordination control of coordinated systems is discussed in Sect. 12.4. Approaches to coordination control are mentioned in Sect. 12.5.

J.H. van Schuppen (✉)
Van Schuppen Control Research, Gouden Leeuw 143,
1103 KB Amsterdam, The Netherlands
e-mail: jan.h.van.schuppen@xs4all.nl

J.H. van Schuppen and T. Villa (eds.), *Coordination Control of Distributed Systems*,
Lecture Notes in Control and Information Sciences 456,
DOI 10.1007/978-3-319-10407-2_12

12.2 Examples of Coordination Control

A list of three examples follows. (1) *Coordination control of underwater vehicles for a mission.* In this case, the coordinator is the mission control or headquarters. The control objectives are to guide the underwater vehicles with their missions and to maintain safety, for example, the vehicles should not collide. The task of the coordinator is to guarantee safety and possibly to assist in optimizing performance.

(2) *Coordination control of a link of a motorway.* Think about a link as a stretch of motorway of about 8 km. The subsystems to be controlled are the sections of the link of the motorway of which, in an 8-km stretch, there are typically 16 sections of 500 m. each. The coordinator is the subsystem responsible for the link. Coordination control by the coordinator for the various subsections is required because otherwise the controller of a subsection would provide access control by ramp-metering causing congestion control in the link.

The control objective of the coordinator is to maintain a steady traffic flow and to limit negative environmental effects. It can do so by communicating to the subsystems of the sections traffic inflows and traffic outflows for their respective sections. From those traffic inflows and outflows, then follow the on-ramp traffic flows of the sections.

(3) *Coordination control of a set of reservoirs with dams.* The subsystems to be controlled are the individual reservoirs. The coordinator is the subsystem of the regional network of reservoirs. The coordinator controls the levels of the successive reservoirs. The control objectives of the coordinator are to generate a steady flow of water and of electricity but also to prevent flooding due to an overflow of water.

Other examples are described in Chap. 1 Case Studies.

12.3 Coordination Control of Linear Systems—Briefly

A brief description of coordinated linear systems follows. Elsewhere in this book a formal definition of such systems is presented.

A *coordinated linear system* consists of the interconnection of three or more subsystems. To simplify the exposition, only three subsystems are considered. The subsystems are distinguished into a coordinator labeled c and the two subsystems which are labeled 1 and 2. The system representation of the coordinator is the linear system,

$$\frac{dx_c(t)}{dt} = A_{cc}x_c(t) + B_{cc}u_c(t), \; x_c(t_0) = x_{c,0}, \tag{12.1}$$

$$y_{1c}(t) = C_{1c}x_c(t) + D_{1c}u_c(t), \tag{12.2}$$

$$y_{2c}(t) = C_{2c}x_c(t) + D_{2c}u_c(t). \tag{12.3}$$

In this system representation, $x_c : T \to \mathbb{R}^{n_c}$ is the state of the coordinator subsystem and u_c is the coordinator input. The coordinator has two outputs, y_{1c} and y_{2c}, which

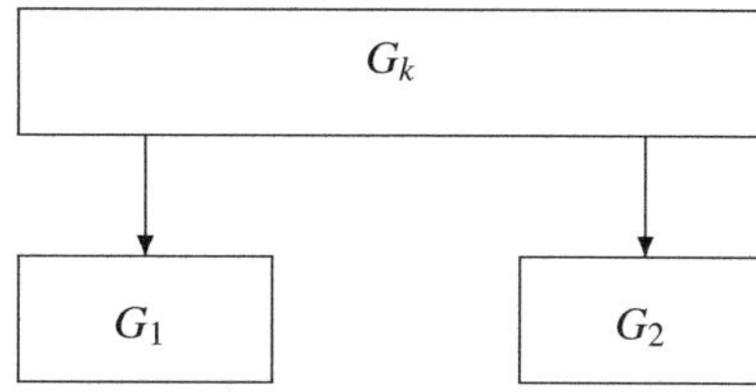

Fig. 12.1 Diagram of a coordinated system

are inputs of the subsystems 1 and 2, respectively. In the examples, these outputs are command signals from the coordinator to the two subsystems. See Fig. 12.1 for a coordinated system with coordinator G_c and subsystems G_1 and G_2.

The system representations of the subsystems 1 and 2 are then, respectively,

$$\frac{d}{dt}x_1(t) = A_{11}x_1(t) + B_{11}u_1(t) + B_{1c}y_{1c}(t), \; x_1(t_0) = x_{1,0}, \tag{12.4}$$

$$\frac{d}{dt}x_2(t) = A_{22}x_2(t) + B_{22}u_2(t) + B_{2c}y_{2c}(t), \; x_2(t_0) = x_{2,0}. \tag{12.5}$$

In this representation, x_1 and x_2 represent the states of the subsystems 1 and 2, respectively, and u_1 and u_2 the inputs of these subsystems, respectively. Note that the output of the coordinator y_{1c} enters subsystem 1 but not subsystem 2 and, correspondingly, y_{2c} enters subsystem 2 but not subsystem 1.

The system consisting of the coordinator and the two subsystems then has the representation

$$\frac{d}{dt}x(t) = \begin{pmatrix} A_{11} & 0 & B_{1c}C_{1c} \\ 0 & A_{22} & B_{2c}C_{2c} \\ 0 & 0 & A_{cc} \end{pmatrix} x(t) + \begin{pmatrix} B_{11} & 0 & B_{1c}D_{1c} \\ 0 & B_{22} & B_{2c}D_{2c} \\ 0 & 0 & B_{cc} \end{pmatrix} u(t), \; x(t_0) = x_0, \tag{12.6}$$

$$x(t) = \begin{pmatrix} x_1(t) \\ x_2(t) \\ x_c(t) \end{pmatrix}. \tag{12.7}$$

Examples of coordination control such as for formation flying may be found in [18].

Many readers who first learn of coordinated linear systems are struck by the fact that there is no feedback from the two subsystems to the coordinator. Such a feedback connection could be an output of the two subsystems, which is an input to the coordinator. The absence of feedback is not an oversight but the outcome of a definition. It is clear that there are coordinated systems which have such a feedback connection. One may define the concept of a mammillary system in which such a feedback connection exists. In the literature on compartmental systems, the concept of a mammillary compartmental systems has been defined (see [10]).

Consider the interconnection of two linear subsystems. It is then possible to decompose the interconnection into a coordinator and two new subsystems such that their joint behavior equals the original interconnection. The system representation which one then obtains equals the coordinated linear system representation formulated above. The decomposition is described in the paper [14].

12.4 Problem of Coordination Control

Consider a distributed control system consisting of two or more subsystems. Consider control objectives of which at least one relates to two or more subsystems. Attention is restricted to the coordination control architecture described above.

Problem 12.1 Consider a distributed system and control objectives. Solve the following subproblems.

1. Formulate the concept of a *coordinated system* in such a way that it applies to many classes of distributed systems, such as linear systems and discrete-event systems. Define the concept of *complexity of a coordinated system*, for example, as the complexity of the state set of the coordinator.
2. Construct, for a distributed system and control objectives, a coordinator subsystem which is a coordinator as defined above. Construct a minimal coordinator according to the concept of complexity defined above.
3. Develop control synthesis for a coordinated system with control objectives. The control synthesis should produce the *control laws* of the coordinator and of each of the subsystems. Note that even though the coordinator is a type of controller, one allows for a further restriction of its behavior by the control law of the coordinator.
4. Evaluate the performance of a controlled coordinated system.

The above problem has been treated extensively for linear systems (see the references below), for discrete-event systems, and for Gaussian control systems.

The concept of a coordinated distributed system developed for linear systems and for discrete-event systems is algebraic and hence can be applied to any class of distributed systems. A coordinated system is defined in terms of a conditional independence relation of subspaces of the state space. This concept is inspired by conditional independence of probability theory. The coordinator makes the behaviors of the other subsystems independent by its restriction on the behaviors of the subsystems. Therefore, control synthesis decomposes into a control synthesis for the coordinator and, separately, for the other subsystems.

The control synthesis problem is then to construct a controller for the coordinator and controllers for each of the subsystems such that the closed-loop system meets the control objectives. The diagram of a closed-loop coordinated system is displayed in Fig. 12.2.

In coordination control, one distinguishes a coordinator which guides all other subsystems in their control tasks. There is only one coordinator in a coordination control of distributed systems. One may also define multilevel coordination control with several groups of subsystems, each with its coordinator. That approach is regarded as control of multilevel system.

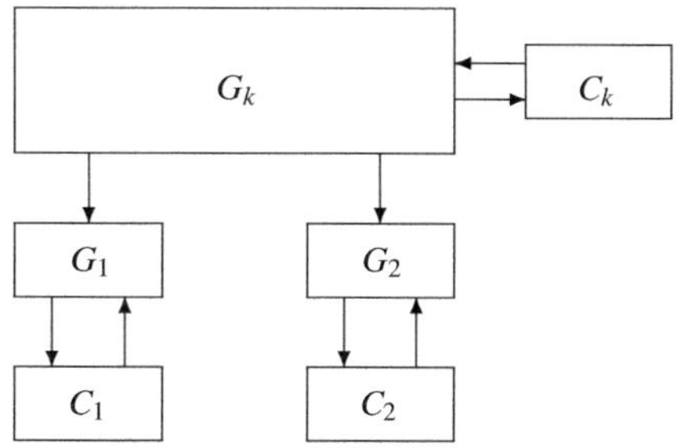

Fig. 12.2 Diagram of a closed-loop coordinated system

12.5 Approaches of Coordination Control

Various approaches to the problem defined above are described.

Several concepts of a coordinated system are discussed in the literature including (1) coordination structure based on engineering modeling; (2) flocking of subsystems; and (3) coordinated systems based on the conditional independence relation. There are likely to be other forms of coordinated systems.

The second subproblem formulated in Sect. 12.4 is the construction of a coordinator for a distributed system. In the reference [14], an algorithm has been proposed to construct a coordinator from the interaction of two or more subsystems. The coordinator consists of the parts of the two subsystems which are observable with respect to the other subsystems. The unobservable parts of the two subsystems then form the particular subsystems of the coordinated system which are directed by the coordinator. If there are three or more subsystems in a distributed system, then one obtains a more complex decomposition.

The next issue is the minimality of a coordinator. It seems a useful principle that the coordinator should have a complexity which is as small as possible. Then, the restriction on the other subsystems is as small as possible, and the remaining subsystems keep a relatively large freedom in their behavior. This is not the only complexity criterion, and one could also limit the complexity of the signals between the coordinator and the subsystems. The question is then: How to construct for a distributed system a coordinator of minimal complexity? Progress on this has been made for linear systems, and the solution involves the concepts of controllability and of observability (see [17]).

Of interest to control synthesis is further the decomposition of a coordinated system in regard to observability and to controllability. Such decompositions have been described for coordinated linear systems (see [15]).

The problem of control synthesis of coordinated systems simplifies because of the concept of coordination. Needed are controllers for the coordinator and for each of the subsystems. The control synthesis separates into (1) control synthesis for the coordinator and (2) separate problems of control synthesis for each of the subsystems after the controlled coordinator subsystem has been fixed. For linear coordinated systems, one can prove that the control synthesis problem (2) is such that it is identical to the control synthesis problem of the subsystems in isolation. The difficulties are thus in the control synthesis of the control law of the coordinator. Both the problems with complete observations and with partial observations are of interest for applications. For coordination control of discrete-event systems, the control synthesis is different due to the fact that the synchronization between the coordinator and the subsystems is more direct.

Control synthesis of coordination control is related to control of leader–follower systems or to control in Stackelberg games. Yet, the problems are often different due to the particularities imposed by the concept of coordination. In the alternatives mentioned, there is only one follower, while in coordination control, there are two or more followers and the coordination of these followers is the main problem.

12.6 Further Research

It should be clear that control engineering requires much more research into coordination control than is available so far. The research topics requiring attention are as follows: (1) the multitude of coordination concepts in the literature and their relation; (2) construction of a coordinated system based on the control objectives and on the minimality of the complexity of a coordinated system; (3) decompositions of coordinated systems; and (4) control synthesis of coordinated systems. Classes of systems for which coordination control may be useful include linear systems, though much work has been done for this class (see the next subsection); particular classes of nonlinear systems such as polynomial and rational systems; and hybrid systems. Experience with still more examples of control engineering of distributed systems is necessary.

12.7 Further Reading

A reader who first learns of coordination control is advised to read the following chapter of the book in which this chapter appears [14]. The other chapters of the part of the book in which this chapter appears are also of interest. An introductory journal article on coordination control of linear systems is [15]. An introductory article on coordination control of discrete-event systems is [22].

Books treating coordination control include [2, 7, 28].

Papers discussing coordination control include [1, 25, 29, 31]. Engineering and life science models in which coordination problems occur include [3–6, 8, 11, 12, 30, 32].

Coordination control of linear systems is treated in [13, 14, 17, 18, 27]. Coordination control of Gaussian systems is found in [26]. Coordination control of discrete-event systems is developed in [19–24]. Coordination in informatics is found in [9, 28].

References

1. Başar T (1980) Optimum coordination of linear interconnected systems. Large Scale Sys 1:17–27
2. Chokshi N, McFarlane D (2008) A distributed coordination approach to reconfigurable process control. Springer series in advanced manufacturing. Springer, London
3. de Sousa JB, Lobo Pereira F, da Silva JE (2009) New problems of optimal path coordination for multi-vehicle systems. In: Proceedings European control conference (ECC.2009). EUCA, European Union Control Association
4. de Sousa JB, Matos A, Pereira FL (2002) Dynamic optimization in the coordination and control of autonomous underwater vehicles. In: Proceedings of decision and control conference, pp 2087–2092. IEEE Press, New York
5. Dimokas N, Katsaros D, Tassiulas L, Manolopoulos Y (2010) High performance, low complexity cooperative caching for wireless sensor networks. Wirel Netw
6. Estrin D, Govindan R, Heidemann J, Kumar S (1999) Next century challenges: scalable coordination in sensor networks. In: Mobicom 1999, ACM, pp 263–270
7. Findeisen W, Bailey FN, Brdys M, Malinowski K, Tatjewski P, Wozniak A (1980) Control and coordination in hierarchical systems. Wiley, Chichester
8. Franco E, Magni L, Parisini T, Polycarpou MM, Raimondo DM (2008) Cooperative constrained control of distributed agents with nonlinear dynamics and delayed information exchange: A stabilizing receding-horizon approach. IEEE Trans Autom Control 53:324–3338
9. Gelernter D, Carriero D (1992) Coordination languages and their significance. Commun ACM 35:97–107
10. Jacquez JA (1985) Compartmental analysis in biology and medicine, 2nd edn. The University of Michigan Press, Ann Arbor
11. Jadbabaie A, Lin J, Morse AS (2003) Coordination of groups of mobile agents using nearest neighbor rules. IEEE Trans Autom Control 48:988–1001
12. Jin U, Liao U, Minai AA, Polycarpou M (2006) Balancing search and target response in cooperative unmanned aerial vehicle (UAV) teams. IEEE Trans SMC: Part B 36:571–587
13. Kempker PL (2012) Coordination control of linear systems. PhD thesis, VU University Amsterdam, Amsterdam
14. Kempker PL, Ran ACM, van Schuppen JH (2009) Construction of a coordinator for coordinated linear systems. In: Proceedings of European control conference (ECC.2009). European Control Association, pp 4979–4984
15. Kempker PL, Ran ACM, van Schuppen JH (2012) Controllability and observability of coordinated linear systems. Linear Algebra Its Appl 437:121–167
16. Kempker PL, Ran ACM, van Schuppen JH (2013) Construction and minimality of coordinated linear systems. Report, VU University Amsterdam, Amsterdam
17. Kempker PL, Ran ACM, van Schuppen JH (2014) Construction and minimality of coordinated linear systems. Linear Algebra Its Appl (submission)
18. Kempker PL, Ran ACM, van Schuppen JH (2014) LQ control for coordinated linear systems. IEEE Trans Autom Control 59:to appear
19. Komenda J, Masopust T, van Schuppen JH (2011) Synthesis of controllable and normal sublanguages for discrete-event systems using a coordinator. Systems Control Lett 60:492–502
20. Komenda J, Masopust T, van Schuppen JH (2012) Supervisory control synthesis of discrete-event systems using a coordination scheme. Automatica 48(2):247–254
21. Komenda J, Masopust T, van Schuppen JH (2012) On algorithms and extensions of coordination control of discrete-event systems. In: Proceedings of WODES pp 245–250
22. Komenda J, Masopust T, van Schuppen JH (2013) Coordination control of distributed discrete-event systems. In: Seatzu C, Silva M, van Schuppen JH (eds) Control of discrete-event systems—automata and petri net perspectives, vol 433. Lecture notes in control and information sciences. Springer Verlag London Ltd., London, pp 147–168

23. Komenda J, Masopust T, van Schuppen JH (2013) Multilevel coordination control of modular DES. In: Proceedings of 6xth IEEE conference on decision and control (CDC.2013), IEEE. IEEE Press, New York
24. Komenda J, van Schuppen JH (2008) Coordination control of discrete-event systems. In: Proceedings of the international workshop on discrete-event systems (WODES.2008), IEEE. IEEE Press, New York, pp 9–15
25. Lin F (1991) Control of large scale discrete event systems: task allocation and coordination. Sys Control Lett 17:169–175
26. Pambakian N (2011) LQG coordination control. Master of science thesis, Delft University of Technology, Delft Center for Systems and Control, Delft, 2011
27. Ran ACM, van Schuppen JH (2008) Control for coordination of linear systems. In:Ball J (ed) Proceedings of international symposium on the mathematical theory of networks andsystems (MTNS.2008; CD-ROM only). Virginia Institute ofTechnology, Blacksburg, VA, U.S.A, p 105
28. Ren W, Cao Y (2011) Distributed coordination of multi-agent networks. Communications and control engineering. Springer, London
29. Speranzon A (2006) Coordination, consensus, and communication in multi-robot control systems. PhD thesis, KTH, Stockholm
30. Wolkenhauer O, Ghosh BK, Cho KH (2004) Control and coordination in biochemical networks. Control Sys Mag 24(4):30–34
31. Wong KC, Murray Wonham W (1998) Modular control and coordination of discrete-event systems. Discrete Event Dyn Sys 8:247–297
32. Yang Y, Polycarpou MM, Minai AA (2007) Multi-UAV cooperative search using an opportunistic learning method. Trans ASME 129:716–728

Chapter 13
Introduction to Coordinated Linear Systems

Pia L. Kempker

13.1 Introduction

The purpose of this chapter is to introduce and motivate the concept of coordinated linear systems, a special class of hierarchical systems. Coordinated linear systems are structured linear systems consisting of one coordinator system and two or more subsystems, each with their own input and output. The coordinator state and input may influence the subsystem states, inputs, and outputs. The state and input of each subsystem, on the other hand, have no influence on the coordinator state, input, or output, and neither can they influence the state, input, or output of the other subsystem. This structure is illustrated in Fig. 13.1.

The concept of a coordinated linear system was first introduced in [1]. In [2], the construction of coordinated linear systems from unstructured linear systems is described, and the concept of a minimal coordinator is introduced. These results are summarized in Chap. 14. The controllability and observability properties of coordinated linear systems are discussed in [3].

Coordinated linear systems can be useful in the study of many applications with an inherent hierarchical structure, but also of applications without a predefined structure, which permit a hierarchical approach. The main motivation for the study of coordinated linear systems is that we expect the structure imposed on the systems to simplify control synthesis: Since the subsystem states and inputs have no influence on the states or outputs of any other part of the system, local control synthesis can be done for each subsystem independently after the control law for the coordinator has been fixed.

An application of coordinated linear systems involving the inherently hierarchical traffic network is described in Chap. 12. Another application, where a hierarchical

P.L. Kempker (✉)
TNO, P.O. Box 5050, 2600 GB Delft, The Netherlands
e-mail: pia.kempker@tno.nl

J.H. van Schuppen and T. Villa (eds.), *Coordination Control of Distributed Systems*,
Lecture Notes in Control and Information Sciences 456,
DOI 10.1007/978-3-319-10407-2_13

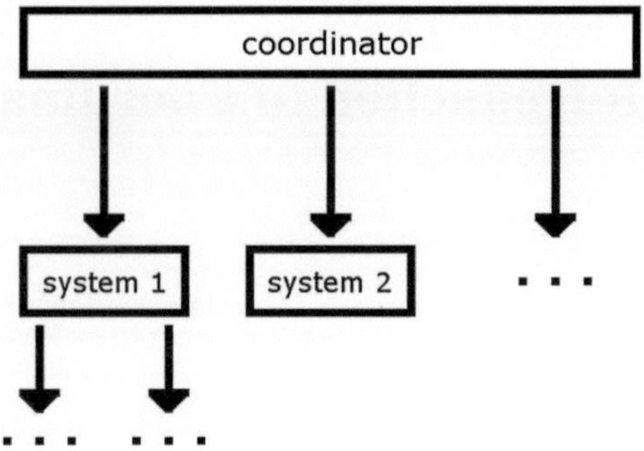

Fig. 13.1 Scheme of a coordinated system

structure is imposed on a group of autonomous underwater vehicles in order to meet the control requirements, can be found in [4].

The principle of a coordinated linear system, with several subsystems and a hierarchical top-to-bottom information structure, is in no way restricted to linear systems. Linear systems are merely one of several classes of systems to be considered; the corresponding problem of coordination control for discrete-event systems is studied in [5].

In the following, coordinated linear systems will be defined, and some of their basic properties will be discussed. For notational simplicity, we restrict attention to two subsystems. Coordinated linear systems with more than two subsystems, and their extension to hierarchical systems with more than two layers, are discussed in [6].

13.2 Definition

In accordance with the geometric framework for linear systems developed in [7], we define coordinated linear systems with inputs and outputs in terms of independence and invariance properties of the state, input, and output spaces:

Definition 13.1 Let a continuous-time, time-invariant linear system with inputs and outputs of the form

$$\begin{aligned}\dot{x}(t) &= Ax(t) + Bu(t),\\ y(t) &= Cx(t)\end{aligned}$$

be given. Moreover, let the state space of the system be decomposed as $X = X_1 \dot{+} X_2 \dot{+} X_c$, and let the input and output spaces be given by $U = U_1 \dot{+} U_2 \dot{+} U_c$ and $Y = Y_1 \dot{+} Y_2 \dot{+} Y_c$. Then, we call the system a **coordinated linear system** if we have that

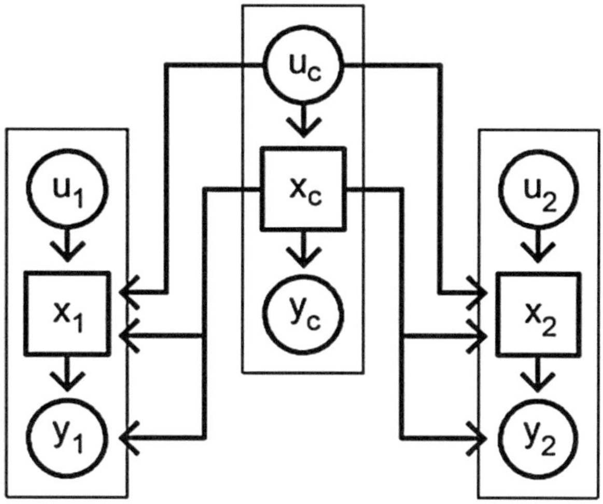

Fig. 13.2 A coordinated linear system with inputs and outputs

1. X_1 and X_2 are A-invariant,
2. $BU_1 \subseteq X_1$ and $BU_2 \subseteq X_2$,
3. and $CX_1 \subseteq Y_1$ and $CX_2 \subseteq Y_2$.

Conditions 1, 2, and 3 in Definition 13.1 imply that the state and input of each subsystem have no influence on the states or the outputs of the coordinator or the other subsystem.

With respect to the decompositions $X = X_1 \dot{+} X_2 \dot{+} X_c$, $U = U_1 \dot{+} U_2 \dot{+} U_c$, and $Y = Y_1 \dot{+} Y_2 \dot{+} Y_c$, the system is then of the form

$$\begin{bmatrix} \dot{x}_1 \\ x_2 \\ x_c \end{bmatrix}(t) = \begin{bmatrix} A_{11} & 0 & A_{1c} \\ 0 & A_{22} & A_{2c} \\ 0 & 0 & A_{cc} \end{bmatrix} \begin{bmatrix} x_1 \\ x_2 \\ x_c \end{bmatrix}(t) + \begin{bmatrix} B_{11} & 0 & B_{1c} \\ 0 & B_{22} & B_{2c} \\ 0 & 0 & B_{cc} \end{bmatrix} \begin{bmatrix} u_1 \\ u_2 \\ u_c \end{bmatrix}(t),$$

$$\begin{bmatrix} y_1 \\ y_2 \\ y_c \end{bmatrix}(t) = \begin{bmatrix} C_{11} & 0 & C_{1c} \\ 0 & C_{22} & C_{2c} \\ 0 & 0 & C_{cc} \end{bmatrix} \begin{bmatrix} x_1 \\ x_2 \\ x_c \end{bmatrix}(t). \tag{13.1}$$

The structure of the system matrices in (13.1) follows directly from conditions 1, 2, and 3 in Definition 13.1.

The interconnections between the different variables of a coordinated linear system are illustrated in Fig. 13.2.

13.3 Basic Properties

The set of coordination-structured matrices

$$\mathbb{R}_{\text{CLS}} = \left\{ \begin{bmatrix} M_{11} & 0 & M_{1c} \\ 0 & M_{22} & M_{2c} \\ 0 & 0 & M_{cc} \end{bmatrix}, \; M_{jj} \in \mathbb{R}^{m_j \times n_j}, \; j = 1, 2, c \right\}$$

forms an algebraic ring (i.e., it is closed with respect to addition and multiplication). In particular, e^M is of the form

$$\exp\left(\begin{bmatrix} M_{11} & 0 & M_{1c} \\ 0 & M_{22} & M_{2c} \\ 0 & 0 & M_{cc} \end{bmatrix} \right) = \begin{bmatrix} e^{M_{11}} & 0 & \star_{1c} \\ 0 & e^{M_{22}} & \star_{2c} \\ 0 & 0 & e^{M_{cc}} \end{bmatrix},$$

where the entries denoted by $\star$ are not specified further. Hence, the information structure imposed by the invariance properties of Definition 13.1 is left unchanged over time by the system dynamics. This means that the conditional independence imposed on subsystems 1 and 2, given the coordinator state and input, is preserved under the system dynamics: No subsystem is ever influenced by the state or input of the other subsystem, not even indirectly via the coordinator.

If $A, B, C, D \in \mathbb{R}_{\text{CLS}}$ are of the appropriate sizes, then the transfer function of the system is given by

$$\hat{G}(z) = D + C(zI - A)^{-1}B = \begin{bmatrix} G_{11}(z) & 0 & \star_{1c} \\ 0 & G_{22}(z) & \star_{2c} \\ 0 & 0 & G_{cc}(z) \end{bmatrix},$$

where $G_{jj}(z) = D_{jj} + C_{jj}(zI - A_{jj})^{-1}B_{jj}$ corresponds to the transfer function of subsystem j for $j = 1, 2, c$ when disregarding the rest of the system, and

$$\star_{ic} = D_{ic} + C_{ii}(zI - A_{ii})^{-1}B_{ic} + (C_{ic} - C_{ii}(zI - A_{ii})^{-1}A_{ic})(zI - A_{cc})^{-1}B_{cc}.$$

Note that the diagonal entries of the linear combination, product, and inverse of matrices in $\mathbb{R}_{\text{CLS}}$ are just the linear combination, product, and inverse of the corresponding diagonal entries of the original matrices, respectively. This means that these operations also preserve the structure of matrices corresponding to more nested hierarchies: If $A \in \mathbb{R}_{\text{CLS}}$ with a diagonal entry $A_{ii} \in \mathbb{R}_{\text{CLS}}$, then operations as above will yield matrices in $\mathbb{R}_{\text{CLS}}$ with the ii-th entry again in $\mathbb{R}_{\text{CLS}}$. Hence, coordinated linear systems can act as building blocks for constructing linear systems with a more complex hierarchical structure: An extension to an arbitrary number of subsystems is straightforward, and nested hierarchies can be modeled by using another coordinated linear system as one of the subsystems of a coordinated linear system. Hierarchical systems that are modeled by such a combination of coordinated linear systems can again be shown to have an information structure that is invariant with respect to the system dynamics.

This invariance property has important consequences for control synthesis: In Chap. 14, it is shown that the problem of stabilizing the overall system via a static state feedback reduces to local stabilization problems for the different parts of the system, and hence can be solved in a fully decentralized manner. LQ optimal control problems for coordinated linear systems decouple partly, allowing for some subproblems to be solved in a decentralized manner (see Chap. 15).

13.4 Concluding Remarks and Further Reading

In this chapter, coordinated linear systems were introduced, and related concepts were summarized. The concept was defined mathematically, and some basic properties were described. An in-depth analysis of coordinated linear systems and related concepts can be found in [6]. Many of the results developed in [1] and [6] are summarized in the following chapters of this part.

References

1. Ran ACM van Schuppen JH (2008) Control for coordination of linear systems. In: Joe Bet al (eds) Proceedings of international symposium on the mathematical theory of networks and systems (MTNS.2008), Virginia Institute of Technology, Blacksburg
2. Kempker PL, Ran ACM, van Schuppen JH (2009) Construction of a coordinator for coordinated linear systems. In: Proceedings of European control conference (ECC.2009), Budapest
3. Kempker PL, Ran ACM, van Schuppen JH (2012) Controllability and observability of coordinated linear systems. Linear Algebra and its Appl 437(1):121–167
4. Kempker PL Ran ACM, van Schuppen JH (2011) A formation flying algorithm for autonomous underwater vehicles. In: Proceedings of 50th IEEE conference on decision and control (CDC.2011), Orlando, FL, USA, pp 1293–1298
5. Komenda J, Masopust T, van Schuppen JH (2010) Synthesis of safe sublanguages satisfying global specification using coordination scheme for discrete-event systems. In: Proceedings of the 10th international workshop on discrete event systems (WODES 2010), Technische Universität Berlin, Berlin, Germany
6. Kempker PL (2012) Coordination control of linear systems. PhD thesis, VU University, Amsterdam, The Netherlands
7. Wonham WM (1979) Linear multivariable control—a geometric approach., Application of mathematicsSpringer, New York

Chapter 14
Coordinated Linear Systems

André C.M. Ran and Jan H. van Schuppen

14.1 Introduction

The problem to be discussed in this essay and the subsequent one is control of linear systems with a modular or distributed structure. The motivation of this research is the frequent occurence of control problems for large-scale linear systems with a clear modular or distributed structure. Frequently, in a distributed system, the *global level* of the coordinator and the *local level* of the subsystems minus the coordinator can be distinguished. In general, the coordinator will consist of parts of the original subsystems but it may also include dynamics of its own.

The concept of a coordinated linear system will be defined. The coordinator subsystem is in its dynamics not affected by the other local subsystems. The dynamics of each subsystem (different from the coordinator) is affected only by its own state and by the state of the coordinator but not by the dynamics of the other subsystems. An equivalent condition for a coordinated linear system is that of conditionally linear independent subspaces of the state space given a coordinator subspace, combined with an invariance condition. The properties of conditionally linearly independent subspaces are investigated.

In the second half of this essay, we discuss control synthesis of distributed systems which admit a coordinated linear control system representation. If the control objectives of a control problem for a distributed system require a tight interaction between the subsystems, then a local control synthesis for each subsystem without

A.C.M. Ran (✉)
Department of Mathematics, VU University Amsterdam, De Boelelaan 1081, 1081 HV Amsterdam, The Netherlands
e-mail: ACM.Ran@few.vu.nl

A.C.M. Ran
Unit for BMI, North-West University, Potchefstroom, South Africa

J.H. van Schuppen
Van Schuppen Control Research, Gouden Leeuw 143, 1103 KB Amsterdam, The Netherlands
e-mail: jan.h.van.schuppen@xs4all.nl

J.H. van Schuppen and T. Villa (eds.), *Coordination Control of Distributed Systems*,
Lecture Notes in Control and Information Sciences 456,
DOI 10.1007/978-3-319-10407-2_14

a coordinator may not meet the control objectives, and hence, coordination control is necessary. The control objectives may then be attained by first carrying out control synthesis of the coordinator and subsequently control synthesis of each of the subsystems.

The equivalent conditions for the control problem are geometric in character and hence are formulated in terms of linear subspaces of the state space. The geometric approach to linear systems was primarily developed by Wonham [13, 14], see the books, see also [12]. The geometric analysis of vector spaces is based on the book [3]. Control of coordination of large-scale and hierarchical systems is treated for example in the book [1].

The usefulness of this essay is that it relates the linear system representation to the concepts of conditional independent subspaces and invariance. This is of interest for the abstract concept of a coordinated system.

14.2 Problem of Coordinated Representations of Linear System Without Inputs by Feedback Equivalence

A linear time-invariant system (see [11]), has a representation

$$\frac{\mathrm{d}x}{\mathrm{d}t} = Ax(t) + Bu(t), \; x(t_0) = x_0,$$
$$y(t) = Cx(t) + Du(t),$$

where T is the time index set, $X = \mathbb{R}^n$ is the state space, $U = \mathbb{R}^m$ is the input space, $Y = \mathbb{R}^p$ is the output space, and $A \in \mathbb{R}^{n\times n}, \; B \in \mathbb{R}^{n\times m}, \; C \in \mathbb{R}^{p\times n}, \; D \in \mathbb{R}^{p\times m}$, while $x : T \to X, \; u : T \to U, \; y : T \to Y$.

The problem to be considered in this essay is as follows. Consider a linear system (without outputs),

$$\frac{\mathrm{d}x}{\mathrm{d}t} = Ax(t) + Bu(t), \; x(t_0) = x_0.$$

Determine a linear control law $u = Fx$ such that, possibly after a state space and an input space transformation, the closed-loop system has the representation

$$\frac{\mathrm{d}x}{\mathrm{d}t} = (A + BF)x(t) = \begin{pmatrix} A_{11} & 0 & A_{1,c} \\ 0 & A_{22} & A_{2,c} \\ 0 & 0 & A_{c,c} \end{pmatrix} x(t),$$
$$x(t_0) = x_0, \; n_1, n_2, n_c \in \mathbb{N}, \; n_1 + n_2 + n_3 = n,$$
$$A_{i,j} \in \mathbb{R}^{n_i \times n_j}, \; \forall i, j \in 1, 2, c.$$

We analyse this problem first for the special case where there is no input, the more general case will be treated afterwards. The corresponding LQ-optimal control problem will be treated in a subsequent essay.

The importance of the special form lies in the following: the state space decomposes into a direct sum of three state spaces, on the first of which the system acts independently of what happens in the second. The dynamics in these first two subsystems is coordinated by the dynamics in the coordinator. Concrete examples of such coordinated linear systems are given for instance in the [7].

14.3 Conditionally Independent Linear Subspaces

In this section, we present auxiliary results from linear algebra.

Consider a linear space X over a field $\mathbb{F}$. The set of linear subspaces of X is denoted by Lat (X). We shall use the notation $X_1 \dot{+} X_2$ to denote the (not necessarily orthogonal) direct sum of two subspaces. Two subspaces $X_1, X_2 \subseteq X$ of a vector space X over a field $\mathbb{F}$ are called *linearly independent subspaces* if $X_1 \cap X_2 = \{0\}$.

Definition 14.1 Consider a linear space X. Two subspaces $X_1, X_2 \in \text{Lat}(X)$ are called *conditionally linear independent given* a subspace $X_c \in \text{Lat}(X)$ if there exist complements $X_{i\backslash c} \subseteq X_i$ of $X_i \cap X_c$, equivalently,

$$X_i = (X_i \cap X_c) \dot{+} X_{i\backslash c}, \; i = 1, 2, \text{ such that,} \tag{14.1}$$

$$X_{1\backslash c}, \; X_{2\backslash c}, \text{ are linearly independent in } X. \tag{14.2}$$

The notation $(X_1, X_2|X_c) \in \text{CILS}$ denotes that the linear subspaces X_1, X_2 are conditionally independent given X_c. In this case, the subspace X_c is called the *coordinator subspace* for X_1 and X_2.

It is immediate from the definition that any two subspaces are conditionally independent given $X_c = X$. However, more interesting is the following observation, which is relatively straightforward.

Theorem 14.1 [10, Lemma 3.2]. *Consider a linear space X and two subspaces $X_1, X_2 \in \text{Lat}(X)$. Define $X_c = X_1 \cap X_2$, and let $X_{i\backslash c}$ be subspaces such that $X_i = (X_i \cap X_c) \dot{+} X_{i\backslash c}$, $i = 1, 2$. Then $X_{1\backslash c} \cap X_{2\backslash c} = \{0\}$. Consequently, $(X_1, X_2|X_1 \cap X_2) \in$ CILS.*

14.4 Coordinated Linear Systems Without Inputs

In this section, we consider linear systems without inputs.

Definition 14.2 Let A be an $n \times n$ matrix. The linear system without input $\frac{dx}{dt} = Ax(t)$ is said to be a *coordinated linear system* (without input) if it has a representation of the form

$$\frac{dx}{dt} = \begin{pmatrix} A_{11} & 0 & A_{1,c} \\ 0 & A_{22} & A_{2,c} \\ 0 & 0 & A_{c,c} \end{pmatrix} x(t), \; x(t_0) = x_0, \tag{14.3}$$

$$n_1, n_2, n_c \in \mathbb{N}, \; n_1 + n_2 + n_c = n, \qquad A_{i,j} \in \mathbb{R}^{n_i \times n_j}, \; \forall i, j \in 1, 2, c.$$

Theorem 14.2 *Let A be an $n \times n$ matrix, and consider the linear system $\frac{dx}{dt} = Ax(t)$, $x(t_0) = x_0$. Consider linear subspaces $X_1, X_2 \in$ Lat (X) with the property that $X_1 \dot{+} X_2 = X$. Define $X_c = X_1 \cap X_2$, and let $X_{i \backslash c}$ be subspaces such that $X_i = X_c \dot{+} X_{i \backslash c}$. Then, it follows from* Theorem 14.1 *that* $(X_1, X_2 | X_c) \in$ CILS.

There exists a basis of X such that with respect to this basis the linear system has the coordinated linear system representation (14.3) *if and only if there are subspaces $X_{1 \backslash c}$ and $X_{2 \backslash c}$ as above such that they are A-invariant:*

$$AX_{1\backslash c} \subseteq X_{1\backslash c}, \; AX_{2\backslash c} \subseteq X_{2\backslash c}. \tag{14.4}$$

Invariant subspaces of matrices are the object of study in [2].

Consider a distributed linear system which consists of an interconnection of two subsystems. Denote the relevant state spaces of these subsystems by X_1 and X_2; suppose that $X = X_1 + X_2$. If the system is not already given in the form of a coordinated system, then it may nevertheless be possible to write the system as a coordinated system with respect to some decomposition. This would have the effect of describing the total system as a combination of two uncoupled system governed by a coordinator. Obviously, it is of interest to select a representation for which the coordinator subspace X_c is, in some way, as small as possible. Denote $X_c = X_1 \cap X_2$. From Theorem 14.1 follows that $(X_1, X_2 | X_c) \in$ CILS. If the invariance condition of Theorem 14.2 holds, then one can choose a basis of X such that the system has a representation as a coordinated linear system. In case the invariance condition (14.4) does not hold it is suggested to extend the coordinator subspace $X_c \subseteq X$ till the invariance condition holds, see [6]. The subspace $X_1 + X_2$ is a coordinator subspace, but there may be smaller subspaces in the range

$$X_1 \cap X_2 \subseteq X_c \subseteq X_1 + X_2.$$

It is then a problem to select a coordinated subspace in this range, possibly of minimal dimension. This is discussed further in Chap. 2 of [4], see also [6].

14.5 Problem of Coordinated Representations of Linear Systems with Inputs by Feedback Equivalence

Consider a time-invariant linear system without outputs, given by the representation $\frac{dx}{dt} = Ax(t) + Bu(t),\ x(t_0) = x_0$. Determine a linear control law $g(x) = Fx$ such that, possibly after a state space and an input space transformation, the closed-loop system has the representation

$$\frac{\mathrm{d}x}{\mathrm{d}t} = (A + BF)x(t) = \begin{pmatrix} A_{11} & 0 & A_{1,c} \\ 0 & A_{22} & A_{2,c} \\ 0 & 0 & A_{c,c} \end{pmatrix} x(t), \tag{14.5}$$

$$x(t_0) = x_0,\ n_1, n_2, n_c \in \mathbb{N},\ n_1 + n_2 + n_3 = n,\quad A_{i,j} \in \mathbb{R}^{n_i \times n_j},\ \forall i, j \in 1, 2, c.$$

14.6 Coordinated Linear Systems with Inputs

Let A be an $n \times n$ real matrix, and let B be an $n \times m$ real matrix.

Definition 14.3 A linear control system with representation

$$\frac{\mathrm{d}x}{\mathrm{d}t} = Ax(t) + Bu(t),\ x(t_0) = x_0, \tag{14.6}$$

is said to be a *coordinated linear control system* if there exists a basis for the state space $X = \mathbb{R}^n$ and for the input space $U = \mathbb{R}^m$ such that with respect to those bases, it has the representation

$$\frac{\mathrm{d}x}{\mathrm{d}t} = \begin{pmatrix} A_{11} & 0 & A_{13} \\ 0 & A_{22} & A_{23} \\ 0 & 0 & A_{33} \end{pmatrix} x(t) + \begin{pmatrix} B_{11} & 0 & B_{13} \\ 0 & B_{22} & B_{23} \\ 0 & 0 & B_{33} \end{pmatrix} u(t),\quad x(t_0) = x_0. \tag{14.7}$$

To avoid trivialities we shall assume from the start that $m \leq n$ and that B has full column rank, i.e. $\ker B = \{0\}$. The latter condition will be dropped later on.

Problem 14.1 The problem we will consider is the following: under what conditions on A and B are there invertible matrices $S \in \mathbb{R}^{n\times n}$ and $T \in \mathbb{R}^{m\times m}$, and a matrix $F \in \mathbb{R}^{m\times n}$ such that

$$S^{-1}(A+BF)S = \begin{pmatrix} A_{11} & 0 & A_{13} \\ 0 & A_{22} & A_{23} \\ 0 & 0 & A_{33} \end{pmatrix}, \qquad S^{-1}BT = \begin{pmatrix} B_{11} & 0 & B_{13} \\ 0 & B_{22} & B_{23} \\ 0 & 0 & B_{33} \end{pmatrix}. \tag{14.8}$$

In other words, in a basis-free formulation, we ask when there are decompositions

$$\mathbb{R}^n = X_1 \dot{+} X_2 \dot{+} X_3, \qquad \mathbb{R}^m = U_1 \dot{+} U_2 \dot{+} U_3,$$

such that with respect to these decompositions, the matrices $A + BF$ and B have the block forms of (14.8). Phrased yet differently, we ask when the pair (A, B) is *feedback equivalent* to a pair of the form (14.8).

To discuss the problem, we need the notion of (A, B)-invariant subspace. Recall, see [2] that a subspace $X_s \subset \mathbb{R}^n$ is called an *(A, B)-invariant subspace* if $AX_s \subset X_s + \text{Im}\, B$.

Theorem 14.3 *Let $A \in \mathbb{R}^{n\times n}$ and $B \in \mathbb{R}^{n\times m}$ with* $\ker B = \{0\}$. *Then there exist invertible matrices $S \in \mathbb{R}^{n\times n}$ and $T \in \mathbb{R}^{m\times m}$, and a feedback matrix $F \in \mathbb{R}^{m\times n}$ such that* (14.8) *holds, if and only if there are (A, B)-invariant subspaces X_1 and X_2 such that $X_1 \cap X_2 = \{0\}$.*

Proof Our first observation is that if such decompositions exist, then $(A + BF)X_i \subset X_i$, $BU_i \subset X_i$ for $i = 1, 2$. Phrased differently, without direct reference to the feedback matrix F: both X_1 and X_2 are (A, B)-invariant subspaces. Furthermore, $X_1 \cap X_2 = \{0\}$.

Conversely, assume that X_1 and X_2 are (A, B)-invariant subspaces such that $X_1 \cap X_2 = \{0\}$. Let X_3 be any complement in $\mathbb{R}^n$ of $X_1 \dot{+} X_2$. Take a basis $\{x_1, \ldots, x_k\}$ in X_1, and a basis $\{x_{k+1}, \ldots, x_l\}$ in X_2. Then, there are vectors u_i, $i = 1, \ldots, l$ in $\mathbb{R}^m$, and vectors $y_1, \ldots, y_k$ in X_1, $y_{k+1}, \ldots, y_l$ in X_2 such that $Ax_i = y_i + Bu_i$. Put $Fx_i = -u_i$. This fixes the action of F on $X_1 \dot{+} X_2$. Let us take $F \mid X_3 = 0$. With this choice of F, we have that $(A + BF)X_i \subset X_i$ for $i = 1, 2$.

It remains to make the decomposition of $\mathbb{R}^m$, and of B. To achieve this, take $U_i = B^{-1}X_i = \{u \mid Bu \in X_i\}$ for $i = 1, 2$. We claim that $U_1 \cap U_2 = \{0\}$. Indeed, assume that $u \in U_1 \cap U_2$. Then, $Bu \in X_1 \cap X_2 = \{0\}$, that is, $u \in \ker B$. Since we assume that $\ker B = \{0\}$, we see that $U_1 \cap U_2 = \{0\}$. We can now take U_3 to be an arbitrary complement of $U_1 \dot{+} U_2$ in $\mathbb{R}^m$. □

Note that it is possible that $U_i = \{0\}$, with $B_{ii} : U_i \to X_i$ being the zero operator.

We can drop the condition that $\ker B = \{0\}$. In this case, the result is as follows. The proof is analogous to the proof of Theorem 14.3.

Theorem 14.4 *Let $A \in \mathbb{R}^{n\times n}$ and $B \in \mathbb{R}^{n\times m}$. Then, there exist an invertible matrix $S \in \mathbb{R}^{n\times n}$, an inverbile matrix $T \in \mathbb{R}^{m\times m}$, and a matrix $F \in \mathbb{R}^{m\times n}$ such that* (14.8) *holds, if and only if there are (A, B)-invariant subspaces X_1 and X_2 of $\mathbb{R}^n$ such that $X_1 \cap X_2 = \{0\}$ and there are subspaces U_1 and U_2 of $B^{-1}X_1$ and $B^{-1}X_2$, respectively, such that $U_1 \cap U_2 = \{0\}$.*

It is of interest to also consider the case of coordinated linear systems in the form above with outputs. There are then three outputs: one of each of the subsystems, and one of the coordinator. The structure of the matrix giving the outputs from the state $x(t)$ is then the same as that of A and B. Thus for the outputs, we have

$$y(t) = \begin{pmatrix} y_1(t) \\ y_2(t) \\ y_3(t) \end{pmatrix} = \begin{pmatrix} C_{11} & 0 & C_{13} \\ 0 & C_{22} & C_{23} \\ 0 & 0 & C_{33} \end{pmatrix} \begin{pmatrix} x_1(t) \\ x_2(t) \\ x_3(t) \end{pmatrix}.$$

The interest of this structured linear system for control synthesis lies in the following idea: first, design a controller for the coordinator and then separately for each of the subsystems, see [5].

14.7 Eigenvalue Assignment of Coordinated Linear Systems

In this section, we apply the results of the previous section to study pole placement for coordinated linear systems.

Proposition 14.1 *Consider a coordinated linear control system with representation* (14.7)*. For any symmetric subset of the complex numbers,* $S_{\text{specification}} \subset \mathbb{C}$ *with at most* n *elements there exist control laws* $g_1(x) = F_{11}x_1 + F_{13}x_3$, $g_2(x) = F_{22}x_2 + F_{23}x_3$, $g_3(x) = F_{33}x_3$, *such that the inputs* $u_1(t) = g_1(x(t))$, $u_2(t) = g_2(x(t))$, $u_3(t) = g_3(x(t))$, *yield a closed-loop linear system with eigenvalues of the system matrix in* $S_{\text{specification}}$ *if and only if the matrix pairs* (A_{11}, B_{11}), (A_{22}, B_{22}), *and*(A_{33}, B_{33}) *are controllable.*

The resulting closed-loop system is then $\frac{dx}{dt} = A_{cl}x(t)$, $x(t_0) = x_0$, *where* A_{cl} *is given by*

$$\begin{pmatrix} A_{11} + B_{11}F_{11} & 0 & A_{13} + B_{11}F_{13} + B_{13}F_{33} \\ 0 & A_{22} + B_{22}F_{22} & A_{23} + B_{22}F_{23} + B_{23}F_{33} \\ 0 & 0 & A_{33} + B_{33}F_{33} \end{pmatrix},$$

which has the required structure specified in (14.5).

The reader can now easily formulate the result corresponding to the above proposition for which only exponential stability of the closed-loop linear system is required in terms of stabilizability.

Note that the control synthesis of a coordinated linear control system proceeds by carrying out first control synthesis of the coordinator and subsequently carrying out independently control synthesis of each of the other subsystems.

14.8 Concluding Remarks

Firstly, the problem of coordination control of distributed linear systems has been discussed. The concept of two subspaces being conditionally linear independent given another subspace has been formulated and the minimal subspace equals the intersection of the two subspaces. The concept of a coordinated linear system has been proposed for the case of a linear time-invariant system without inputs. A linear system without inputs admits a decomposition as a coordinated linear system if an invariance condition for two subspaces holds.

Secondly, the problem of coordination control of distributed linear systems has been discussed. The concept of a coordinated linear system has been proposed for the case with inputs. A linear control system admits a decomposition as a coordinated linear control system if there are two (A, B)-invariant subspaces of the state space that are independent. In that case, the coordinator system has a state space which is a complement to the direct sum of these two (A, B)-invariant subspaces. Control synthesis for such a system can be carried out by first doing control synthesis for the coordinator and then for the two subsystems.

The current essay is based on [10]. For further developments, see [4, 5].

Control for coordination of discrete event systems has been developed in the paper [9].

14.9 Further Reading

The next essay in this book will discuss LQ-control of coordinated systems, see also [7]. The PhD thesis [4] contains much of the relevant developments concerning coordinated systems. Compare also [1].

References

1. Findeisen W, Bailey FN, Brdys M, Malinowski K, Tatjewski P, Wozniak A (1980) Control and coordination in hierarchical systems. Wiley, Chichester
2. Gohberg I, Lancaster P, Rodman L (1986) Invariant subspaces of matrices with applications, Canadian mathematical society series of monographs and advanced texts. Wiley, New York
3. Halmos PR (1993) Finite dimensional vector spaces. Springer, New York
4. Kempker PL (2012) Coordination control of linear systems. PhD thesis, VU University
5. Kempker PL, Ran ACM, van Schuppen JH (2012) Controllability and observability of coordinated linear systems. Linear Algebra Appl 437:121–167
6. Kempker PL, Ran ACM, van Schuppen JH (2009) Construction of a coordinator for coordinated linear systems. In: Proceedings of the European control conference (ECC 2009), Budapest
7. Kempker PL, Ran ACM, van Schuppen JH (2014) LQ control for coordinated linear systems. IEEE Trans Autom Control 59:851–862. doi:10.1109/TAC.2013.2293412
8. Kempker PL, Ran ACM, van Schuppen JH (2014) Construction and minimality of coordinated linear systems. Linear Algebra Appl 452:202–236
9. Komenda J, van Schuppen JH (2008) Coordination control of discrete-event systems. In: Proceedings of the international workshop on discrete-event systems (WODES 2008). IEEE Press, New York

10. Ran ACM, van Schuppen JH (2008) Control for coordination of linear systems. In: Ball J et al (ed) Proceedings of the international symposium on the mathematical theory of networks and systems (MTNS 2008; CD-ROM only), Virginia Institute of Technology, Blacksburg
11. Sontag ED (1998) Mathematical control theory: deterministic finite dimensional systems, 2nd edn., Graduate text in applied mathematicsSpringer, New York
12. Trentelman HL, Stoorvogel AA, Hautus MLJ (2001) Control theory for linear systems. Springer, London
13. Wonham WM (1974) Linear multivariable control: a geometric approach. Lecture notes in economics and mathematical systems, vol 101. Springer, Berlin
14. Wonham WM (1979) Linear multivariable control: a geometric approach. Springer, Berlin

Chapter 15
LQ Optimal Control for Coordinated Linear Systems

Pia L. Kempker

15.1 Introduction

Coordinated linear systems were introduced in Chap. 13. This chapter deals with LQ optimal control for coordinated linear systems: In order to preserve the hierarchical information structure underlying coordinated linear systems, the solution method used for unstructured linear systems has to be adjusted, and some properties of the optimal solution in the unstructured case are lost.

For unstructured linear systems of the form $\dot{x} = Ax + Bu$, with $x(t_0) = x_0$, the infinite-horizon LQ optimal control problem is

$$\min_{\{u(\cdot):[t_0,\infty)\to U\}} \int_{t_0}^{\infty} \left(x^T Qx + u^T Ru\right) \mathrm{d}t,$$

with Q positive semidefinite and R positive definite. If (A, B) is a stabilizable pair and (Q, A) is a detectable pair, then this problem is solved by $u^* = -R^{-1}B^T Xx$, where X is the unique stabilizing solution of the algebraic Riccati equation

$$XBR^{-1}B^T X - XA - A^T X - Q = 0.$$

This solution is independent of the initial condition x_0; the total cost $J(x_0, u^*) = x_0^T Xx_0$, however, does depend on x_0.

This approach cannot be directly applied to coordinated linear systems: In most cases, the feedback matrix $F = -R^{-1}B^T X$ will not be a coordination-structured matrix, and applying this feedback would hence destroy the top-to-bottom informa-

P.L. Kempker (✉)
TNO, P.O. Box 5050, 2600 GB Delft, The Netherlands
e-mail: pia.kempker@tno.nl

J.H. van Schuppen and T. Villa (eds.), *Coordination Control of Distributed Systems*, Lecture Notes in Control and Information Sciences 456,
DOI 10.1007/978-3-319-10407-2_15

tion structure imposed on coordinated linear systems. This means that we have to restrict the admissible inputs to those inputs that preserve the coordinated structure:

Problem 15.1 We consider the restricted LQ optimal control problem

$$\min_{\{u(\cdot)\in U_{\mathrm{CLS}}\}} \int_{t_0}^{\infty} \left(x^T \begin{bmatrix} Q_{11} & 0 & 0 \\ 0 & Q_{22} & 0 \\ 0 & 0 & Q_{cc} \end{bmatrix} x + u^T \begin{bmatrix} R_{11} & 0 & 0 \\ 0 & R_{22} & 0 \\ 0 & 0 & R_{cc} \end{bmatrix} u \right) dt,$$

where the set of admissible input trajectories is given by

$$U_{\mathrm{CLS}} = \left\{ Fx, \text{ where } F = \begin{bmatrix} F_{11} & 0 & F_{1c} \\ 0 & F_{22} & F_{2c} \\ 0 & 0 & F_{cc} \end{bmatrix}, \text{ and } \sigma(A+BF) \subset \mathbb{C}^- \right\},$$

and the state x has dynamics $\dot{x} = \begin{bmatrix} A_{11} & 0 & A_{1c} \\ 0 & A_{22} & A_{2c} \\ 0 & 0 & A_{cc} \end{bmatrix} x + \begin{bmatrix} B_{11} & 0 & B_{1c} \\ 0 & B_{22} & B_{2c} \\ 0 & 0 & B_{cc} \end{bmatrix} u$, with initial state $x(t_0) = x_0$.

Note that for simplicity we have restricted attention to linear state feedbacks: In the unstructured case, the optimal input trajectory turns out to be of the form $u = Fx$; however, there is no indication that this will also be the case for coordinated linear systems.

15.2 Conditionally Optimal Control, Given F_{cc}

Our main result concerning possible solutions to Problem 15.1 states that if the coordinator feedback is fixed (either it was assigned by a higher-level controller, or it is used as a parameter), then the standard procedure of solving an algebraic Riccati equation will produce an optimal feedback in the class of coordination-structured matrices:

Proposition 15.1 *For a given matrix F_{cc} (such that $A_{cc} + B_{cc}F_{cc}$ is stable), define*

$$\widetilde{A} = \begin{bmatrix} A_{11} & 0 & A_{1c} + B_{1c}F_{cc} \\ 0 & A_{22} & A_{2c} + B_{2c}F_{cc} \\ 0 & 0 & A_{cc} + B_{cc}F_{cc} \end{bmatrix}, \quad \widetilde{B} = \begin{bmatrix} B_{11} & 0 \\ 0 & B_{22} \\ 0 & 0 \end{bmatrix}, \tag{15.1}$$

$$\widetilde{Q} = \begin{bmatrix} Q_{11} & 0 & 0 \\ 0 & Q_{22} & 0 \\ 0 & 0 & 0 \end{bmatrix}, \quad \widetilde{R} = \begin{bmatrix} R_{11} & 0 \\ 0 & R_{22} \end{bmatrix}. \tag{15.2}$$

Then, the stabilizing solution of the Riccati equation

$$\widetilde{X}\widetilde{B}\widetilde{R}^{-1}\widetilde{B}^T\widetilde{X} - \widetilde{X}\widetilde{A} - \widetilde{A}\widetilde{X} - \widetilde{Q} = 0 \tag{15.3}$$

is of the form

$$\widetilde{X}(F_{cc}) = \begin{bmatrix} X_{11} & 0 & X_{1c}(F_{cc}) \\ 0 & X_{22} & X_{2c}(F_{cc}) \\ X_{1c}^T(F_{cc}) & X_{2c}^T(F_{cc}) & X_{cc}(F_{cc}) \end{bmatrix}. \tag{15.4}$$

Moreover, let $Y(F_{cc})$ *be the solution of the Lyapunov equation*

$$(A_{cc} + B_{cc}F_{cc})^\top Y(F_{cc}) + Y(F_{cc})(A_{cc} + B_{cc}F_{cc}) = (Q_{cc} + F_{cc}^\top R_{cc}F_{cc}). \tag{15.5}$$

Then, the conditionally optimal feedback for Problem 15.1, *conditioned on* F_{cc}, *is given by*

$$F^* = \begin{bmatrix} -R_{11}^{-1}B_{11}^TX_{11} & 0 & -R_{11}^{-1}B_{11}^TX_{1c}(F_{cc}) \\ 0 & -R_{22}^{-1}B_{22}^TX_{22} & -R_{22}^{-1}B_{22}^TX_{2c}(F_{cc}) \\ 0 & 0 & F_{cc} \end{bmatrix}, \tag{15.6}$$

with cost

$$J(x_0, F^*x) = x_0^T \begin{bmatrix} X_{11} & 0 & X_{1c}(F_{cc}) \\ 0 & X_{22} & X_{2c}(F_{cc}) \\ X_{1c}^T(F_{cc}) & X_{2c}^T(F_{cc}) & X_{cc}(F_{cc}) - Y(F_{cc}) \end{bmatrix} x_0. \tag{15.7}$$

From the proof, which will not be given here, it also follows that the Riccati equation in Proposition 15.1 decouples into several smaller Riccati-type and Lyapunov-type equations and that the submatrices X_{11} and X_{22} are independent of F_{cc}.

Applying the optimal feedback then leads to a closed-loop system of the form

$$\dot{x} = \begin{bmatrix} A_{11} - B_{11}R_{11}^{-1}B_{11}^TX_{11} & 0 & A_{1c} - B_{11}R_{11}^{-1}B_{11}^TX_{1c}(F_{cc}) + B_{1c}F_{cc} \\ 0 & A_{22} - B_{22}R_{22}^{-1}B_{22}^TX_{22} & A_{2c} - B_{22}R_{22}^{-1}B_{22}^TX_{2c}(F_{cc}) + B_{2c}F_{cc} \\ 0 & 0 & A_{cc} + B_{cc}F_{cc} \end{bmatrix} x,$$

which is again a coordinated linear system.

15.3 Possible Choices of F_{cc}

In the case that the coordinator feedback F_{cc} was not assigned by a higher-level controller, the choice of F_{cc} is part of the optimal control problem: Given the conditionally optimal solution $\widetilde{X}(F_{cc})$ given in Proposition 15.1, Problem 15.1 can be rewritten as

$$\min_{\{\text{stabilizing } F_{cc}\}} x_0^T\widetilde{X}(F_{cc})x_0.$$

Since we have no explicit expression for $\widetilde{X}(F_{cc})$ in terms of F_{cc}, we have no general solution for this problem. From examples, we do, however, know that different initial conditions lead to different optimal coordinator feedbacks, and hence, the solution to the restricted LQ optimal control problem considered here does depend on the initial state.

One admissible (but in general not optimal) choice for F_{cc} is $F_{cc} = -R_{cc}^{-1} B_{cc}^T X_{cc}$, where X_{cc} is the unique stabilizing solution of the algebraic Riccati equation

$$X_{cc} B_{cc} R_{cc}^{-1} B_{cc}^T X_{cc} - X_{cc} A_{cc} - A_{cc}^T X_{cc} - Q_{cc} = 0.$$

This solution corresponds to the local LQ optimal control feedback of the coordinator when the subsystems are ignored. The example below shows that this feedback can perform arbitrarily worse than the optimal feedback in the sense of Problem 15.1.

15.4 Example

We illustrate some properties of the solution of the LQ problem for coordinated linear systems in the following example, which was worked out in [9]:

Consider a coordinated linear system with only one subsystem s and coordinator c. Let the system and cost matrices be given by

$$A = \begin{bmatrix} 1 & 0 \\ 0 & -1 \end{bmatrix}, \quad B = \begin{bmatrix} 1 & 1 \\ 0 & 0 \end{bmatrix}, \quad Q = \begin{bmatrix} 1 & 0 \\ 0 & 0 \end{bmatrix}, \quad R = \begin{bmatrix} \alpha_s & 0 \\ 0 & \alpha_c \end{bmatrix},$$

with parameters $\alpha_s, \alpha_c > 0$. The coordinator system is not controllable in this example; however, the subsystem is both locally controllable and controllable via the coordinator input. Hence, the LQ problem for this example reduces to the problem in which input (or combination of inputs) should be used to stabilize the subsystem state, taking into account that different costs (α_s and α_c) are associated with the different inputs. Using Theorem 15.1, we get

$$X = \begin{bmatrix} \alpha_s \left(1 + \sqrt{1 + \frac{1}{\alpha_s}}\right) & \alpha_s f_{cc} \\ \alpha_s f_{cc} & \frac{1}{2} (\alpha_s + \alpha_c) f_{cc}^2 \end{bmatrix}.$$

With initial state $x_0 = \begin{bmatrix} x_{0s} \\ x_{0c} \end{bmatrix}$, the cost is given by

$$J(x_0, Fx(\cdot)) = x_0^T X x_0 = \alpha_s \left(1 + \sqrt{1 + \frac{1}{\alpha_s}}\right) x_{0s}^2 + 2\alpha_s f_{cc} x_{0s} x_{0c} + \frac{1}{2} (\alpha_s + \alpha_c) f_{cc}^2 x_{0c}^2.$$

The unique minimizer of this cost is given by

$$f_{cc}^* = -\frac{2\alpha_s x_{0s}}{(\alpha_s + \alpha_c)\, x_{0c}}$$

and the corresponding minimal cost is

$$J\left(x_0, F^* x(\cdot)\right) = \alpha_s \left(1 + \sqrt{1 + \frac{1}{\alpha_s}} - \frac{2\alpha_s}{\alpha_s + \alpha_c}\right) x_{0s}^2.$$

For comparison, we also give the centralized optimum:

$$G^* = -\begin{bmatrix} \frac{\alpha_c\left(1+\sqrt{1+\frac{1}{\alpha_s}+\frac{1}{\alpha_c}}\right)}{\alpha_s+\alpha_c} & 0 \\ \frac{\alpha_s\left(1+\sqrt{1+\frac{1}{\alpha_s}+\frac{1}{\alpha_c}}\right)}{\alpha_s+\alpha_c} & 0 \end{bmatrix}, \quad J(x_0, G^* x(\cdot)) = \frac{\alpha_s \alpha_c \left(1 + \sqrt{1 + \frac{1}{\alpha_s} + \frac{1}{\alpha_c}}\right)}{\alpha_s + \alpha_c} x_{0s}^2.$$

From this, we can derive the following properties:

- The optimal coordinator feedback f_{cc}^* depends on x_0, while G^* does not.
- For $\alpha_c \to \infty$, both $J(x_0, F^* x(\cdot))$ and $J(x_0, G^* x(\cdot))$ approach $\alpha_s \left(1 + \sqrt{1 + \frac{1}{\alpha_s}}\right) x_{0s}^2$.
- For $\alpha_c \to 0$, we have that $J(x_0, F^* x(\cdot)) \to \alpha_s \left(\sqrt{1 + \frac{1}{\alpha_s}} - 1\right) x_{0s}^2$, but $J(x_0, G^* x(\cdot))$ approaches 0.

The fact that f_{cc}^* depends on the initial state is a major drawback of the LQ coordination control problem considered in this chapter and also implies that a closed-loop solution for f_{cc}^* cannot be derived directly from the matrix equations characterizing the cost, as in the centralized case.

For very large α_c, using the coordinator input for stabilizing the subsystem state is very costly compared to the local input, and hence, both the centralized control law and the coordination control law converge to a local feedback law for the subsystem state. If α_c is very small, then using the coordinator input is very cheap compared to the local input, and the relative cost difference $\frac{J(x_0, F^* x(\cdot)) - J(x_0, G^* x(\cdot))}{J(x_0, G^* x(\cdot))}$ of applying the coordination control law instead of the centralized control law approaches ∞.

15.5 Concluding Remarks

In this chapter, our main results and considerations concerning LQ optimal control for coordinated linear systems were summarized. Further research should concentrate on the impact of different methods of finding admissible coordinator feedbacks F_{cc} on the total cost. Moreover, we want to find conditions on the system and cost matrices which would characterize the following special cases:

- The optimal F_{cc} when ignoring the subsystems is also optimal for the overall control problem,
- The solution of Problem 15.1 is also optimal within the larger class of nonlinear structure-preserving control laws, and
- The solution of Problem 15.1 is also optimal within the larger class of unstructured feedback matrices.

LQ optimal control for coordinated linear systems was applied to the problem of formation flying for autonomous underwater vehicles in [4].

15.6 Further Reading

This chapter summarizes some of the results of [1]. The problem of finding the optimal F_{cc} numerically was addressed in [9]. Note that in this chapter we restricted attention to structure-preserving *static* state feedbacks: The related problem of finding the optimal structure-preserving *dynamic* state feedback for LQ coordination control problems in the frequency domain is discussed in [7, 8]. Other relevant references are [2, 3, 5, 6, 10].

References

1. Kempker PL, Ran ACM, van Schuppen JH (2014) LQ control for coordinated linear systems. IEEE Trans Autom Control 59(4):851–862
2. Kempker PL, Ran ACM, van Schuppen JH (2009) Construction of a coordinator for coordinated linear systems. In: Proceedings of European control conference (ECC.2009), Budapest
3. Kempker PL, Ran ACM, van Schuppen JH (2012) Controllability and observability of coordinated linear systems. Linear Algebra Appl 437:121–167
4. Kempker PL, Ran ACM, van Schuppen JH (2011) A formation flying algorithm for autonomous underwater vehicles. In: Proceedings of 50th IEEE conference on decision and control (CDC.2011), Orlando, FL, USA
5. Komenda J, Masopust T, van Schuppen JH (2010) Synthesis of safe sublanguages satisfying global specification using coordination scheme for discrete-event systems. In: Proceedings of the 10th international workshop on discrete event systems (WODES 2010), Technische Universität Berlin, Berlin, Germany
6. Ran ACM, van Schuppen JH (2008) Control for coordination of linear systems. In: Ball J et al (ed) Proceedings of international symposium on the mathematical theory of networks and systems (MTNS.2008), Virginia Institute of Technology, Blacksburg
7. Shah P, Parrilo PA (2008) A partial order approach to decentralized control. In: Proceedings of 47th IEEE conference on decision and control (CDC.2008)
8. Shah P, Parrilo PA (2010) H_2-Optimal decentralized control over posets: a state-space solution for state-feedback. In: Proceedings of 49th IEEE conference on decision and control (CDC.2010)
9. Willems R (2012) A numerical approach to LQ control for coordinated linear systems. Master's thesis, Vrije Universiteit Amsterdam
10. Wonham WM (1979) Linear multivariable control—a geometric approach. Application of mathematics, 2nd edn., vol 10. Springer, New York

Chapter 16
Supervisory Control of Discrete-Event Systems

Jan Komenda and Tomáš Masopust

16.1 Motivation

In our daily lives, we are confronted with many different machines that can be viewed as instances of discrete-event systems. Probably, the most popular examples are personal computers, laptops, tablets, mobile phones, ATM machines, beverage machines, copy machines, medical scanners, and others. Many large engineering systems, such as manufacturing or transportation systems, contain discrete-event components, for instance conveyor belts, robots, storage capacities, on-board computers in vehicles, etc. Large complex engineering systems are often built out of the smaller ones as their synchronous or asynchronous compositions. Notice that the composition of discrete-event systems results in a discrete-event system again.

Discrete-event systems composed of two or more subsystems are called distributed. As the complexity of distributed systems grows, human operators are not able to design a controller by hand, and a formal approach to design a controller or supervisor is needed. The task of a supervisor is to impose a given requirement on the behavior of the system that is usually referred to as control specification. Traditionally, control specifications are formulated only informally and software engineers translate them into a control software manually. This is a time-consuming and error-prone process; moreover, the produced software needs to be verified using model-checking techniques.

Supervisory control is a formal method providing a theory to design supervisors for discrete-event systems. Using supervisory control theory, the control design process becomes fully automated. Typical examples of control specifications are to avoid dangerous states in the systems such as overflows or underflows of buffers in manufacturing systems, to avoid collision of vehicles in transportation systems or to

J. Komenda (✉) · T. Masopust
Institute of Mathematics, Academy of Sciences of the Czech Republic, Žižkova 22, 61662 Brno, Czech Republic
e-mail: komenda@ipm.cz

T. Masopust
e-mail: masopust@math.cas.cz

J.H. van Schuppen and T. Villa (eds.), *Coordination Control of Distributed Systems*,
Lecture Notes in Control and Information Sciences 456,
DOI 10.1007/978-3-319-10407-2_16

control access to databases with many users, where a new user may enter only after all previous users have completed their tasks (writing to the database). Supervisory control theory is also applicable to continuous-time and hybrid systems (those composed of a discrete and continuous components) after these systems are abstracted into discrete-event systems.

The aim of this essay is to acquaint the reader with the basic notions, concepts, and results of discrete-event systems and supervisory control theory. It is an introduction to the subsequent essay on coordination control, Chap. 17. For further details, the reader is referred to [2, 9, 12]. The pioneering work on this topic is described in the papers [6, 7].

16.2 Concepts

This chapter differs from the other chapters of this book in the considered model. The model used in this chapter is a finite automaton, also called a finite-state machine. In supervisory control theory, automata are usually called generators. An introduction to automata theory and formal languages can be found in [4, 10]. Here, we recall only the notions necessary for the presented theory.

Consider a discrete-event system. Each action of such a system is described by an event. Let Σ denote a finite nonempty set of *events*. The behavior of a system is then a finite sequence of actions; hence, we can see it as a finite sequence of events. Such a finite sequence is called a *word* over the event set Σ. The set of all words over the event set Σ is denoted by Σ^*. The initial behavior of a system, where no action has yet been performed, is described by the *empty word*, denoted by ε. For instance, for an event set $\Sigma = \{e, h, l, o\}$, the word *hello* is an example of a word over Σ.

As already mentioned, the basic model of discrete-event systems used here is a generator. A *generator* is a quintuple $G = (Q, \Sigma, f, q_0, Q_m)$, where Q is a finite nonempty set of *states*, Σ is a finite set of events (*event set*), $f: Q \times \Sigma \to Q$ is a *partial transition function*, $q_0 \in Q$ is the *initial state*, and $Q_m \subseteq Q$ is a set of *marked states*. For instance, let $G = (\{1, 2, 3\}, \{\$1, \$2, tea, coffee\}, f, 1, \{1\})$ be a generator, modeling a simple beverage machine. A graphical representation of generators uses labeled graphs as demonstrated in Fig. 16.1. States of the generator are drawn as circles, a transition $f(q, a) = p$ is depicted as a labeled arrow from state q to state p labeled by event a, the initial state is denoted by an incoming arrow that does not come from any other state, and the marked states are drawn as double circles.

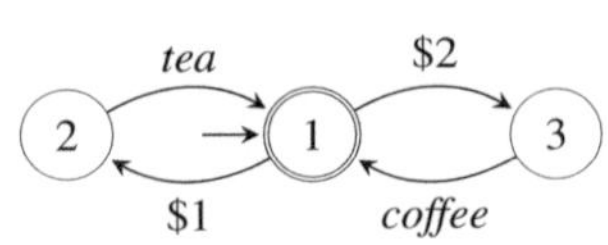

Fig. 16.1 A graphical representation of the generator G of a simple beverage machine; tea costs \$1, while we need to pay \$2 to get coffee

The transition function f can be extended from events to words as the function $\hat{f}:Q \times \Sigma^* \to Q$ so that $\hat{f}(q, \varepsilon) = q$ and $\hat{f}(q, aw) = \hat{f}(f(q, a), w)$, for $a \in \Sigma$ and $w \in \Sigma^*$. The behavior of a generator G is described in terms of languages. The language *generated* by G is the set $L(G) = \{s \in \Sigma^* \mid \hat{f}(q_0, s) \in Q\}$, and the language *marked* by G is the set $L_m(G) = \{s \in \Sigma^* \mid \hat{f}(q_0, s) \in Q_m\}$. Obviously, $L_m(G) \subseteq L(G)$. Thus, for our example, we have $L(G) = \{\varepsilon, \$1, \$1\ tea, \$1\ tea\ \$2, \$1\ tea\ \$2\ coffee, \ldots\}$ and $L_m(G) = \{\varepsilon, \$1\ tea, \$1\ tea\ \$2\ coffee, \ldots\}$.

The set of behaviors of a generator is called a (regular) language. On the other hand, a *(regular) language* L over an event set Σ is a set $L \subseteq \Sigma^*$ such that there exists a generator G with $L_m(G) = L$. The prefix closure $\overline{L}$ of a language L over an event set Σ is the set of all prefixes of all its words, that is, $\overline{L} = \{w \in \Sigma^* \mid$ there exists $u \in \Sigma^*$ such that $wu \in L\}$; language L is prefix-closed if $L = \overline{L}$.

To control a system, we need to specify which events can be controlled, that is, disabled by a supervisor. This is done by the notion of a controlled generator. A *controlled generator* over an event set Σ is a triple (G, Σ_c, Γ), where G is a generator over $\Sigma = \Sigma_c \cup \Sigma_u$, where Σ_c is the set of *controllable events*, $\Sigma_u = \Sigma \setminus \Sigma_c$ is the set of *uncontrollable events*, and $\Gamma = \{\gamma \subseteq \Sigma \mid \Sigma_u \subseteq \gamma\}$ is the *set of control patterns*. Only controllable events can be disabled by the supervisor, while uncontrollable events cannot be prevented from happening. Typical instances of uncontrollable events are fault events, ticks of clocks, high priority events, unpreventable events due to hardware or actuation limitations, events that should not be disabled, and so on. Supervisors choose events among those from control patterns.

Example 16.1 Let $G = (\{1, 2, 3\}, \{a, b, c, d\}, f, 1, \{1\})$ be a generator depicted in Fig. 16.2 with the set of controllable events $\Sigma_c = \{a, c\}$. Then, $\Sigma_u = \{b, d\}$ are uncontrollable events (whose transitions are depicted by dashed arrows) and $\Gamma = \{\{b, d\}, \{a, b, d\}, \{b, c, d\}, \{a, b, c, d\}\}$ is the set of control patterns. One can think of this example as a simple manufacturing system, where a single resource (e.g., a machine) is shared by two manufacturing lines: one line is represented by operations (event sequences) $(ab)^*$ and the other one by operations $(cd)^*$. All possible schedules are considered, that is, the resource can be attributed to an arbitrary sequence of both lines.

A *supervisor* for the controlled generator (G, Σ_c, Γ) is a map $S:L(G) \to \Gamma$. The meaning of a supervisor is that for any state of the system, i.e., a word w generated by the generator, $S(w)$ defines the set of events that are enabled in the system after w is generated. By definition of Γ, only controllable events can be disabled.

Applying a supervisor on a (controlled) generator results in the closed-loop system. The *closed-loop system* associated with the controlled generator (G, Σ_c, Γ)

Fig. 16.2 Generator G

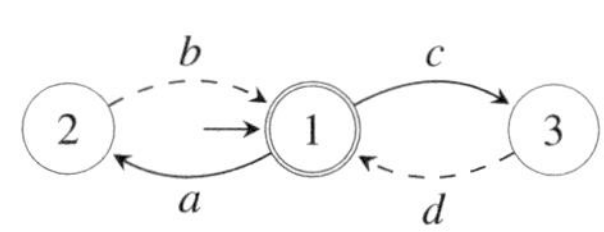

and the supervisor S is the resulting (supervised/controlled) system defined as the minimal language $L(S/G) \subseteq \Sigma^*$ such that

1. $\varepsilon \in L(S/G)$ and
2. if $s \in L(S/G)$, $a \in S(s)$, and $sa \in L(G)$, then $sa \in L(S/G)$.

The intuition is that the supervisor disables some transitions of the generator G, but it can never disable any transition under an uncontrollable event. The marked language of the closed-loop system is defined as $L_m(S/G) = L(S/G) \cap L_m(G)$, meaning that in the closed-loop system, the states are marked in the same way as in G.

If the closed-loop system is nonblocking, that is, $\overline{L_m(S/G)} = L(S/G)$, then the supervisor S is called *nonblocking*.

Example 16.2 Consider the controlled generator from Example 16.1. Our goal is to define a supervisor that disables events c and a in an alternating way when the plant is back in the initial state. More precisely, c is disabled in the initial state at the beginning of the work of the system (that is, the generated word is ε), a is disabled in the state after ab is generated, c is disabled after $abcd$ is generated and so forth. We define the supervisor S as follows. For $k \geq 0$,

- $S((abcd)^k) = \{a, b, d\}$,
- $S((abcd)^k ab) = \{b, c, d\}$,
- for all other words w, $S(w) = \{a, b, c, d\}$.

The closed-loop system is then $L(S/G) = \overline{L_m(S/G)} = \overline{\{(abcd)^k \mid k \geq 0\}}$, so the supervisor is nonblocking.

The following two concepts play a central role in supervisory control [12].

Definition 16.1 (*Controllable language*) Let G be a generator over an event set Σ. A language $K \subseteq L(G)$ is *controllable* with respect to $L(G)$ and Σ_u if

$$\overline{K}\Sigma_u \cap L(G) \subseteq \overline{K}\,.$$

The concept of controllability of a specification language in supervisory control theory differs from controllability in classical control theory of linear or nonlinear systems. It is, however, closely related to the concept of invariant spaces from geometrical control theory, because it requires that one cannot exit from the specification by an uncontrollable transition while staying within the plant language.

Definition 16.2 (*$L_m(G)$-closed language*) Let G be a generator. A nonempty language $K \subseteq L_m(G)$ is *$L_m(G)$-closed* if

$$K = \overline{K} \cap L_m(G)\,.$$

Example 16.3 Consider the generator G defined in Example 16.1, and define a specification language K as the language of the generator depicted in Fig. 16.3. Note that

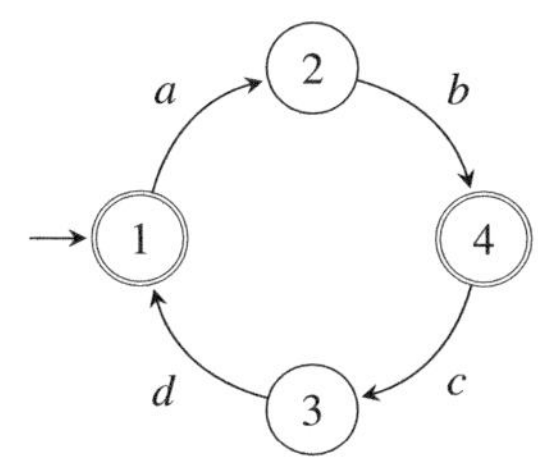

Fig. 16.3 Generator of the specification K

the specification restricts the behavior of the manufacturing system by imposing a particular schedule so that the resource is attributed in an alternating way to both manufacturing lines. One can verify that K is controllable with respect to $L(G)$ and Σ_u, and $L_m(G)$-closed.

16.3 Supervisory Control Problem

In this section, we formally define the supervisory control problem. Let K be a specification language, and let G be a plant (generator). The control objective of supervisory control is to find a nonblocking supervisor S (if possible) such that the closed-loop system satisfies the specification, that is,

$$L(S/G) = \overline{K} \qquad \text{and} \qquad L_m(S/G) = K\,.$$

It cannot be satisfied if $K \not\subseteq L_m(G)$; therefore, we can assume that $K \subseteq L_m(G)$.

16.4 Supervisory Control Theory

The supervisory control problem is to find conditions that are equivalent to the existence of a supervisor that achieves a specification. Two conditions defined above, *controllability* and $L_m(G)$-*closedness*, are necessary and sufficient for the existence of a nonblocking supervisor that achieves the specification, see [2, 12] for the proofs of the following theorems.

Theorem 16.1 *Consider the problem specified above. There exists a nonblocking supervisor S solving the problem if and only if the specification language K is both controllable with respect to $L(G)$ and Σ_u, and $L_m(G)$-closed.*

Example 16.4 Consider the plant and the specification of Example 16.3. By Theorem 16.1, there exists a supervisor S solving the supervisory control problem. This supervisor is described in Example 16.2. Note that the supervisor can be represented as an automaton. Moreover, if the specification K is controllable with respect to the plant language $L(G)$ and Σ_u, the generator for the specification is precisely the automaton representation of the supervisor. Thus, the automaton representation of the supervisor S is depicted in Fig. 16.3.

It remains to explain what to do if the specification language is not controllable (in some sense, the $L_m(G)$-closedness is not an issue, because if K is not $L_m(G)$-closed, then $\overline{K} \cap L_m(G)$ is considered as a new specification, cf. [2]). For uncontrollable specification languages, controllable sublanguages of the specification are considered instead. Note that control specifications are most often safety specifications expressed by a language inclusion and, therefore, it is reasonable to compute a controllable sublanguage of a specification, if the specification fails to be controllable. The notation $C(K, L(G), \Sigma_u)$ stands for the set of controllable sublanguages of the specification K with respect to $L(G)$ and Σ_u. It is not hard to check that controllability is preserved by language unions. Consequently, there always exists the supremal controllable sublanguage of the specification language among the controllable sublanguages, denoted by $\sup C(K, L(G), \Sigma_u)$, see [2, 12].

Theorem 16.2 *The supremal controllable sublanguage of a specification language always exists and is equal to the union of all controllable sublanguages of the specification.*

The supremal controllable sublanguage is an important concept, because it represents the maximally permissive (or, equivalently, minimally restrictive) solution to the supervisory control problem.

16.5 Nonblockingness in Distributed Systems

In this section, we study the blocking issue that can appear in distributed discrete-event systems. A *distributed discrete-event system* with synchronous communication is a concurrent system formed by the synchronous product of several local subsystems. The engineering relevance of distributed systems modeled by discrete-event systems can be justified by showing that supervisory control for the following systems has been investigated in the literature: control of a rapid thermal multiprocessor [1], databases [2], chemical plants [8], feature interaction in telephone networks [11], theme park vehicles [3], a controller for traffic lights [5].

Synchronous product is a standard way of constructing large-scale systems as a composition of potentially a large number of smaller systems. Formally, a synchronous product of languages $L_1 \subseteq \Sigma_1^*$ and $L_2 \subseteq \Sigma_2^*$ is the language

$$L_1 \| L_2 = P_1^{-1}(L_1) \cap P_2^{-1}(L_2) \subseteq \Sigma^* ,$$

where $P_i : \Sigma^* \to \Sigma_i^*$ are natural projections, for $i = 1, 2$. A (*natural projection*) $P : \Sigma^* \to \Sigma_0^*$, where $\Sigma_0 \subseteq \Sigma$, is a homomorphism defined so that $P(a) = \varepsilon$ for $a \in \Sigma \setminus \Sigma_0$, and $P(a) = a$ for $a \in \Sigma_0$. The projection of a word is thus uniquely determined by projections of its letters. The *inverse image* of P is denoted by $P^{-1}:\Sigma_0^* \to 2^{\Sigma^*}$.

Example 16.5 Let $P:\{a,b,c\}^* \to \{a,b\}^*$ be a projection. Then, the projection of a word $abcba$ is $P(abcba) = abba$. On the other hand, the inverse image of $abba$ is $P^{-1}(abba) = \{c^i ac^j bc^k bc^\ell ac^m \mid i,j,k,\ell,m \geq 0\}$.

For two generators, $G_1 = (Q_1, \Sigma_1, f_1, q_{01}, Q_{m1})$ and $G_2 = (Q_2, \Sigma_2, f_2, q_{02}, Q_{m2})$, the generator $G_1 \| G_2$ is defined as the accessible part (i.e., the part of the state set which can be reached from the initial state) of the generator $(Q_1 \times Q_2, \Sigma_1 \cup \Sigma_2, f, (q_{01}, q_{02}), Q_{m1} \times Q_{m2})$, where

$$f((x,y),e) = \begin{cases} (f_1(x,e), f_2(y,e)), & \text{if } f_1(x,e) \in Q_1 \text{ and } f_2(y,e) \in Q_2, \\ (f_1(x,e), y), & \text{if } f_1(x,e) \in Q_1 \text{ and } e \notin \Sigma_2, \\ (x, f_2(y,e)), & \text{if } e \notin \Sigma_1 \text{ and } f_2(y,e) \in Q_2, \\ \text{undefined}, & \text{otherwise.} \end{cases}$$

It is known that the relation to the synchronous product of languages is as follows: $L(G_1 \| G_2) = L(G_1) \| L(G_2)$ and $L_m(G_1 \| G_2) = L_m(G_1) \| L_m(G_2)$.

Example 16.6 Consider two generators G_1 and G_2 depicted in Fig. 16.4. Note that the only event shared by the generators is the event a and that events b and d are private in the respective generators. Then, the synchronous product (also called parallel composition) $G_1 \| G_2$ is depicted in Fig. 16.5.

Recall that a generator G is nonblocking if $\overline{L_m(G)} = L(G)$, that is, if every generated word from $L(G)$ can be extended to a marked word from $L_m(G)$. Otherwise, we say that the system is blocking, which typically arises in discrete-event systems formed by the synchronous product. It is well known that the synchronous product of two nonblocking generators G_1 and G_2 can be blocking.

Fig. 16.4 Generators G_1 and G_2

Fig. 16.5 Synchronous product $G_1 \| G_2$

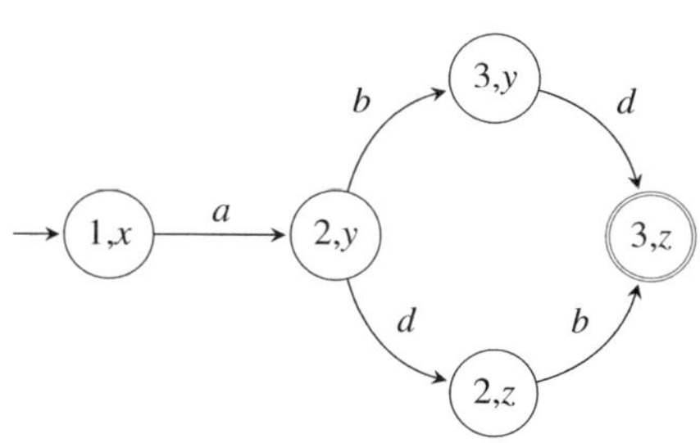

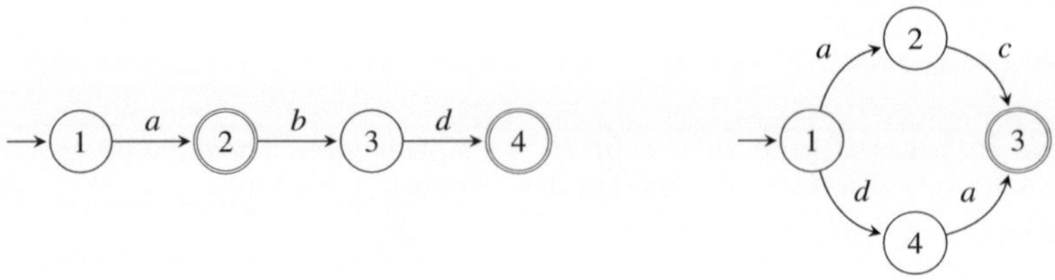

Fig. 16.6 Generators G_1 and G_2

Fig. 16.7 Synchronous product $G_1 \| G_2$

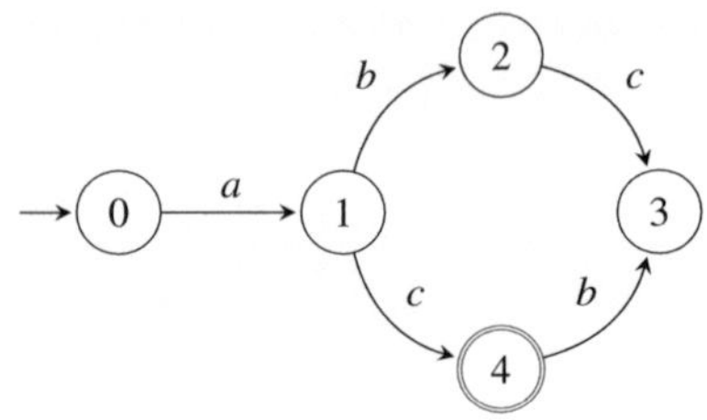

Example 16.7 Consider nonblocking generators G_1 and G_2 depicted in Fig. 16.6. Their synchronous product, depicted in Fig. 16.7, is blocking because no marked state is accessible from state 3.

References

1. Balemi S, Hoffmann GJ, Gyugi P, Wong-Toi H, Franklin GF (1993) Supervisory control of a rapid thermal multiprocessor. IEEE Trans Automat Control 38(7):1040–1059
2. Cassandras CG, Lafortune S (2008) Introduction to discrete event systems. Springer, Berlin
3. Forschelen STJ, Mortel-Fronczak JM, Su R, Rooda JE (2012) Application of supervisory control theory to theme park vehicles. Discrete Event Dyn Syst 22(4):511–540
4. Hopcroft JE, Motwani R, Ullman JD (2006) Introduction to automata theory, languages, and computation. Addison-Wesley, Boston
5. Kurshan RP (1994) Computer-aided verification of coordinating processes: the automata-theoretic approach. Princeton University Press, Princeton
6. Ramadge PJ, Wonham WM (1987) Supervisory control of a class of discrete event processes. SIAM J Control Optim 25(1):206–230
7. Ramadge PJ, Wonham WM (1989) The control of discrete event systems. Proc of IEEE 77(1):81–98
8. Sang-Heon L (1998) Structural decentralised control of concurrent discrete-event systems. PhD thesis, Australian National University, Canberra
9. Seatzu C, Silva M, van Schuppen JH (eds) (2013) Control of discrete-event systems, vol 433 of LNCIS. Springer, London
10. Sipser M (2005) Introduction to the theory of computation. Course Technology, Boston
11. Thistle JG, Malhamé RP, Hoang HH, Lafortune S (1997) Feature interaction modelling, detection and resolution: a supervisory control approach. In FIW, UK, pp 93–107
12. Wonham WM (2009) Supervisory control of discrete-event systems. Department of electrical and computer engineering, University of Toronto, Lecture notes

Chapter 17
Coordination Control of Distributed Discrete-Event Systems

Jan Komenda and Tomáš Masopust

17.1 Motivation

Supervisory control of distributed discrete-event systems with synchronous communication and a global specification is a difficult problem. Control synthesis relying on the composition of all subsystems is not feasible in general because of its exponential complexity in the number of subsystems. Hence, a local control synthesis is preferred. However, controllers based only on the local control synthesis may be blocking, and, if not blocking, they may not reach the performance of the global control synthesis. Therefore, a form of coordination between the subsystems is needed.

The coordination control architecture proposed in [11] is applicable in general and deals with a control synthesis for distributed systems with a global specification. Coordination control was first developed for prefix-closed languages in [10] and then further extended to partial observations in [6]. A nonprefix-closed extension is discussed in [7]. The approaches for prefix-closed languages are implemented in the software library libFAUDES [14].

For simplicity, the theoretical development considers the special case of two subsystems. However, the extension to more subsystems is straightforward as demonstrated in the example consisting of three subsystems.

17.2 Problem

Consider a system given by a composition of generators G_1 and G_2 over the event sets Σ_1 and Σ_2, respectively. Let G_k be a coordinator over an event set Σ_k such

J. Komenda (✉) · T. Masopust
Institute of Mathematics, Academy of Sciences of the Czech Republic, Žižkova 22, 61662 Brno, Czech Republic
e-mail: komenda@ipm.cz

T. Masopust
e-mail: masopust@math.cas.cz

J.H. van Schuppen and T. Villa (eds.), *Coordination Control of Distributed Systems*,
Lecture Notes in Control and Information Sciences 456,
DOI 10.1007/978-3-319-10407-2_17

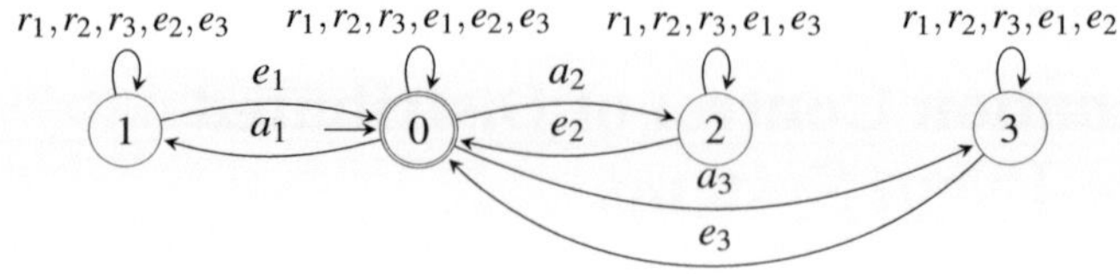

Fig. 17.1 Specification K

that $\Sigma_k \supseteq \Sigma_1 \cap \Sigma_2$. Assume that the specification $K \subseteq L_m(G_1 \| G_2 \| G_k)$ and its prefix-closure $\overline{K}$ are conditionally decomposable with respect to event sets Σ_1, Σ_2, and Σ_k (see Sect. 17.3). The aim of the coordination control synthesis is to determine nonblocking supervisors S_1, S_2, S_k for respective generators such that

$$L_m(S_k/G_k) \subseteq P_k(K) \text{ and } L_m(S_i/[G_i \parallel (S_k/G_k)]) \subseteq P_{i+k}(K),$$

for $i = 1, 2$, and the closed-loop system with the coordinator satisfies

$$L_m(S_1/[G_1 \parallel (S_k/G_k)]) \parallel L_m(S_2/[G_2 \parallel (S_k/G_k)]) = K. \qquad \diamond$$

Note that one could expect that the equality $L(S_1/[G_1 \parallel (S_k/G_k)]) \parallel L(S_2/[G_2 \parallel (S_k/G_k)]) = \overline{K}$ for prefix-closed languages should also be required in the statement of the problem, but it is sufficient to require the equality for marked languages since it implies that

$$\begin{aligned}
\overline{K} &= \overline{L_m(S_1/[G_1 \parallel (S_k/G_k)]) \parallel L_m(S_2/[G_2 \parallel (S_k/G_k)])} \\
&\subseteq \overline{L_m(S_1/[G_1 \parallel (S_k/G_k)])} \parallel \overline{L_m(S_2/[G_2 \parallel (S_k/G_k)])} \\
&\subseteq \overline{P_{1+k}(K)} \parallel \overline{P_{2+k}(K)} \\
&= \overline{K}.
\end{aligned}$$

If such supervisors exist, their synchronous product is a nonblocking supervisor for the global plant, cf. [5].

Example 17.1 Database transactions are examples of discrete-event systems that should be controlled to avoid incorrect behaviors. Transactions are modeled by a sequence of request (r), access (a), and exit (e) operations. Often, several users access the database, which can lead to inconsistencies when executed concurrently, because not all interleavings of operations give a correct behavior. Consider three users with events $\mathrm{r}_i, \mathrm{a}_i, \mathrm{e}_i$, where $i = 1, 2, 3$. All possible schedules are described by the behavior of the plant $G_1 \| G_2 \| G_3$, where G_1, G_2, G_3 are nonblocking generators with $L_m(G_i) = \{(\mathrm{r}_i\mathrm{a}_i\mathrm{e}_i)^j \mid j \in \mathbb{Z}_+\}$, which is also denoted as $(\mathrm{r}_i\mathrm{a}_i\mathrm{e}_i)^*$, and the set of controllable events is $\Sigma_c = \{\mathrm{a}_i \mid i = 1, 2, 3\}$. The specification K (Fig. 17.1) describes the correct behavior consisting in finishing the transaction in the exit stage before another transaction can proceed to the exit phase.

17.3 Concepts

The reader is referred to Chap. 16 for the basic notions and concepts of discrete-event systems and supervisory control. Having a global specification, the first step we need to do is to identify the right parts of the specification corresponding to each of the respective subsystems.

A language K is *conditionally decomposable* with respect to event sets Σ_1, Σ_2, Σ_k, where $\Sigma_1 \cap \Sigma_2 \subseteq \Sigma_k \subseteq \Sigma_1 \cup \Sigma_2$, if

$$K = P_{1+k}(K) \parallel P_{2+k}(K),$$

where $P_{i+k}: (\Sigma_1 \cup \Sigma_2)^* \to (\Sigma_i \cup \Sigma_k)^*$ is a projection, for $i = 1, 2$.

There always exists an extension of Σ_k that satisfies this condition; $\Sigma_k = \Sigma_1 \cup \Sigma_2$ is such a trivial example. A polynomial algorithm to check whether the condition is satisfied and, if not, to extend the event set Σ_k so that it becomes satisfied can be found in [9]. The question which extension is the most appropriate requires further investigation. To find the minimal extension with respect to set inclusion is an NP-hard problem [8].

For event sets $\Sigma_i, \Sigma_j, \Sigma_\ell \subseteq \Sigma$, in what follows we use the notation P_ℓ^{i+j} to denote the projection from $(\Sigma_i \cup \Sigma_j)^*$ to Σ_ℓ^*. If $\Sigma_i \cup \Sigma_j = \Sigma$, we simplify the notation to P_ℓ. Moreover, $\Sigma_{i,u} = \Sigma_i \cap \Sigma_u$ denotes the set of locally uncontrollable events of the event set Σ_i.

Languages K and L are *synchronously nonconflicting* if $\overline{K \parallel L} = \overline{K} \parallel \overline{L}$.

Lemma 17.1 ([8]) *Let K be a language. If the language $\overline{K}$ is conditionally decomposable, then the languages $P_{1+k}(K)$ and $P_{2+k}(K)$ are synchronously nonconflicting.*

17.4 Construction of a Coordinator

In the statement of the problem above, we have mentioned the notion of a coordinator. The fundamental problem, however, is the construction of such a coordinator. We now discuss one of the possible constructions of a suitable coordinator.

Algorithm 1 (*Construction of a coordinator*) Consider two subsystems G_1 and G_2 over the event sets Σ_1 and Σ_2, respectively, and let K be a specification language. Construct an event set Σ_k and a coordinator G_k as follows:

1. Set $\Sigma_k = \Sigma_1 \cap \Sigma_2$ to be the set of all shared events.
2. Extend Σ_k with events of $\Sigma_1 \cup \Sigma_2$ so that K and $\overline{K}$ are conditionally decomposable (for instance using a method described in [9]).
3. Set the coordinator $G_k = P_k(G_1) \parallel P_k(G_2)$; for a generator G and a projection P, $P(G)$ is a generator whose behavior satisfies $L(P(G)) = P(L(G))$ and $L_m(P(G)) = P(L_m(G))$.

Example 17.2 Consider the statement of Example 17.1. We can verify that for $\Sigma_k = \{a_1, a_2, a_3\}$, the specification language K and its prefix-closure $\overline{K}$ are conditionally decomposable with respect to Σ_1, Σ_2, Σ_3 and Σ_k. The coordinator is then computed as $G_k = P_k(G_1)\|P_k(G_2)\|P_k(G_3)$.

From the complexity viewpoint, the problem is that the projected generator $P_k(G_i)$ can have exponential number of states compared to the generator G_i. So far, the only known condition ensuring that the projected generator is smaller (in the number of states) than the original one is the observer property (see Definition 17.1 below). Therefore, we might need to add step (2b) to further extend Σ_k so that the projection P_k is an $L(G_i)$-observer, for $i = 1, 2$. A polynomial algorithm how to do this can be found in [2, 16].

Definition 17.1 (*Observer property*) Let $\Sigma_k \subseteq \Sigma$. The projection $P_k\colon \Sigma^* \to \Sigma_k^*$ is an *L-observer* for a language $L \subseteq \Sigma^*$ if for every $t \in P(L)$ and $s \in \overline{L}$, if $P(s)$ is a prefix of t, then there exists $u \in \Sigma^*$ such that $su \in L$ and $P(su) = t$, cf. Fig. 17.2.

Example 17.3 The projection P_k from Example 17.2 is a K-observer, but it is not an $L_m(G_i)$-observer for $i = 1, 2, 3$. However, the projected generators $P_k(G_i)$, $i = 1, 2, 3$, have only one state.

Theorem 17.1 *If a projection P is an $L(G)$-observer, for a generator G, then the minimal generator for the language $P(L(G))$ has no more states than G.*

Based on this result, the coordinator G_k is expected to be quite small compared to the global plant $G_1\|G_2$.

17.5 Theory

The theory presented here is based on the latest results that can be found in [7] (see also [8]), together with the results from [10].

Let G_1 and G_2 be two generators over Σ_1 and Σ_2, respectively, and let G_k be a coordinator over Σ_k. A language $K \subseteq L(G_1 \parallel G_2 \parallel G_k)$ is *conditionally controllable* for generators G_1, G_2, G_k and uncontrollable event sets $\Sigma_{1,u}$, $\Sigma_{2,u}$, $\Sigma_{k,u}$ if

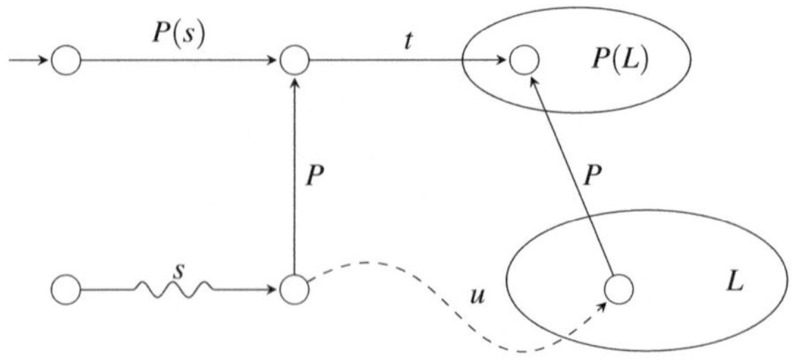

Fig. 17.2 Demonstration of the observer property

1. $P_k(K)$ is controllable with respect to $L(G_k)$ and $\Sigma_{k,u}$,
2. $P_{1+k}(K)$ is controllable with respect to $L(G_1) \parallel \overline{P_k(K)}$ and $\Sigma_{1+k,u}$,
3. $P_{2+k}(K)$ is controllable with respect to $L(G_2) \parallel \overline{P_k(K)}$ and $\Sigma_{2+k,u}$,

where $\Sigma_{i+k,u} = (\Sigma_i \cup \Sigma_k) \cap \Sigma_u$, for $i = 1, 2$.

Example 17.4 Consider Example 17.2. It can be verified that $P_k(K) = \{a_1, a_2, a_3\}^*$ is controllable with respect to $L(G_k) = P_k(K)$ and $\Sigma_{k,u} = \emptyset$. This does not hold for $P_{i+k}(K)$ because the language is not included in $L(G_i)\|\overline{P_k(K)}$, for $i = 1, 2, 3$.

As in the monolithic case, we need a notion similar to $L_m(G)$-closedness. A nonempty language $K \subseteq \Sigma^*$ is *conditionally closed* for generators G_1, G_2, G_k if

1. $P_k(K)$ is $L_m(G_k)$-closed,
2. $P_{1+k}(K)$ is $L_m(G_1) \parallel P_k(K)$-closed,
3. $P_{2+k}(K)$ is $L_m(G_2) \parallel P_k(K)$-closed.

Example 17.5 Consider Example 17.2. It can be verified that $P_k(K)$ is $L_m(G_k)$-closed, but $P_{i+k}(K)$ is not $L_m(G_i)\|P_k(K)$-closed, for $i = 1, 2, 3$.

If the specification K is conditionally closed and conditionally controllable, then there exists a nonblocking supervisor S_k such that $L_m(S_k/G_k) = P_k(K)$, which follows from the basic theorem of supervisory control applied to languages $P_k(K)$ and $L(G_k)$, see [1] or Chap. 16.

Theorem 17.2 *Consider the problem specified above. There exist nonblocking supervisors S_1, S_2, S_k solving the problem if and only if the specification language K is both conditionally controllable with respect to G_1, G_2, G_k and $\Sigma_{1,u}$, $\Sigma_{2,u}$, $\Sigma_{k,u}$, and conditionally closed with respect to G_1, G_2, G_k.*

Example 17.6 Consider Example 17.2. According to Examples 17.4 and 17.5, there do not exist such supervisors that would reach the specification K.

If the specification is not conditionally controllable, we can compute the supremal conditionally controllable sublanguage.

Theorem 17.3 *The supremal conditionally controllable sublanguage of a specification language always exists and is equal to the union of all conditionally controllable sublanguages of the specification.*

Consider the problem specified above and define the languages

$$\begin{aligned}
\sup C_k &= \sup C(P_k(K), L(G_k), \Sigma_{k,u}),\\
\sup C_{1+k} &= \sup C(P_{1+k}(K), L(G_1)\|\overline{\sup C_k}, \Sigma_{1+k,u}),\\
\sup C_{2+k} &= \sup C(P_{2+k}(K), L(G_2)\|\overline{\sup C_k}, \Sigma_{2+k,u}).
\end{aligned} \tag{17.1}$$

Example 17.7 Consider Example 17.2. We can compute $\sup C_k$ (Fig. 17.4) and $\sup C_{1+k}$, $\sup C_{2+k}$, $\sup C_{3+k}$ depicted in Fig. 17.3.

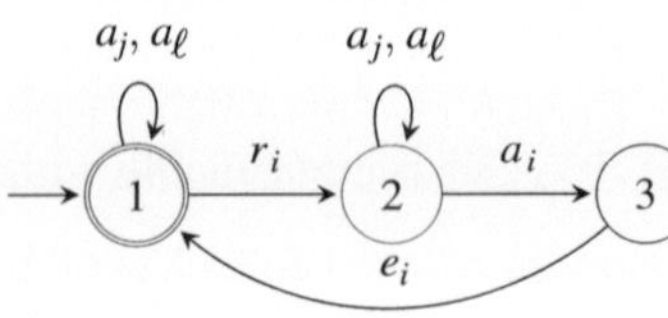

Fig. 17.3 Supervisor $\sup C_{i+k}$, $\{i, j, \ell\} = \{1, 2, 3\}$

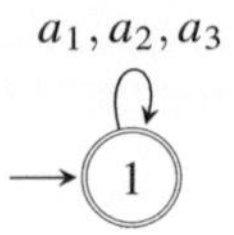

Fig. 17.4 Coordinator

For the languages defined in (17.1), it always holds that $P_k(\sup C_{i+k}) \subseteq \sup C_k$, for $i = 1, 2$. If the converse inclusion also holds, we obtain the supremal conditionally controllable sublanguage.

Theorem 17.4 *Consider the languages defined in* (17.1). *If* $\sup C_k \subseteq P_k(\sup C_{i+k})$, *for* $i = 1, 2$, *then the language* $\sup C_{1+k} \| \sup C_{2+k}$ *is the supremal conditionally controllable sublanguage of* K.

Example 17.8 Consider the coordinator and supervisors computed in Example 17.7. We can verify that the assumptions of Theorem 17.4 are satisfied. As the language $\sup C_k$ is $L_m(G_k)$-closed and $\sup C_{i+k}$ is $L_m(G_i) \| \sup C_k$-closed, for $i = 1, 2, 3$, they form a solution for the database problem by Theorems 17.4 and 17.2.

17.6 Coordinator for Nonblockingness

In this section, we discuss the coordinator for nonblockingness in the coordination control framework. Recall first that a generator G is nonblocking if $\overline{L_m(G)} = L(G)$.

Theorem 17.5 *Consider languages* L_1 *over* Σ_1 *and* L_2 *over* Σ_2, *and let the projection* $P_0 \colon (\Sigma_1 \cup \Sigma_2)^* \to \Sigma_0^*$, *with* $\Sigma_1 \cap \Sigma_2 \subseteq \Sigma_0$, *be an* L_i*-observer, for* $i = 1, 2$. *Let* G_0 *be a nonblocking generator with* $L_m(G_0) = P_0(L_1) \parallel P_0(L_2)$. *Then the language* $L_1 \| L_2 \| L_m(G_0)$ *is nonblocking, that is,* $\overline{L_1 \| L_2 \| L_m(G_0)} = \overline{L_1} \| \overline{L_2} \| \overline{L_m(G_0)}$.

This result is used in the coordination control synthesis as follows. Local supervisors $\sup C_{1+k}$ and $\sup C_{2+k}$ are computed as in (17.1), and the properties of Theorem 17.4 are verified. If they are satisfied, the computed supervisors are the solution of the problem. However, they can still be blocking. In such a case, we can choose the language $L_C = P_0(\sup C_{1+k}) \parallel P_0(\sup C_{2+k})$, where the projection P_0 is a $\sup C_{i+k}$-observer, for $i = 1, 2$, (actually, we take the supremal controllable sublanguage of $P_0(\sup C_{1+k}) \parallel P_0(\sup C_{2+k})$, cf. [8]) and obtain that the equality

$$
\begin{aligned}
\overline{\sup C_{1+k} \parallel \sup C_{2+k} \parallel L_C} &= \overline{\sup C_{1+k} \parallel \sup C_{2+k}} \\
&= \overline{\sup C_{1+k}} \parallel \overline{\sup C_{2+k}} \parallel \overline{L_C}
\end{aligned}
$$

holds by Theorem 17.5. In other words, L_C is the behavior of a nonblocking coordinator. This gives the following algorithm.

Algorithm 2 (*Coordinator for nonblockingness*) Consider the notation above.

1. Compute $\sup C_{1+k}$ and $\sup C_{2+k}$ as defined in (17.1).
2. Let $\Sigma_0 := \Sigma_k$ and $P_0 := P_k$.
3. Extend Σ_0 so that the projection P_0 is both a $\sup C_{1+k}$- and a $\sup C_{2+k}$-observer.
4. Define the coordinator C as the minimal nonblocking generator such that $L_m(C) = \sup C(P_0(\sup C_{1+k}) \parallel P_0(\sup C_{2+k}),\ \overline{P_0(\sup C_{1+k})} \parallel \overline{P_0(\sup C_{2+k})},\ \Sigma_{0,u})$.

Example 17.9 Consider the solution of the database problem computed in Example 17.7. It can be verified that the language $\sup C_{1+k} \| \sup C_{2+k} \| \sup C_{3+k}$ is nonblocking; hence, we do not need a coordinator for nonblockingness in this example.

17.7 Prefix-Closed Languages

Here, we assume that the specification is prefix-closed. The following notion is required. More details, an explanation and examples can be found in [16].

Definition 17.2 (*Local control consistency*) Let L be a prefix-closed language over Σ, and let $\Sigma_0 \subseteq \Sigma$. The projection $P_0 \colon \Sigma^* \to \Sigma_0^*$ is *locally control consistent* (LCC) with respect to $s \in L$ if for all $\sigma_u \in \Sigma_0 \cap \Sigma_u$ such that $P_0(s)\sigma_u \in P_0(L)$, it holds that either there does not exist any $u \in (\Sigma \setminus \Sigma_0)^*$ such that $su\sigma_u \in L$, or there exists $u \in (\Sigma_u \setminus \Sigma_0)^*$ such that $su\sigma_u \in L$. The projection P_0 is LCC with respect to a language L if P_0 is LCC for all words of L.

Consider generators G_1, G_2, G_k, and denote $L_i = L(G_i)$, for $i = 1, 2, k$. There is not yet a general procedure to compute the supremal conditional controllable sublanguage. However, there is a procedure for prefix-closed specifications.

Theorem 17.6 *Let $K \subseteq L_1 \| L_2 \| L_k$ be a prefix-closed language over $\Sigma_1 \cup \Sigma_2 \cup \Sigma_k$, where $L_i = \overline{L_i} \subseteq \Sigma_i^*$, $i = 1, 2, k$. Assume that the language K is conditionally decomposable and consider the languages defined in* (17.1). *Let the projection P_k^{i+k} be an $(P_i^{i+k})^{-1}(L_i)$-observer and LCC for $(P_i^{i+k})^{-1}(L_i)$, for $i = 1, 2$. Then, $\sup C_{1+k} \| \sup C_{2+k}$ is the supremal conditionally controllable sublanguage of K.*

The following corollary explains the relation to the notion of controllability of the monolithic case.

Corollary 17.1 *In the setting of* Theorem 17.6, *the supremal conditionally controllable sublanguage of K is controllable with respect to $L_1 \| L_2 \| L_k$ and Σ_u.*

Finally, the last theorem states the conditions under which the solution is optimal.

Theorem 17.7 *Consider the setting of* Theorem 17.6. *If, in addition, $L_k \subseteq P_k(L)$ and P_{i+k} is LCC for $P_{i+k}^{-1}(L_i \| L_k)$, for $i = 1, 2$, then $\sup C(K, L_1 \| L_2 \| L_k, \Sigma_u)$ is the supremal conditionally controllable sublanguage of K.*

17.8 Further Reading

The theory presented here is based on paper [7]. This topic is still under investigation. For other structural conditions on local plants under which it is possible to synthesize the supervisors locally, but which are quite restrictive, see [3, 12]. Among the most successful approaches to supervisory control of distributed discrete-event systems are those that combine distributed and hierarchical control [16, 17], or the approach based on interfaces [13]. For coordination control of linear or stochastic systems, the reader is referred to [4, 15].

References

1. Cassandras CG, Lafortune S (2008) Introduction to discrete event systems. Springer, Berlin
2. Feng L, Wonham WM (2010) On the computation of natural observers in discrete-event systems. Discrete Event Dyn Syst 20:63–102
3. Gaudin B, Marchand H (2004) Supervisory control of product and hierarchical discrete event systems. Eur J Control 10(2):131–145
4. Kempker PL, Ran ACM, van Schuppen JH (2009) Construction of a coordinator for coordinated linear systems. In: ECC 2009, pp 4979–4984
5. Komenda J, Masopust T, van Schuppen JH (2011) Coordinated control of discrete event systems with nonprefix-closed languages. In: IFAC World Congress, pp 6982–6987
6. Komenda J, Masopust T, van Schuppen JH (2011) Synthesis of controllable and normal sublanguages for discrete-event systems using a coordinator. Syst Control Lett 60(7):492–502
7. Komenda J, Masopust T, van Schuppen JH (2012) On algorithms and extensions of coordination control of discrete-event systems. In: WODES, pp 245–250
8. Komenda J, Masopust T, van Schuppen JH (2013) Coordination control of discrete-event systems revisited. Extended version of [7]. Submitted for publication
9. Komenda J, Masopust T, van Schuppen JH (2012) On conditional decomposability. Syst Control Lett 61(12):1260–1268
10. Komenda J, Masopust T, van Schuppen JH (2012) Supervisory control synthesis of discrete-event systems using a coordination scheme. Automatica 48(2):247–254
11. Komenda J, van Schuppen JH (2008) Coordination control of discrete event systems. In: WODES, pp 9–15
12. Komenda J, van Schuppen JH, Gaudin B, Marchand H (2008) Supervisory control of modular systems with global specification languages. Automatica 44(4):1127–1134
13. Leduc RJ, Pengcheng D, Raoguang S (2009) Synthesis method for hierarchical interface-based supervisory control. IEEE Trans Automat Control 54(7):1548–1560
14. Moor Th et al. (2012) libFAUDES—a discrete event systems library [online]. Available at http://www.rt.eei.uni-erlangen.de/FGdes/faudes/
15. Ran ACM, van Schuppen JH (2008) Control for coordination of linear systems. In: MTNS 2008, Blacksburg, USA
16. Schmidt K, Breindl C (2011) Maximally permissive hierarchical control of decentralized discrete event systems. IEEE Trans Autom Control 56(4):723–737
17. Schmidt K, Moor Th, Perk S (2008) Nonblocking hierarchical control of decentralized discrete event systems. IEEE Trans Autom Control 53(10):2252–2265
18. Wonham WM (2009) Supervisory control of discrete-event systems. Department of Electrical and Computer Engineering, University of Toronto, Lecture notes

Part IV
Distributed Control

Distributed control of distributed systems was defined in Chap. 11. In this part, distributed control is discussed. Chapter 18 provides an introduction to team theory. The next chapter presents research results on team theory for distributed stochastic systems. Signaling of information in distributed systems is described in Chap. 20. Control of distributed manufacturing systems is treated in Chap. 21.

Chapter 18
What Is Team Theory?

Jan H. van Schuppen

18.1 Motivation

Team theory is a mathematical formalism for a stochastic decision problem in which a team, consisting of two or more team members, cooperates to achieve a common control objective. The information of any team member about the underlying state of the problem is in general different from that of the other teams. If the observations were identical, then there is only an optimization problem.

Team theory developed out of the need for a mathematical model of cooperating teams or groups within an organization in which all team members have the same objective yet different information. Distributed dynamic systems are a special case of this. Team theory, inspired by game theory, has adopted the concept of person-by-person equilibrium which leads to a set of optimization problems with only one decision maker and are hence solvable by available theory. The motivation of the originators of team theory, J. Marschak and R. Radner, comes from cooperating teams or departments within an organization.

The purpose of this article is to provide a brief introduction to the problem of teams and the main result of the topic. The reader is advised to read also about team theory for decentralized control in Chap. 19.

18.2 The Team Theory Framework

A team problem is a special case of an optimization problem or a decision problem. It can also be regarded as a special case of a game problem. The research area of decision problems is based on probability theory and developed differently from optimization though both topics are quite related.

J.H. van Schuppen (✉)
Van Schuppen Control Research, Gouden Leeuw 143,
1103 KB Amsterdam, The Netherlands
e-mail: jan.h.van.schuppen@xs4all.nl

J.H. van Schuppen and T. Villa (eds.), *Coordination Control of Distributed Systems*,
Lecture Notes in Control and Information Sciences 456,
DOI 10.1007/978-3-319-10407-2_18

Definition 18.1 The *team theory setting* is characterized by the following properties of a *team problem*:

- there are two or more decision makers, an essential condition;
- it is a decision problem such that each decision maker takes only one decision;
- the observations available to any tuple of decision makers differ in general, which is also an essential characteristic;
- all decision makers have the same control objective, they strive to achieve the same goal;
- one distinguishes: (1) the case without communication between decision makers and (2) the case with communication between decision makers; a further extension in the latter case is to allow a sequence of steps of communication exchanges.

The last item of the above setting requires comments. In the original formulation, there was no communication between the various decision makers at all. H. Witsenhausen and Y.C. Ho have written about an extension of the problem where at first one or more decision makers exchange information they have observed of the state of the model and then each of them separately reach a decision. This allows the inclusion into team theory of the problem of communication for which situation there is a need for better understanding and concepts. This extension is not further analyzed in the chapter though it is discussed at Further Reading.

The team problem can be formulated either as a deterministic problem or as a stochastic problem, the latter seems more realistic. Below a formulation as a stochastic team problem will be used.

The exposition below is borrowed from the papers [10, 11].

Definition 18.2 *Team problem.* A *team* is a set with as elements *team members*, with $p \in \mathbb{Z}_+$, $p \geq 2$, the *number of team members*, which have a common *team objective*. Each team member has an observation about the uncertainty of the model and the observations of any tuple of team members differ. A *team decision rule* is a set of functions, one for each team member, mapping the member's observation to his/her action or input. The problem is to determine an *optimal team decision rule* according to a *cost function* based on the team objective.

The mathematical definition is as follows. Consider a probability space (Ω, F, P) consisting of a probability space Ω which is a complete separable metric space, a σ-algebra F, and a probability measure P. The initial state is random and is modeled by the probability space. Each team member has partial observations of the initial state. The observation of team member $i \in \mathbb{Z}_p$ is denoted by $y_i(\omega) = h_i(\omega)$ for a measurable function $h_i \colon \Omega \to \mathbb{R}^{r_i}$. Denote $r = \sum_{i=1}^{p} r_i$, $h \colon \Omega \to \mathbb{R}^r$, and y as the vector of the y_i so that then $y = h(\omega)$. Denote the σ-algebra generated by y_i by F^{y_i}. The observation y_i induces a σ-algebra on the range space denoted by $H^{y_i} \subseteq B(\mathbb{R}^{r_i})$. A notation involving an unobserved state random variable would allow for a more realistic exposition.

Definition 18.3 A *team decision rule*, or a *team strategy*, or a *control law* is a set of measurable functions, of which the function for team member i is denoted by $g_i \colon \mathbb{R}^{r_i} \to \mathbb{R}^{m_i}$. Define $m = \sum_{i=1}^{p} m_i$. Denote the sets of team decision rules by

$$G_i = \{g_i: \mathbb{R}^{r_i} \to \mathbb{R}^{m_i} | g_i \text{ is } H^{y_i} \text{ measurable}\},$$
$$G = G_1 \times G_2 \times \cdots \times G_p.$$

Denote for any team member $i \in \mathbb{Z}_p = \{1, 2, \ldots, p\}$; the team decision rule of all members except member i by $g_{-i} = g \backslash g_i \in G_{-i}$.

Definition 18.4 Finally, define the *team cost* and the *cost function* as the measurable functions,

$$C: \mathbb{R}^m \times \Omega \to \mathbb{R}, \quad J: G \to \mathbb{R}, \tag{18.1}$$
$$J(g) = E[C(g(h(\omega)), \omega)]. \tag{18.2}$$

18.3 Team Problem

Problem 18.1 The *team problem* in mathematical form is then to solve the optimization problem,

$$\inf_{g \in G} J(g), \tag{18.3}$$
$$g^* \in G, \text{ is called an optimal team decision rule if} \tag{18.4}$$
$$J(g^*) = \inf_{g \in G} J(g). \tag{18.5}$$

The team problem is a special case of an optimization problem though over a function space and as such can be solved by known methods. However, the novel concept is that of a person-by-person equilibrium which decentralizes or distributes the optimization problem from the global level of all team members to an optimization problem of each team member separately.

18.4 Concepts of Team Theory

Definition 18.5 Consider the Team Problem 18.1. A team decision rule $g^* \in G$ is said to be a *person-by-person equilibrium* (alternative terms of the literature: *player-by-player equilibrium* or *member-by-member equilibrium*) if

$$J(\{g_i^*, g_{-i}^*\}) \leq J(\{g, g_{-i}^*\}), \quad \forall\, g_i \in G_i, \forall\, i \in \mathbb{Z}_p. \tag{18.6}$$

The concept of a person-by-person equilibrium is a specialization of the concept of the Nash equilibrium for a game problem first proposed by J. Nash to a team problem.

The following two major questions are now of interest: (1) Is a team decision rule which is a person-by-person equilibrium also an optimal team decision rule? (2)

How to compute a person-by-person equilibrium? The answers to these questions form team theory and are summarized below.

There follow two more concepts.

Definition 18.6 Consider the Team Problem 18.1. Call a team decision rule $g \in G$ *stationary* if the following two conditions both hold:

$$J(g) = E[C(g(h(\omega)), \omega)] < \infty, \tag{18.7}$$

$$\frac{\mathrm{d}}{\mathrm{d}u_j} E[C(\hat{g}^j(h(\omega), u_j), \omega)|F^{y_j}]|_{u_j = g_j(h_j(\omega))} = 0, \tag{18.8}$$

$$\text{a.s. } P, \forall\, j \in \mathbb{Z}_p, \quad \text{where,}$$

$$\hat{g}_i^j(h(\omega), u_j) = \begin{cases} g_i(h_i(\omega)), & i \neq j, \\ u_j, & i = j, \end{cases} \quad \forall\, j, i \in \mathbb{Z}_p. \tag{18.9}$$

Stationarity is the first-order condition of person-by-person optimality. An optimal team decision rule is a stationary team decision rule.

The next condition is needed so that the cost function is locally well defined at a particular control law.

Definition 18.7 Consider Team Problem 18.1. The cost function J is called *locally finite* at the team decision rule $g \in G$ if the following two conditions both hold,

$$|J(g)| < \infty, \tag{18.10}$$

$$\forall\, \delta \in \mathbb{R}^m, \text{ such that } |J(g+\delta)| < \infty, \ \exists\, k_1, \ldots, k_p \in \mathbb{R}_+ \text{ such that}$$
$$|J(g_1 + h_1\delta_1, \ldots, g_p + h_p\delta_p)| < \infty, \tag{18.11}$$
$$\forall\, h_1 \in U_1 = \mathbb{R}^{m_1}, \ldots, h_p \in U_p = \mathbb{R}^{m_p}, \ \|h_1\| \leq k_1, \ldots, \|h_p\| \leq k_p.$$

18.5 Team Theory

Theorem 18.1 [18]. *Consider Team* Problem 18.1. *If there exists a team decision rule $g^* \in G$ such that*

1. *the team cost function $C: \mathbb{R}^m \times \Omega \to \mathbb{R}$ is convex and differentiable in $\mathbb{R}^m$ almost everywhere on Ω;*
2. $\inf_{g \in G} J(g) > -\infty$;
3. *the cost function J is localy finite at $g^* \in G$; and*
4. *$g^* \in G$ is stationary;*

then, $g^* \in G$ is a person-by-person equilibrium and an optimal team decision rule.

The main interest of the above theorem is that it provides a sufficient condition with respect to which a person-by-person equilibrium is an optimal decision rule. In fact, the two conditions are equivalent because, if the first three conditions are assumed, then an optimal decision rule is a person-by-person equilibrium.

There exists an example of a team problem which does not satisfy the conditions of the above theorem, see [10].

Theorem 18.2 [10, Th. 3]. *Consider Team* Problem 18.1. *Assume that there exists a team decision rule $g^* \in G$ such that:*

1. $\forall\, \omega \in \Omega$, *the function* $C(., \omega)\colon \mathbb{R}^m \to \mathbb{R}$ *is convex and differentiable;*
2. $\inf_{g \in G} J(g) > -\infty$;
3. g^* *is a stationary team decision rule; and*
4. *for all* $g \in G$ *such that* $E[C(g(h(\omega), \omega)] < \infty$ *the following two expressions are well defined,*

$$E[(D_{u_i} C((u_i, u_{-i}), \omega)|_{u=g^*(h(\omega))}(g_i - g_i^*)],$$

$$\sum_{i=1}^{p} E[(D_{u_i} C((u_i, u_{-i}), \omega)|_{u=g^*(h(\omega))}(g_i - g_i^*)], \quad \forall\, i \in \mathbb{Z}_p.$$

(a) *Then,* $g^* \in G$ *is a person-by-person equilibrium and an optimal team decision rule.*

(b) *If, in addition, the team cost function* $C(., .)$ *is strictly convex for almost all* $\omega \in \Omega$, *then* g^* *is the unique optimal decision rule a.s. P.*

The hypotheses of Theorem 18.1 imply those of Theorem 18.2. There exists an example which satisfies the conditions of Theorem 18.2 but not the conditions of Theorem 18.1; hence, the first-mentioned result is a strict generalization of the second-mentioned theorem. The authors of the paper [11] conjecture that if the team cost function is a convex function, then any stationary team decision rule is also an optimal team decision rule.

The paper [11] provides a procedure on how to compute an optimal person-by-person equilibrium of a team problem via an iteration as for a fixed point problem. If the conditions of the theorem hold, then the stationarity conditions lead to a set of equations which have to be solved.

An example of a team problem is the case in which the unknown random variable has a Gaussian probability distribution, the observations are linear functions of the unknown random variable hence also Gaussian, and the cost function is quadratic. See [13, 18] for the formulas.

18.6 Further Research

Though team theory has been known for a several decades, there are still several problem issues which require attention. Several of these research issues also arise in the distributed control of distributed systems. A list of open control issues is:

1. What are *sufficient and necessary conditions* for the implication that a person-by-person optimal decision rule is a team optimal decision rule? For the case of continuous input spaces, the convexity of the cost function is a sufficient condition. For the case of discrete or mixed continuous-discrete input spaces, the conditions are not known. However, see the paper [2] for a sufficient condition.

2. *Extension from team theory to decentralized control*? Can sufficient and necessary conditions for the team problems as discussed in the previous item be extended to optimal decentralized control problems, thus for decentralized control systems and cost functions on a time axis? The convexity condition imposed in the team problem described would then have to be imposed on the Hamiltonian of the optimal decentralized control problem. Theory for decentralized control based on this equivalence would be a significant advance for control theory.
3. *What is the structure of the minimal team decision rules*? One expects that the teams each contribute to particular aspects of the team problem based on their observations and on their effect on the cost function. For the case of a team problem with Gaussian distributionsk, one expects that a decomposition of the model with respect to the effects on the cost function and observations will play a role. What can a decision maker determine about the observations of other decision makers and how can this information be used to his/her advantage? These questions remain to be explored. The results could be very useful for control of decentralized systems.
4. *How to compute a set of team decision rules which are a person-by-person equilibrium*? Note that the problem asks for a decision rule which is function, it is not an optimization over a finite-dimensional spaces. In practice, a sequence of sets of team decision rules is constructed, but then it has to be established that that sequence converges to a person-by-person equilibrium.
5. The case of *team problems with communication between decision makers* requires attention because the related problems are very relevant for distributed control with communication. What is the control law of a team member for requesting information of other team members and how should the information received be processed? See the essay on distributed control with direct communication between controllers, Chap. 22.

For further research, the author favors an investigation into the relation of the various inputs of the team members on the cost, which amounts to an investigation of the interactions of the input variables and the controlled outputs. The author also favors for decentralized control problems a control theoretic formulation based on the theory of dynamic games. He does not favor the approach called dynamic team theory, see below for references.

18.7 Further Reading

To the reader who wants to learn team theory without prior knowledge, the author recommends the paper [4] and the book by Marschak and Radner, [13].

The reader is advised to continue after the reading of this chapter with the reading of Chap. 19.

The first journal papers on the subject are by Marschak and Radner, see [12, 18]. The motivation of Marschak was apparently microeconomics and the economics of firms, see the book made for his commemmoration with his list of publications, [16, pp. 337–341]. A major theoretical development is due to Krainak, Marcus, and Speyer, see [10, 11]. An extension to discontinuous action spaces is presented in [2]. A recent summary may be found in [23, Sect. 4].

The team problem can be regarded as a statistical decision problem for which references are [3, 14, 17].

The relation of team theory and communication is explored in the papers [6, 7].

An extension is that each team has the choice to communicate part of its observation or part of its decision to other teams. See for example [5]. The teams must then process the received information before taking a decision. They may also have different a priori beliefs about the probability distribution of the unknown state variable, see [22].

A further extension is that there are communications of team members but that the information of the communication exchanges depend on the decisions taken by team members who have already submitted a decision. This is called a *dynamic team problem* though in control theory it would be considered a decentralized control problem. This case can be distinguished into a *sequential decision problem* and a *non-sequential decision problem* depending on whether an order in the decisions can be constructed. See for references of the latter cases Chaps. 24 and 25.

H. Witsenhausen has shown that a control problem of centralized or of decentralized control with two or more controllers can be transformed into a team problem as considered in the body of this chapter. This can be done by considering the controller at any particular time to be a different team member. See [24–26].

Team theory has been applied to management and to economic theory, see [8, 9, 20]. Applications of the team theory are described in [1, 13, 15, 16, 19, 21].

References

1. Beckmann M (1958) Decision and team problems in airline reservations. Econometrica 26:134–145
2. de Waal PR, van Schuppen JH (2000) A class of team problems with discrete action spaces: optimality conditions based on multimodularity. SIAM J Control Opt 38:875–892
3. Ferguson TS (1967) Mathematical statistics: a decision theoretic approach. Academic Press, New York
4. Ho YC (1980) Team decision theory and information structures. Proc IEEE 68:644–654
5. Ho YC, Chu KC (1972) Team decision theory and information structures in optimal control problems—parts I and II. IEEE Trans Control 17:15–28
6. Ho YC, Karstner MP (1978) Market signalling: an example of a two person decision problem with a dynamic information structure. IEEE Trans Autom Control 23:350–361
7. Ho YC, Karstner P, Wong E (1978) Teams, signaling and information theory. IEEE Trans Autom Control 23:305–311
8. Hurwicz L, Radner R, Reiter S (1975) A stochastic decentralized resource allocation process I. Econometrica 43:187–221
9. Hurwicz L, Radner R, Reiter S (1975) A stochastic decentralized resource allocation process II. Econometrica 43:363–393

10. Krainak J, Speyer JL, Marcus SI (1982) Static team problems—part I: sufficient conditions and the exponential cost criterion. IEEE Trans Autom Control 27:839–848
11. Krainak J, Speyer JL, Marcus SI (1982) Static team problems—part II: affine control laws, projection algorithms, and the LEGT problem. IEEE Trans Autom Control 27:848–859
12. Marschak J (1955) Elements for a theory of teams. Manage Sci 1:127–137
13. Marschak J, Radner R (1972) Economic theory of teams. Yale University Press, New Haven
14. Mas-Colell A, Whinston MD, Green JR (1995) Microeconomic theory. Oxford University Press, Oxford
15. McGuire CB (1961) Some team models of a sales organization. Manage Sci 7:101–130
16. McGuire CB, Radner R (1972) Decision and organization, volume 12 of studies in mathematical and managerial Economics. North-Holland Publishing Company, Amsterdam
17. Pratt JW, Raiffa H, Schlaifer R (2008) Introduction to statistical decision theory. MIT Press, Cambridge
18. Radner R (1962) Team decision problems. Ann Math Statist 33:857–881
19. Radner R (1972) Allocation of a scarce resource under uncertainty: an example of a team. In: McGuire CB, Radner R (eds) Decision and organization. North-Holland Publishing Company, Amsterdam, pp 217–236
20. Radner R, Arrow KJ (1979) Allocation of resources in large teams. Econometrica 47:361–392
21. Schoute FC (1978) Symmetric team problems and multi access wire communication. Autom J IFAC 14:255–269
22. Teneketzis D, Varaiya P (1984) Consensus in distributed estimation with inconsistent beliefs. Syst Control Lett 4:217–221
23. van Schuppen J H (2011) Control of distributed stochastic systems—introduction, problems, and approaches. In: Proceedings of IFAC World Congress, pp 6029–6035
24. Witsenhausen HS (1971) On information structures, feedback and causality. SIAM J Control 9:149–160
25. Standard A (1973) Form for sequential stochastic control. Math Syst Theor 7:5–11
26. Witsenhausen HS (1988) Equivalent stochastic control problems. Math Control Signals Syst 1:3–11

Chapter 19
Team Theory and Information Structures of Stochastic Dynamic Decentralized Decision

C.D. Charalambous and N.U. Ahmed

19.1 Motivation

Static team theory is a mathematical formalism of decision problems with multiple decision makers acting on different information affecting a single payoff [1]. Its generalization to dynamic team theory has far reaching implications in all human activity including science and engineering systems. In general, decentralized systems consist of multiple local observation posts and decision or control stations, in which the actions applied at the decision stations are calculated using different information, that is, the arguments of the decision strategies are different. We call, as usual, "information structures or patterns" the information available for decisions at the decision stations to implement their actions, and we call such informations "decentralized information structures" if the information available for decisions at the various decisions stations are not identical to all stations. Moreover, we call an information structure "classical" if all decision stations have the same information structures, and the information structure at the decision stations is nested, also called perfect recall, (i.e., a decision station that at some time has available certain information will have available the same information at any subsequent times), otherwise we call it "nonclassical." Early work discussing the importance of information structures in decision making and its applications is found in [2, 3], while more recent one is found in [4–6]. The most general examples with nested information structures admitting closed form solutions appear to be the ones in [7] .

Stochastic discrete-time dynamic decision problems with nonclassical information structures are often formulated using team theory; the two methods proposed

C.D. Charalambous (✉)
University of Cyprus, 75 Kallipoleos, P.O. Box 20537, 1678 Nicosia, Cyprus
e-mail: chadcha@ucy.ac.cy

N.U. Ahmed
University of Ottawa, School of Engineering and Computer Science, 161 Louis Pasteur, Ottawa, Canada
e-mail: ahmed@site.uottawa.ca

J.H. van Schuppen and T. Villa (eds.), *Coordination Control of Distributed Systems*,
Lecture Notes in Control and Information Sciences 456,
DOI 10.1007/978-3-319-10407-2_19

over the years are based on identifying conditions so that discrete-time stochastic dynamic team problems can be equivalently reduced to static team problems [5] and dynamic programming [6].

In this chapter, we discuss recent results by the authors in [8–12] on stochastic dynamic decision problems with nonclassical information structures, formulated using dynamic team theory. For an introduction to static team theory and related literature, we suggest the paper by Jan van Schuppen in this book.

Our objectives are the following.

(1) Apply Girsanov's change of probability measure to transform stochastic dynamic team problems to equivalent problems in which the state and/or the observations and information structures are not affected by any of the team decisions;
(2) apply stochastic maximum principle to derive necessary and sufficient team and PbP optimality conditions.

We illustrate the importance of Girsanov's change of probability measure [13] in generalizing Witsenhausen's [5] notion on equivalence between discrete-time stochastic dynamic team problems which can be transformed to equivalent static team problems, to continuous-time Itô stochastic nonlinear differential decentralized decision problems, and to general classes of discrete-time models. The optimal strategies of Witsenhausen's counterexample [2] can be derived using this method. We also invoke the stochastic maximum principe to present necessary and sufficient team and PbP optimality conditions described in terms of conditional variational inequalities with respect to the information structures and BSDEs.

19.2 Team Theory of Stochastic Dynamic Systems

Given the information structure $\{I^k(t) : t \in [0, T]\}$ available at the kth decision station and the corresponding admissible regular strategies $\mathbb{U}^k_{reg}[0, T]$ for $k = 1, \dots, K$, the team decision problem is defined as follows.

$$\inf\{J(u^1, \dots, u^K) : (u^1, \dots, u^K) \in \times_{k=1}^{K} \mathbb{U}^k_{reg}[0, T]\} \tag{19.1}$$

$$J(u^1, \dots, u^K) = \mathbf{E}\Big\{ \int_0^T \ell(t, x(t), u^1(t, I^1), \dots, u^K(t, I^K))\mathrm{d}t + \varphi(x(T))\Big\}, \tag{19.2}$$

subject to stochastic dynamics with state $x(\cdot)$ and noisy observations at the observation posts $\{y^i(\cdot) : i = 1, \dots, M\}$, which are solutions of the Itô differential equations

$$\begin{aligned} dx(t) =& f(t, x(t), u^1(t, I^1), \dots, u^K(t, I^K))\mathrm{d}t \\ &+ \sigma(t, x(t), u^1(t, I^1), \dots, u^K(t, I^K))dW(t), \quad x(0) = x_0, \end{aligned} \tag{19.3}$$

$$dy^m(t) = h^m(t, x(t), u^1(t, I^1), \dots, u^K(t, I^K))\mathrm{d}t + D^{m,\frac{1}{2}}(t)dB^m(t), \; m = 1, \dots, M. \tag{19.4}$$

Here, $u^k(t, I^k) \in \mathbb{A}^k \subseteq \mathbb{R}^{d_k}$ and $\{W(t) : t \in [0,T]\}, \{B^m(t) : t \in [0,T]\}$ are independent Brownian motion (BM) processes. The stochastic system (19.3) can be specialized to any interconnected subsystems architectures. For simplicity, we assume $K = M \equiv N$ and we consider the following information structures. Denote the observation posts which communicate to the ith decision station by $\mathscr{O}(i) \subset \{1, 2, \ldots, i-1, i+1, \ldots, N\} \subset \mathbb{Z}_N \stackrel{\triangle}{=} \{1, 2, \ldots, N\}, i = 1, \ldots, N$.

Nonclassical Information Structures. The decision applied at the ith decision station, $u_t^i \equiv u^i(t, I^i)$ at time $t \in [0,T]$ is a nonanticipative measurable function of $I^i(s) \stackrel{\triangle}{=} \{y^i(s), y^j(s - \varepsilon_j) : \varepsilon_j > 0, j \in \mathscr{O}(i)\}, 0 \leq s \leq T, i = 1, \ldots, N$.

Letting $\mathscr{G}_{0,t}^{I^i} \stackrel{\triangle}{=} \sigma\Big\{I^i(s) : 0 \leq s \leq t\Big\}$ denote the minimal $\sigma-$algebra generated by $\{I^i(s) : 0 \leq s \leq t\}, t \in [0,T]$, then $\{\mathscr{G}_{0,t}^{I^i} : i = 1, \ldots N\}$ are nonclassical because they are different .

Restricting $\mathscr{G}_{0,t}^{I^i}$ to $\mathscr{G}^{I^i(t)} \stackrel{\triangle}{=} \sigma\big\{I^i(t)\}$, then $\{\mathscr{G}^{I^i(t)} : i = 1, \ldots, N\}$ are nonclassical because they are different, and nonnested, because $\mathscr{G}^{I^i(t)} \not\subseteq \mathscr{G}^{I^i(\tau)}, \forall \tau > t, i = 1, \ldots, N$.

A generic information structure is denoted by $\mathscr{G}_T^i \stackrel{\triangle}{=} \{\mathscr{G}_t^i : t \in [0,T]\}$, which can be specialized to $\{\mathscr{G}_{0,t}^{y^i} : t \in [0,T]\}, \{\mathscr{G}_{0,t}^{I^i} : t \in [0,T]\}, \{\mathscr{G}^{I^i(t)} : t \in [0,T]\}$.

Next, we describe the set of admissible relaxed strategies which are conditional distributions, since regular strategies are special cases of delta measures [9, 11].

Definition 19.1 (The Admissible Relaxed Strategies) The strategy applied at ith decision station is a conditional distribution defined by

$$u_t^i(\Gamma) = q_t^i(\Gamma|\mathscr{G}_t^i), \qquad \text{for } \ t \in [0,T], \quad \text{and} \qquad \forall\, \Gamma \in \mathscr{B}(\mathbb{A}^i), \quad i = 1, \ldots, N.$$

Each strategy is member of an appropriate function space [9, 11] denoted by $\mathbb{U}_{rel}^i[0,T], i = 1, \ldots, N$. An N tuple of relaxed strategies is $\mathbb{U}_{rel}^{(N)}[0,T] \stackrel{\triangle}{=} \times_{i=1}^N \mathbb{U}_{rel}^i[0,T]$.

The notation for relaxed strategies $u \in \mathbb{U}_{rel}^{(N)}[0,T]$ is $f(t,x,u_t) \stackrel{\triangle}{=} \int (f(t,x,\xi^1, \ldots, \xi^N)) \times {}_{i=1}^N u_t^i(d\xi^i)$ and similarly for $\{\sigma, h, \ell\}$ in (19.1)–(19.4).

Problem 19.1 (Team and PbP Optimality) A relaxed strategy $u^o \in \mathbb{U}_{rel}^{(N)}[0,T]$ for (19.1)–(19.4) is called team optimal if

$$J(u^{1,o}, u^{2,o}, \ldots, u^{N,o}) \leq J(u^1, u^2, \ldots, u^N), \qquad \forall u \stackrel{\triangle}{=} (u^1, u^2, \ldots, u^N) \in \mathbb{U}_{rel}^{(N)}[0,T],$$

and it is called PbP optimal if it satisfies $\tilde{J}(u^{i,o}, u^{-i,o}) \leq \tilde{J}(u^i, u^{-i,o}), \forall u^i \in \mathbb{U}_{rel}^i[0,T], \forall i \in \mathbb{Z}_N, \tilde{J}(v, u^{-i}) \stackrel{\triangle}{=} J(u^1, u^2, \ldots, u^{i-1}, v, u^{i+1}, \ldots, u^N)$.

Notation. $C([0,T], \mathbb{R}^n)$: space of continuous $\mathbb{R}^n-$valued functions defined on $[0,T]$.

$L^2_{\mathbb{F}_T}([0,T],\mathbb{R}^n)$: space of $\{\mathbb{F}_{0,t} : t \in [0,T]\}-$adapted $\mathbb{R}^n-$valued random processes $\{z(t) : t \in [0,T]\}$ such that $\mathbb{E}\int_{[0,T]} |z(t)|^2_{\mathbb{R}^n} \mathrm{d}t < \infty$.
$L^2_{\mathbb{F}_T}([0,T],\mathcal{L}(\mathbb{R}^m,\mathbb{R}^n))$: space of $\{\mathbb{F}_{0,t} : t \in [0,T]\}-$adapted $n \times m$ matrix valued random processes $\{\Sigma(t) : t \in [0,T]\}$ such that $\mathbb{E}\int_{[0,T]} |\Sigma(t)|^2_{\mathcal{L}(\mathbb{R}^m,\mathbb{R}^n)} \mathrm{d}t < \infty$.
$B^\infty_{\mathbb{F}_T}([0,T],L^2(\Omega,\mathbb{R}^n))$: space of $\{\mathbb{F}_{0,t} : t \in [0,T]\}$-adapted $\mathbb{R}^n-$ valued second-order random processes endowed with the norm topology $\| x \|^2 \triangleq \sup_{t\in[0,T]} \mathbb{E}|x(t)|^2_{\mathbb{R}^n}$.

19.3 Equivalent Stochastic Dynamic Team Problems

In this section, we invoke Girsanov's theorem to transform the original stochastic dynamic decision problem to an equivalent decision problem with corresponding observations and information structures which are independent and independent of any of the team decisions. Consider

(WP1) $x(0) = x_0$: an $\mathbb{R}^n$-valued random variable with distribution $\Pi_0(dx)$;
(WP2) $\{W(t) : t \in [0,T]\}$: an $\mathbb{R}^m$-valued standard BM, independent of $x(0)$;
(WP3) $\{y^i(t) \triangleq \int_0^t D^{i,\frac{1}{2}}(s) dB^i(s) : t \in [0,T]\}$: $\mathbb{R}^{k_i}$-valued, $i = 1,\ldots,N$, mutually independent BMs, independent of $\{W(t) : t \in [0,T]\}$.

where $W(\cdot) \in C([0,T],\mathbb{R}^m)$, with Borel $\sigma-$algebra $\mathcal{F}^W_{0,T} \triangleq \mathcal{B}(C[0,T],\mathbb{R}^m))$ and $\mathbb{P}^W$ its Wiener measure on it, similarly for $y^i(\cdot)$, with $\mathcal{F}^{y^i}_{0,T} \triangleq \mathcal{B}(C(0,T],\mathbb{R}^{k_i}))$, $\mathbb{P}^{y^i}$, $i = 1,\ldots,N$, and $\mathcal{B}(C(0,T],\mathbb{R}^k)) \triangleq \otimes^N_{i=1}\mathcal{B}(C(0,T],\mathbb{R}^{k_i}))$, $k = \sum^N_{i=1} k_i$, $\mathbb{P}^y \triangleq \times^N_{i=1}\mathbb{P}^{y^i}$. Then, we define the reference probability space $(\Omega,\mathbb{F},\{\mathbb{F}_{0,t} : t \in [0,T]\},\mathbb{P})$, by $\Omega \triangleq \mathbb{R}^n \times C([0,T],\Re^m) \times C([0,T],\mathbb{R}^k)$, $\mathbb{F} \triangleq \mathcal{B}(\mathbb{R}^n) \otimes \mathcal{B}(C([0,T],\mathbb{R}^m) \otimes \mathcal{B}(C([0,T],\mathbb{R}^k)$, $\mathbb{F}_{0,t} \triangleq \mathcal{B}(\mathbb{R}^n) \otimes \mathcal{F}^W_{0,t} \otimes \mathcal{G}^y_{0,t}$, $t \in [0,T]$, $\mathbb{P} \triangleq \Pi_0 \times \mathbb{P}^W \times \mathbb{P}^y$, the independent observations (WP3), which are independent of the decisions, and the state process by

$$dx^u(t) = f(t,x^u(t),u_t)\mathrm{d}t + \sigma(t,x^u(t),u_t)dW(t), \qquad x(0) = x_0, \quad t \in (0,T]. \tag{19.5}$$

Then, we introduce the exponential functions for $i = 1,\ldots,N$:

$$\Lambda^{i,u}(t) \triangleq \exp\Big\{\int_0^t h^{i,*}(s,x(s),u_s)D^{i,-1}(s)dy^i(s) - \frac{1}{2}\int_0^t h^{i,*}(s,x(s),u_s)D^{i,-1}(s)h^i(s,x(s),u_s)ds\Big\}, \quad \Lambda^u(t) \triangleq \prod_{i=1}^N \Lambda^{i,u}(t),$$

$$d\Lambda^u(t) = \Lambda^u(t)\sum_{i=1}^N h^{i,*}(t,x(t),u_t)D^{i,-1}(t)dy^i(t), \qquad \Lambda^u(0) = 1, \quad t \in [0,T]. \tag{19.6}$$

For $u \in \mathbb{U}_{rel}^{(N)}[0, T]$, the payoff under the reference probability space $(\Omega, \mathbb{F}, \mathbb{P})$ is

$$J(u) \stackrel{\triangle}{=} \mathbb{E}\Big\{ \int_0^T \Lambda^u(t)\ell(t, x(t), u_t)\mathrm{d}t + \Lambda^u(T)\varphi(x(T)\Big\}. \tag{19.7}$$

Under the reference probability measure $\mathbb{P}$, the payoff (19.7) with $\Lambda^u(\cdot)$ given by (19.6), and the state process satisfying (19.5) is a transformed problem with observations which are not affected by any of the team decisions. If $\{\Lambda^u(t) : t \in [0, T]\}$ is an $(\{\mathbb{F}_{0,t} : t \in [0, T]\}, \mathbb{P})$-martingale, then, we define a probability measure by setting

$$\frac{d\mathbb{P}^u}{d\mathbb{P}}\Big|_{\mathbb{F}_T} = \Lambda^u(T). \tag{19.8}$$

Under the probability measure $(\Omega, \mathbb{F}, \mathbb{P}^u)$, the observations, state, and payoff are defined by (19.2)–(19.3), with $B^i(\cdot)$ replaced by $B^{i,u}(\cdot)$, $i = 1, \ldots, N$, and $\mathbf{E}$ by $\mathbb{E}^u$.

Fact 1. There formulation of the stochastic dynamic team problem under probability space $(\Omega, \mathbb{F}, \{\mathbb{F}_{0,t} : t \in [0, T]\}, \mathbb{P}^u)$, (19.1)–(19.4), is equivalent to that under the reference probability space $(\Omega, \mathbb{F}, \{\mathbb{F}_{0,t} : t \in [0, T]\}, \mathbb{P})$, (19.5)–(19.7), in which $\{y^i(t) : t \in [0, T]\}$, $i = 1, \ldots, N$ are Brownian motions, and hence, the information structures are independent of the team decisions.

Fact 2. Girsanov's approach implies that the probability measure $\mathbb{P}^u$ and the Brownian motions $\{B^{i,u}(t) : t \in [0, T]\}$ depend on u but $\{y^i(t) : t \in [0, T]\}$ and $\{\mathscr{G}_{0,t}^{y^i} : t \in [0, T]\}$, $i = 1, \ldots, N$, are fixed â priori and independent of u. When the information structures are functionals of the state process $\{x(t) : t \in [0, T]\}$, then for σ independent of u, we can also make $\{x(t) : t \in [0, T]\}$ to be independent of u by replacing (19.5) by $dx(t) = \sigma(t, x(t))dW(t)$ [12]. We can derive a BSDE relating the value process and PbP optimality [12].

Fact 3. Girsanov's measure transformation generalizes and makes precise Witsenhausen's [5] equivalence of discrete-time stochastic dynamic decision problems to static team problems. The "common denominator condition" and "change of variables" described in [5] are equivalent to the "change of probability measure" via (19.8) and the associated Bayes' theorem of expressing $J(u)$ via (19.7) [11, 12].

19.4 Team and PbP Optimality Conditions

In this section, we describe some of the consequences of **Fact 1, Fact 2, Fact 3**, in deriving necessary and sufficient team and PbP optimality conditions.

19.4.1 Discrete-Time Equivalence of Static and Dynamic Team Problems

Consider a discrete-time team problem on the probability space $(\Omega, \mathbb{F}, \{\mathbb{F}_{0,t} : t \in \mathbb{N}_0^T\}, \mathbb{P}^u)$, $\mathbb{N}_0^T \triangleq \{0, 1, \ldots, T\}$, $\mathbb{N}_1^T \triangleq \{1, 2, \ldots, T\}$ described by

$$x(t+1) = f(t, x(0), \ldots, x(t), u^1(t), \ldots, u^N(t)) + G(t)w^u(t+1), \quad t \in \mathbb{N}_0^{T-1}, \tag{19.9}$$

$$y^m(t) = h^m(t, x(t), u^1(t), \ldots, u^N(t)) + D^{\frac{1}{2},m}(t)b^{m,u}(t), \ t \in \mathbb{N}_0^T, \ \forall m \in \mathbb{Z}_N, \tag{19.10}$$

$$J(u) = \mathbb{E}^u\{\sum_{t=0}^{T-1} \ell(t, x(t), u^1(t), \ldots, u^N(t)) + \varphi(x(T))\}, \tag{19.11}$$

where $\{x(0), w^u(\cdot), b^{m,u}(\cdot)\}$ are independent, distributed according to $\Pi_0(dx)$, $\{w(t) \sim \zeta_t(\cdot) \triangleq \text{Gaussian}(0, I_{n\times n}) : t \in \mathbb{N}_1^T\}$, $\{b^{m,u}(t) \sim \lambda_t^m(\cdot) \triangleq \text{Gaussian}(0, I_{k_m\times k_m}) : t \in \mathbb{N}_0^T\}$, and $G(\cdot)$, $D^{\frac{1}{2},m}(\cdot)$ invertible, for $m = 1, \ldots, N$.

Next, we specify the information structures. For $t \in \{0, 1, \ldots, T\}$, let $\mathscr{Y}_t \triangleq \{(\tau, m) \in \{0, 1, \ldots, t\} \times \{1, 2, \ldots, N\}\}$. A data basis at time $t \in \{0, 1, \ldots, T\}$ for the kth decision station is a subset $\mathscr{Y}_{t,k} \subseteq \mathscr{Y}_t$. The interpretation is that the decision applied by the kth station at time t is based on $\{y^\mu(\tau) : (\tau, \mu) \in \mathscr{Y}_{t,k}\}$, i.e., $u^k(t) \equiv \gamma_t^k(\{y^\mu(\tau) : (\tau, \mu) \in \mathscr{Y}_{t,k}\}, t \in \mathbb{N}_0^T, k = 1, \ldots, N$.

We introduce Girsanov's measure transformation via the following quantity.

$$\Theta_{0,t}^u \triangleq \prod_{\tau=1}^{t} \frac{\zeta_\tau(G^{-1}(\tau-1)(x(\tau) - f(\tau-1, x(0), \ldots, x(\tau-1), u^1(\tau-1), \ldots, u^N(\tau-1))))}{|G(\tau-1)|\zeta_\tau(x(\tau))}$$
$$\cdot \frac{\lambda_\tau(D^{-\frac{1}{2}}(\tau)(y(\tau) - h(\tau, x(\tau), u^1(\tau), \ldots, u^N(\tau))))}{|D^{\frac{1}{2}}(\tau)|\lambda_\tau(y(\tau))}, \quad t \in \mathbb{N}_1^T, \tag{19.12}$$

where $\Theta_{0,0}^u \triangleq \frac{\lambda_0(D^{-\frac{1}{2}}(0)(y(0)-h(0,x(0),u^1(0),\ldots,u^N(0))))}{|D^{\frac{1}{2}}(0)|\lambda_0(y(0))}$. Under $(\Omega, \mathbb{F}, \{\mathbb{F}_{0,t} : t \in \mathbb{N}_0^T\}, \mathbb{P})$, $\{(x(t), y^m(t)) : t \in \mathbb{N}_0^T\}$ are independent, with $x(0)$ having distribution $\Pi_0(dx)$, $\{x(t) \sim \zeta_t(\cdot) : t \in \mathbb{N}_1^T\}$, and $\{y^m(t) \sim \lambda_t^m(\cdot) : t \in \mathbb{N}_0^T\}$, for $m = 1, \ldots, N$, unaffected by the team decisions, and the discrete-time team payoff is given by

$$J(u) = \int \Big\{\Theta_{0,T}^u(x(0), u^1(0), \ldots, u^N(0), y(0), \ldots, x(T), u^1(T), \ldots, u^N(T), y(T))$$
$$\cdot\Big(\sum_{t=1}^{T-1} \ell(t, x(t), u^1(t), \ldots, u^N(t)) + \varphi(x(T))\Big)\Big\}$$
$$.\Pi_0(dx(0)).\lambda_0(y(0)).\prod_{t=1}^{T} \zeta_t(x(t)).\lambda_t(y(t))dx(t).dy(t), \tag{19.13}$$

$$J(\gamma_{[0,T]}^*) = \inf\Big\{J(\gamma_{[0,T]}) : \gamma_{[0,T]} \in \mathbb{U}^{(N)}[0,T], \quad J(\cdot) \equiv (19.3)\Big\}. \tag{19.14}$$

This is the transformed equivalent stochastic team problem.

Fact 4. Dynamic team (19.9)–(19.11) is equivalent to the static team (19.13), (19.14), and hence, optimality conditions in [4] are applicable to (19.14). This is applied in [11] to compute the optimal strategies of Witsenhausen's [2] counterexample.

19.4.2 Continuous-Time Stochastic Maximum Principle for Team Optimality

Next, we present the optimality conditions for Problem 19.1, derived in [11]. Consider the equivalent team problem under the reference probability space $(\Omega, \mathbb{F}, \{\mathbb{F}_{0,t} : t \in [0, T]\}, \mathbb{P})$, described by $\{\Lambda, x\}$ satisfying (19.6), (19.5), and reward (19.7). Let $X \triangleq Vector\{\Lambda, x\} \in \mathbb{R} \times \mathbb{R}^n$, $\overline{W}(\cdot) \triangleq Vector\{\int_0^\cdot D^{\frac{1}{2}}(s)dB(s), W(\cdot)\} \in \mathbb{R}^{k+m}$, $h(t, x, u) \triangleq Vector\{h^1(t, x, u), \ldots, h^N(t, x, u)\}$, $L(t, X, u) \triangleq \Lambda\ell(t, x, u)$, $\Phi(X) \triangleq \Lambda\varphi(x)$, then

$$dX(t) = F(t, X(t), u_t)dt + G(t, X(t), u_t)d\overline{W}(t), \; X(0) = X_0, \; t \in (0, T]. \tag{19.15}$$

$$J(u) = \mathbb{E}\{\int_0^T L(t, X(t), u_t)dt + \Phi(X(T))\}. \tag{19.16}$$

The Hamiltonian $\mathcal{H} : [0, T] \times \mathbb{R}^{n+1} \times \mathbb{R}^{n+1} \times \mathcal{L}(\mathbb{R}^{m+k}, \mathbb{R}^{n+1}) \times \mathcal{M}_1(\mathbb{A}^{(N)}) \to \mathbb{R}$ is

$$\mathcal{H}(t, X, \Psi, Q, u) \triangleq \langle F(t, X, u), \Psi\rangle + tr(Q^*G(t, X, u)) + L(t, X, u). \tag{19.17}$$

For any $u \in \mathbb{U}_{rel}^{(N)}[0, T]$, the state process satisfies (19.15), the adjoint process $(\Psi, Q) \in L^2_{\mathbb{F}_T}([0, T], \mathbb{R}^{n+1}) \times L^2_{\mathbb{F}_T}([0, T], \mathcal{L}(\mathbb{R}^{m+k}, \mathbb{R}^{n+1}))$ and it satisfies the BSDE

$$d\Psi(t) = -\mathcal{H}_X(t, X(t), \Psi(t), Q(t), u_t)dt + Q(t)d\overline{W}(t), \; \Psi(T) = \Phi_X(X(T)). \tag{19.18}$$

Theorem 19.1 ([11] Team Optimality Conditions. Necessary Conditions) *For an element $u^o \in \mathbb{U}_{rel}^{(N)}[0, T]$ with the corresponding solution $X^o \in B^\infty_{\mathbb{F}_T}([0, T], L^2(\Omega, \mathbb{R}^{n+1}))$ to be team optimal, it is necessary that*
(1) *There exists $(\Psi^o, Q^o) \in L^2_{\mathbb{F}_T}([0, T], \mathbb{R}^{n+1}) \times L^2_{\mathbb{F}_T}([0, T], \mathcal{L}(\mathbb{R}^{m+k}, \mathbb{R}^{n+1}))$.*
(2) *The variational inequality is satisfied:*

$$\sum_{i=1}^N \mathbb{E}\Big\{\int_0^T \mathcal{H}(t, X^o(t), \Psi^o(t), Q^o(t), u_t^{-i,o}, u_t^i - u_t^{i,o})dt\Big\} \geq 0, \quad \forall u \in \mathbb{U}_{rel}^{(N)}[0, T].$$

(3) *(Ψ^o, Q^o) is a unique solution of the BSDE (19.18) with $u^o \in \mathbb{U}_{rel}^{(N)}[0, T]$ satisfying*

$$\mathbb{E}\Big\{\mathscr{H}(t, X^o(t), \Psi^o(t), Q^o(t), u_t^{-i,o}, v^i)|\mathscr{G}_t^i\Big\} \geq \mathbb{E}\Big\{\mathscr{H}(t, X^o(t), \Psi^o(t), Q^o(t), u_t^o)|\mathscr{G}_t^i\Big\},$$
$$\forall v^i \in \mathscr{M}_1(\mathbb{A}^i), a.e. t \in [0,T], \mathbb{P}|_{\mathscr{G}_t^i} - a.s., i \in \mathbb{Z}_N \quad (19.19)$$

Sufficient Conditions. *Let $(X^o(\cdot), u^o(\cdot))$ denote an admissible state and decision pair and let $\Psi^o(\cdot)$ the corresponding adjoint processes and assume*

(C) $\mathscr{H}(t, \cdot, \Psi, Q, v)$ is convex in $X \in \mathbb{R}^{n+1}$; $\Phi(\cdot)$ is convex in $X \in \mathbb{R}^{n+1}$. Then, $(X^o(\cdot), u^o(\cdot))$ is optimal if it satisfies (19.19).

Fact 5. The necessary conditions for team optimality are equivalent to those of PbP optimality, and under (C), PbP optimality implies team optimality. Since regular strategies $\mathbb{U}_{reg}^{(N)}[0,T]$ are embedded into relaxed strategies$\mathbb{U}_{rel}^{(N)}[0,T]$, from Theorem 19.1, we obtain the optimality conditions for regular strategies [8, 11]. Applications of (19.19) lead to fixed point-type equations. Example can be carried out as in [9].

19.5 Additional Results and Open Issues

For noiseless and noisy nonclassical information structures, related results and examples, without invoking Girsanov's measure transformation, are found in [9, 10].

Extensions to a stochastic differential equation driven by continuous and jump martingales can be derived from those in [8].

Extensions of the stochastic maximum principle to discrete-time dynamic systems can be derived from those in [9–11].

Extensions to minimax or Nash-equilibrium strategies are still open problems.

References

1. Marschak J, Radner R (1972) Economic theory of teams. Yale University Press, London
2. Witsenhausen HS (1968) A counter example in stochastic optimum control. SIAM J Control Optim 6(1):131–147
3. Witsenhausen HS (1971) Separation of estimation and control for discrete time systems. Proc IEEE, pp 1557–1566
4. Krainak J, Speyer JL, Marcus SI (1982) Static team problems-part I: sufficient conditions and the exponential cost criterion. IEEE Trans Autom Control 27(4):839–848
5. Witsenhausen H (1988) Equivalent stochastic control problems. Math Control Signals Systems (MCSS) 1:3–11
6. Nayyar A, Mahajan A, Teneketzis D (2013) Decentralized stochastic control with partial history sharing: a common information approach. IEEE Trans Autom Control 58(7):1644–1658
7. Kumar PR, van Schuppen JH (1980) On nash equilibrium solutions in stochastic differential games. IEEE Trans Autom Control 25(6):1146–1149
8. Ahmed NU, Charalambous CD (2013) Stochastic minimum principle for partially observed systems subject to continuous and jump diffusion processes and driven by relaxed controls. SIAM J Control Optim 51(4):3235–3257
9. Charalambous CD, Ahmed NU (2013) Centralized versus decentralized team optimality of distributed stochastic differential decision systems with noiseless information structures-Part

I: General theory, Submitted to IEEE Transactions on Automatic Control, p. 39, February 2013. [Online]. Available: http://arxiv.org/abs/1302.3452, http://arxiv.org/abs/1302.3416
10. Charalambous CD, Ahmed NU (2013) Team games optimality conditions of distributed stochastic differential decision systems with decentralized noisy information structures, Submitted to IEEE Transactions on Automatic Control, p 41, April 2013. (CDC2013) [Online]. Available: http://arxiv.org/abs/1304.3246
11. Charalambous CD, Ahmed NU (2013) Dynamic team theory of stochastic differential decision dystems with decentralized noisy information structures via Girsanov's measure transformation, Submitted to MCSS, p 50, November 2013. [Online]. Available: http://arxiv.org/abs/1309.1913
12. Charalambous CD (2013) Dynamic team theory of stochastic differential decision systems with decentralized noiseless feedback information structures via Girsanov's measure transformation, Submitted to MCSS, p 36, September 2013. [Online]. Available: http://arxiv.org/abs/1310.1488
13. Beneš VE (1971) Existence of optimal stochastic control laws. SIAM J Control Optim 9(3):446–475

Chapter 20
Signaling of Information

César A. Uribe and Jan H. van Schuppen

20.1 Motivation

In decentralized control, a set of controllers works simultaneously toward the minimization of a certain cost function. Different controllers have, in general, different information available. Each controller has partial observations of the system, and it bases its decisions on local information only. The construction of a common system state estimate might not be possible. The information available to each controller is defined by its *Information Pattern* or *Information Structure*, see Chap. 24. Control problems with several kinds of information structures have been studied in the past [1–3]. Under perfect communication of information (i.e., classical information patterns), all controllers have access to a common set of information resulting in a control problem with partial observations for each of the controllers. Nonetheless, among all information patterns, control theory with non-classical information structures is the least understood. Non-classical information structures impose strong constraints for the communication of information. For example, this is the case when no communication between controllers exists or measurements are delayed by more than one-time step [4, 5].

A solution for such communication constraints can be proposed by redefining the control task. Each controller should not only work toward performance improvement, but also toward sending information to other controllers so they have sufficient information to take coordinated decisions. A controller could use the plant to store information and to communicate with itself or other controllers at future instances

C.A. Uribe (✉)
Coordinated Science Laboratory,University of Illinois at Urbana–Champaign, 1308 W Main St, Urbana, IL 61801, USA
e-mail: cauribe2@illinois.edu

J.H. van Schuppen
Van Schuppen Control Research, Gouden Leeuw 143, 1103 KB Amsterdam, The Netherlands
e-mail: jan.h.van.schuppen@xs4all.nl

J.H. van Schuppen and T. Villa (eds.), *Coordination Control of Distributed Systems*, Lecture Notes in Control and Information Sciences 456,
DOI 10.1007/978-3-319-10407-2_20

(thus signaling). Then, the question arises whether *signaling* can be done and how to encode and decode the information.

20.2 The Third Role of Control

Traditionally, control strategies have been assumed to accomplish two main tasks:

- **Control the system (Performance):** Modify the system states such that they meet certain design specifications. Stabilization around operating points and minimization of cost functions are classical examples of such objectives.
- **Information reconstruction (Estimation):** Modify the available information such that the decision maker has a more complete knowledge about the system, which facilitates the definition of optimal actions.

An additional control task has been identified as well, namely the *Communication Task*. The control actions not only strive for performance or estimation but also for communicating information to other controllers. Given that no explicit communication channels exist between the controllers, the system can be used as the communication channel. A dynamical system working as a communication channel presents new challenges from the information theoretic prospective. It can be viewed as a channel with memory, inter-symbol interference, delays, and input-memory costs. The study of such channels is still very limited, and further research is required. Definition of channel capacity and other information theoretic aspects for dynamical systems channels are still yet to be made. There is the problem of the design of the encoder and the decoder once the decentralized control problem and the channel are fixed. The design is nonstandard because the channel has dynamics and feedback. The design methods available include the framework of classical encoder design or decentralized control with a tuple of encoder–decoder laws which achieves a Nash equilibrium. These issues require further study, see [6].

A classic example for the third role of control is the Witsenhausen's Counterexample [3]. Witsenhausen showed that optimal control laws for a system with non-classical information patterns do not satisfy the initial assumptions about the separation principle or linearity. The form of the proposed nonlinear control law suggests the controller does not only consider performance or estimation but communication as well, namely *signaling* [7]. One of the controllers law resembles an encoder, while the other controller acts as an n-bit quantizer. Despite the fact that the implications of the third task have been known since the early 1970s, this aspect of control is not well understood yet.

20.3 Signaling: Implicit and Explicit Communications

Signaling can be understood as the ability of a tuple of controllers to convey information from one to another by using the dynamic system as a communication channel [8]. Signaling was initially described in [9] based on Kuhn-type extensive games. A controller could transmit the necessary information to other controllers by modifying the evolution of the state under specific control inputs. Therefore, a controller can obtain information from other controllers by observing the dynamics of the system. This idea was further used to define controllability under non-classical information structure [10]. Even if the system is not controllable by individual controllers, signaling local state estimates expands the controllability space. Different definitions of signaling and its effects have been identified in the literature: learning the dynamics of the system, communicating a controllers' beliefs about the system state to other controllers, communicating a controller' beliefs about other controllers, or communicating a controller' beliefs about their own future actions.

In [11], the author shows how a controller can add to a regular input signal another small input signal encoding its observations for deterministic systems. Nonetheless, signaling in distributed systems gets more complicated if there are noise effects. The signal could not be seen effectively because of the presence of noise in the observations. Nevertheless, signaling seems possible in decentralized stochastic control systems. Signaling in stochastic systems is discussed in [12, 13]. In [14], the authors define control, signaling and sensing schemes to minimize the sum rate required for decentralized stabilizability in a multicontroller system structure based on signaling strategies introduced in [10]. In [15], the authors extended this approach to stochastic systems in the presence of erasure links. Recently, in [16], the author provides a deep analysis of the signaling phenomena by developing fundamental results toward an implicit communications theory. However, a clear analysis of examples is still missing.

In [16], the author introduced an interpretation of the Witsenhausen Counterexample. Two controllers or decision makers act on a mass to move it from one position to another. One of them has perfect observations of the state of the system, (i.e., noiseless channel) but its actions are heavily penalized. Conversely, the second controller has much influential power in the position of the mass due to the costless actions, but its measurements are noisy. The control objective is to define the actions of both controllers such that the state of the system is as close to zero as possible. The question is as follows: How to communicate the perfect observations of the first controller to the second controller? This situation is common in decentralized stochastic control problems, but no complete solution toward the synthesis of communication protocols for signaling has been fully developed [17].

The fundamental difference between explicit and implicit communication can be defined as follows. Given a decentralized control system composed of several controllers, explicit communication refers to the existence of external communication channels connecting several of the controllers (allowing them to share parts of their local information). This produces a classical information pattern, for which centralized control synthesis tools can be applied. On the other hand, implicit communication refers to the identification of the dynamic system as a channel. Therefore, changing the control objective by introducing additional communication requirements to convey information from one controller to another.

In [18], authors show that if propagation delays in the system dynamics are slower than transmission delays on an external channel, there is no incentive to do signaling. Furthermore, the resulting problem can be formulated as a convex optimization problem. This is a special case of their criterion of quadratic invariance. This result is based on a clear decomposition of the stochastic system into several subsystems with well-defined communication channels between them.

Although extensive use of communication networks enriches the information structures of a certain problem, explicit communication channels do not always imply an increase of performance. An incentive for signaling is present as long as the external communication links are imperfect [19]. There are cases where implicit communication is better than explicit communication channels or at least it performs better under a wide range of signal-to-noise ratio. The impact of imperfect channels in control is of great real-application interest.

One of the most interesting results about the usefulness of implicit communications was presented in [19]. The author evaluates how a real external channel (with finite bandwidth) might improve the performance of a decentralized system. Comparison between the Witsenhausen counterexample against a similar structure with an additional communication channel between the controllers was analyzed. The related cost when using the external communication channel was computed numerically and compared with four different nonlinear control laws developed for the original Witsenhausen counterexample. The performance of the linear strategy with the Gaussian communication channel is better than the selected nonlinear strategies only for signal-to-noise ratio as high as ten. This proves that even when communication channels are available, the use of nonlinear strategies and signaling can induce a relevant increase of performance. For specific characteristics of the compared algorithms, see [19].

In addition to recognizing the importance of the control tasks under limited communication channels, there is still a lack of general understanding of the signaling aspect of control. There is even more pronounced distance with engineering applications of signaling due to its inherent complexity compared with traditional decentralized control approaches.

20.4 An Academic Example of Signaling

In [20], the authors introduce a discrete-time finite-space stochastic system with a non-classical information pattern. The system structure is denoted as the *observer-controller decentralized control system*, and it provides an academic example of signaling.

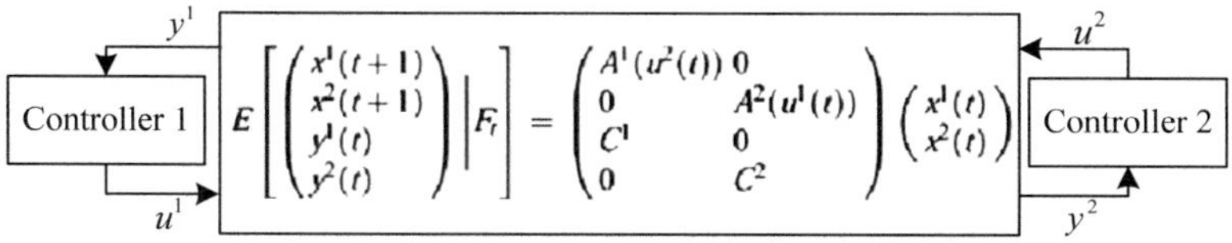

Fig. 20.1 The observer-controller system

Definition 20.1 Define a *observer-controller decentralized control system* as the mathematical structure,

$$\{(\Omega, P), T, X^{1,2}, Y^{1,2}, U^{1,2}, x^{1,2}, y^{1,2}, u^{1,2}, (A^{1,2}, C^{1,2})\}, \text{ where,}$$

$$T = \{t_0, t_0 + 1, \ldots, t_1\}, X^1 = \{\overline{x}_1, \ldots, \overline{x}_{n_1}\}, X^2 = \{\overline{x}_1, \ldots, \overline{x}_{n_2}\}, Y^1 = \{\overline{y}_1^1, \ldots, \overline{y}_{p_1}^1\},$$
$$Y^2 = \{\overline{y}_1^2, \ldots, \overline{y}_{p_2}^2\}, U^1 = \{\{\overline{u}_1^1, \overline{u}_2^1, \ldots, \overline{u}_{m_1}^1\}, U^2 = \{\overline{u}_1^2, \overline{u}_2^2, \ldots, \overline{u}_{m_2}^2\}.$$

where X is called the state set and Y and U are the outputs and inputs sets, respectively. The dynamics of the system by the conditional measure for the one-step transition of the state and for the two outputs are given by:

$$E\left[\begin{pmatrix} x^1(t+1) \\ x^2(t+1) \\ y^1(t) \\ y^2(t) \end{pmatrix} \middle| F_t\right] = \begin{pmatrix} A^1(u^2(t)) \; 0 & \\ 0 & A^2(u^1(t)) \\ C^1 & 0 \\ 0 & C^2 \end{pmatrix} \begin{pmatrix} x^1(t) \\ x^2(t) \end{pmatrix}. \tag{20.1}$$

Define the stochastic process and the function, $\tilde{x}^i : \Omega \times T \to X^i$ for $i = 1, 2$ with its indicator representation $x^i : \Omega \times T \to \mathbb{R}_+^{n_i}$, called the *state process*, $\tilde{y}^i : \Omega \times T \to Y^i$ for $i = 1, 2$ with its indicator representations $y^i : \Omega \times T \to \mathbb{R}_+^{p_i}$, called the *output process* of the system for Controller i, $\tilde{u}^i : \Omega \times T \to U^i$ for $i = 1, 2$ with its indicator representations $u^i : \Omega \times T \to \mathbb{R}_+^{m_i}$, called the *input process* of Controller i, $A^1 : \mathbb{R}_+^{m_2} \to \mathbb{R}_+^{n\times n}$, $A^2 : \mathbb{R}_+^{m_1} \to \mathbb{R}_+^{n\times n}$, called the *input-dependent state transition matrices*, $C^1 : \mathbb{R}^{p_1}$, $C^2 : \mathbb{R}^{p_2}$, called the *input-dependent output matrices*. In order to simplify the notation, the σ-algebra family $\{F_t \subseteq F, \forall\, t \in T\}$ is used, to which all processes are assumed to be measurable for all $t \in T$, $y(t)$ is F_t measurable, $x(t+1)$ is F_t measurable, and $u(t)$ is F_{t-1} measurable. Additionally, the event $\{x_i(\omega, t) = 1\}$ is denoted as $\{x_i(t)\}$. Define the state transition matrix and the observation map as:

$$A_{i,j}^1(e_m) = P\left(x_i^1(t+1)|x_j^1(t), u^2(t)_m\right), A_{i,j}^2(e_k) = P\left(x_i^2(t+1)|x_j^2(t), u^1(t)_k\right),$$
$$C_{i,j}^1 = P\left(y_i^1(t+1)|x_j^2(t)\right), C_{i,j}^1 = P\left(y_i^2(t+1)|x_j^2(t)\right).$$

The system consists of two control systems, with state x_1 and x_2, respectively, connected in a loop. Controller 1 partially observes the state x^1 and influences the dynamics of state x^2 and Controller 2 partially observes the state x_2 and influences the dynamics of state x^1, see Fig. 20.1.

Definition 20.2 Define the *class of time-varying (deterministic) control laws based on partial observations* as the functions and sets,

$$y^1(t_0:t) = \{y^1(t_0), y^1(t_0+1), \ldots, y^1(t)\}, u^1(t_0:t) = \{u^1(t_0), u^1(t_0+1), \ldots, u^1(t)\},$$
$$g^1(t) = g^1(t, y^1(t_0:t-1), u^1(t_0:t-1)) : Y_1^{t-t_0} \times U_1^{t-t_0} \to U_1,\ \forall\, t \in T,$$

and similarly for Controller 2. Denote the corresponding sets of control laws by,

$$G_{\text{clpo},i} = \{g_i\},\ i = 1, 2,\ \ G_{\text{clpo}} = G_{\text{clpo},1} \times G_{rmclpo,2}.$$

The *closed-loop system* thus can be defined as the time-varying stochastic system with representation,

$$E\left[\left(\begin{matrix} x_g^1(t+1) \\ x_g^2(t+1) \\ y^1(t) \\ y^2(t) \end{matrix}\right) \middle| F_t \right] = \left(\begin{matrix} A^1(g^2(t,.)) & 0 \\ 0 & A^2(g^1(t,.)) \\ C^1 & 0 \\ 0 & C^2 \end{matrix}\right) \left(\begin{matrix} x_g^1(t) \\ x_g^2(t) \end{matrix}\right). \tag{20.2}$$

Definition 20.3 Define the cost function on a finite horizon as:

$$J: G_{\text{clpo}} \to \mathbb{R}_+,\ b_{cl}: T \to \mathbb{R}_+^{n_1+n_2},\ b_1 \in \mathbb{R}_+^{n_1+n_2}.$$

$$J(g^1, g^2) = E\left[\sum_{s=t_0}^{t_1-1} b_{cl}(s)^T \left[x_{g^2}^1(s)\ \ x_{g^1}^2(s)\right]^T + b_1^T \left[x_{g^2}^1(s)\ \ x_{g^1}^2(s)\right]^T\right], \tag{20.3}$$

The problem of *decentralized control* for the *observer-controller decentralized control system* on a finite horizon is to solve the infimization problem, including the determination of the value J^* and of the optimal control law $(g^1, g^2)^* \in G$,

$$J^* = \min_{g^1, g^2 \in G} J(g^1, g^2) = J(g^{1,*}, g^{2,*}). \tag{20.4}$$

The above-defined optimal stochastic control problem has a non-classical information structure. The two controllers have different partial observations of the plant. In this case, the *signaling* phenomenon may occur in which any controller can signal to another controller part of its partial observations via the control system. The problem defined above is currently investigated for concepts of synthesis and design of signaling.

20.5 The Problem of Signaling

The problem of signaling in decentralized control can be defined as: Synthesize signaling laws for decentralized control of stochastic systems and analyze the effects with the questions: What? When? To whom? How to send information? In addition, how to strike a balance between the signaling and the other control objectives?

20.6 Research Issues

The following issues require attention of control researchers.

1. Find and analyze examples of decentralized control systems in which signaling takes place. For example, investigate whether signaling is always present in control with non-classical information patterns.
2. Analyze the information theoretic problem of communication of information via a dynamic system. The use of communication and information theory is needed for this issue. The concept of directed information has to be used because of the overall feedback structure, see Chap. 35.
3. What information needs to be signaled? The concepts of common and private information of a set of controllers have to be formulated and developed. For the definitions of common and private information with Gaussian observation processes, see Chap. 26. With this concept in mind, it is clear that common information of two controllers never needs to be communicated, while private is best communicated. The effects of noise and the cost function should also play a role.
4. When should the information be signaled? It has to be determined when the information is useful for the control purpose. For a particular form of signaling in the form of requests, see [10].
5. To whom should the information be communicated? A conjecture is that it should only be communicated to one's nearest neighbors in the network of the decentralized systems. The conjecture depends on the amount of interaction in a distributed system and on the graph structure of the system.
6. How to synthesize signaling laws? The theory for this has to wait until more has been investigated for the above questions.
7. How to strike a balance between the overall control objectives and the signaling objectives?

20.7 Most Relevant Literature

A reader who is introduced to the topic for the first time is recommended to read the papers [10, 12]. Books and theses of interest include [16, 21, 22].

References

1. Andersland M, Teneketzis D (1992) Information structures, causality, and nonsequential stochastic control I: design-independent properties. SIAM J Control Optim 30(6):1447–1475
2. Ho YC (1980) Team decision theory and information structures. Proc IEEE 68:644–654
3. Witsenhausen HS (1971) On information structures, feedback and causality. SIAM J Control 9(2):149–160

4. Nayyar A, Mahajan A, Teneketzis D (2013) Decentralized stochastic control with partial history sharing: a common information approach. IEEE Trans Autom Control (in print)
5. Varaiya P, Walrand J (1978) On delayed sharing patterns. IEEE Trans Autom Control 23(3):443–445
6. Charalambous CD, Kourtellaris CK, Hadjicostis C (2011) Optimal encoder and control strategies in stochastic control subject to rate constraints for channels with memory and feedback. In: Proceedings of 50th IEEE conference on decision and control (CDC-ECC)
7. Mitter S, Sahai A (1998) Information and control: Witsenhausen revisited. Lecture notes in control and information sciences, pp 281–293
8. Grover P, Sahai A (2010) Is Witsenhausen's counterexample a relevant toy? In: Proceedings of 49th IEEE conference on decision and control (CDC), pp 585–590
9. Sandell NR, Athans M (1975) A finite-state, finite-memory minimum principle. In: Proceedings of IEEE conference on decision and control, vol 14, pp 637–644, Dec 1975
10. Kobayashi H, Hanafusa H, Yoshikawa T (1978) Controllability under decentralized information structure. IEEE Trans Autom Control 23(2):182–188
11. Bismut J (1973) An example of interaction between information and control: the transparency of a game. IEEE Trans Autom Control 18(5):518–522
12. Ho Y-C (1980) Team decision theory and information structures. Proc IEEE 68(6):644–654
13. Ho Y-C, Kastner MP, Wong E (1978) Teams, signaling, and information theory. IEEE Trans Autom Control 23(2):305–312
14. Yüksel S, Başar T (2007) Optimal signaling policies for decentralized multicontroller stabilizability over communication channels. IEEE Trans Autom Control 52(10):1969–1974
15. Liu J, Gupta V (2012) Decentralized control over analog erasure links. In: Proceedings of 51st IEEE conference on decision and control (CDC), pp 6926–6931, Dec 2012
16. Grover P (2011) Actions can speak more clearly than words. PhD thesis, Electrical Engineering and Computer Sciences, University of California at Berkeley
17. van Schuppen J (2011) Control of distributed stochastic systems —introduction, problems, and approaches. In: Proceedings of IFAC World Congress
18. Rotkowitz M, Lall S (2006) A characterization of convex problems in decentralized control. IEEE Trans Autom Control 51(2):274–286
19. Martins NC (2006) Witsenhausen's counter example holds in the presence of side information. In: Proceedings of 45th IEEE conference on decision and control (CDC), pp 1111–1116
20. Uribe CA, van Schuppen JH (2013) Analysis of signaling in a finite stochastic system motivated by decentralized control. In: Proceedings of 52nd IEEE conference on decision and control (CDC), To appear
21. Spence M (1974) Market signalling: information transfer in hiring and related screening processes. Harvard University Press, Cambridge
22. Yüksel S, Başar T (2013) Stochastic networked control systems: stabilization and optimization under information constraints. Springer, Berlin

Chapter 21
Distributed Control of Manufacturing Networks: Analysis of Performance

Konstantin K. Starkov, Alexander Y. Pogromsky and Jacobus E. Rooda

21.1 Motivation

Nowadays, distributed controllers are widely used for manufacturing networks. Valuable manufacturing control approaches were developed using queuing theory, Petri nets, dynamic and linear programming, and hybrid systems. Though some of these approaches are widely used in industry, the performance of their operation still remains as a challenging problem.

21.2 Problem

In our research, we tackle the problem of performance analysis of manufacturing networks operated under distributed surplus-based control (see, e.g., [3–5] and references therein). In the surplus-based control, decisions are made based on the production demand tracking error which is the difference between the cumulative demand and the cumulative output of a system. Commonly, the efficiency in coordination and distribution of products throughout a manufacturing networks is evaluated by means of simulation and statistical analysis. Our goal is to contribute to these means by using classical tools from control theory.

K.K. Starkov (✉)
NSPYRE B.V., Utrecht, The Netherlands
e-mail: kkstarkov@gmail.com

A.Y. Pogromsky · J.E. Rooda
Eindhoven University of Technology, Eindhoven, The Netherlands
e-mail: a.pogromsky@tue.nl

J.E. Rooda
e-mail: J.E.Rooda@tue.nl

J.H. van Schuppen and T. Villa (eds.), *Coordination Control of Distributed Systems*,
Lecture Notes in Control and Information Sciences 456,
DOI 10.1007/978-3-319-10407-2_21

21.3 Theory and concepts

Let us first introduce the reader to an optimal control strategy for a simple case of a manufacturing network consisting of one machine.

A. Description of a model (One Machine)

In discrete time, a cumulative number of produced products in time k for a single manufacturing machine can be described as a sum of its production rates at each time step until time k. Thus, the flow model of a manufacturing machine in discrete time is defined as

$$y(k+1) = y(k) + u(k) + f(k), \tag{21.1}$$

where $y(k) \in \mathbb{R}$ is the cumulative output of the machine in time k, $u(k) \in \mathbb{R}$ is the control signal, and $f(k) \in \mathbb{R}$ is an unknown external disturbance.

Under the assumption that there is always sufficient quantity of the raw material to feed the machine, the control aim is to track the non-decreasing cumulative production demand. We define the production demand by using $y_d(k) \in \mathbb{R}$ given by $y_d(k) = y_{d0} + v_d k + \varphi(k)$ where y_{d0} is a positive constant that represents the initial production demand, v_d is a positive constant that defines the average desired demand rate, and $\varphi(k) \in \mathbb{R}$ is the bounded fluctuation that is imposed on the linear demand $v_d k$. Specifically, the problem is to minimize the output tracking error $\varepsilon(k) = y_d(k) - y(k)$ in the class of control strategies fed by available data:

$$u(k) = U_k[y(0), \ldots, y(k), y_d(0), \ldots, y_d(k)] \in \{0; 1\}.$$

This means that in time step k, the control input u is limited to taking the value of 1 when the machine is required to produce and taking the value of 0 when no production is required. Here, it is considered that the production speed of the machine is of 1 lot per time unit.

The increment of $\varepsilon(k)$ is given by:

$$\varepsilon(k+1) = \varepsilon(k) - u(k) + v_d + \Delta\varphi(k) - f(k). \tag{21.2}$$

The external disturbance $f(k)$ and the fluctuation $\varphi(k)$ in (21.2) are bounded

$$\alpha_1 < \Delta\varphi(k) - f(k) < \alpha_2 \qquad \forall k \in \mathbb{N} \tag{21.3}$$

where $\Delta\varphi(k) = \varphi(k+1) - \varphi(k)$ and α_1, α_2 are unknown constants that obey the following bounds $\alpha_2 < 1 - v_d$, and $\alpha_1 > -v_d$.

These conditions imply that the machine can never produce products faster than its maximal speed and that considering the presence of perturbations bounded by (α_1, α_2), the demand rate can only be positive, respectively. Note that in practice, it can be rather unnatural to set bounds on market fluctuations together with a machine perturbations. Thus, in our further research given in ([10] Chap. 4), we evaluated system perturbations independently from the demand rate fluctuations. Accordingly, the results of this chapter can be re-adjusted to more practical bounds.

Thus, from the above-mentioned inequalities, the following condition (also known as capacity condition) holds

$$0 < \xi(k) < 1 \qquad \forall k, \tag{21.4}$$

where for the notation $\xi(k) := v_d + \Delta\varphi(k) - f(k)$ is used.

B. Results on performance (One Machine)

Here, we provide the obtained results on optimal surplus-based controller for one machine. It has been shown in [6] that the following control strategy

$$u(k) = \text{sign}_+\big[\varepsilon(k)\big] \quad \text{where} \quad \text{sign}_+(\varepsilon) := \begin{cases} 1 & \text{if } \varepsilon > 0 \\ 0 & \text{if } \varepsilon < 0 \\ 0\ldots 1 & \text{if } \varepsilon = 0 \end{cases}, \tag{21.5}$$

is optimal with respect to the performance index

$$J_T = \sup_{\xi(0),\ldots,\xi(T-1)} \sum_{k=0}^{T} |\varepsilon(k)|^p \to \min_U, \tag{21.6}$$

for any given T, as well as with respect to the performance criterion

$$J_\infty = \lim_{k\to\infty} \sup \sup_{\xi(\cdot)} |\varepsilon(k)|^p \to \min_U . \tag{21.7}$$

This is true irrespective of the choice of $p \in [1, +\infty)$. The sup is taken over all $\xi(\cdot)$ satisfying (21.4), $U = \{U_k(\cdot)\}_{k=0}^{\infty}$ is the control strategy. Formula (21.6) deals with a finite and given time horizon T of the experiment, whereas it is infinite in (21.7), and $p \in [1, +\infty)$ is a given parameter.

In this section, we have shown the basic model of a single manufacturing machine and described the optimality of its surplus-based controller. Next, we extend our analysis to a line of N manufacturing machines operated under decentralized surplus-based control.

C. Manufacturing line with intermediate buffers of limited capacity.

Figure (21.1) presents a diagram of a line of N manufacturing machines with machines M_j, buffers B_j, and infinite product supply. Here, the control strategy for one machine is modified with respect to the number of buffers and machines present in the line. New limitations such as desired buffer content and buffer capacity restriction are considered in the model.

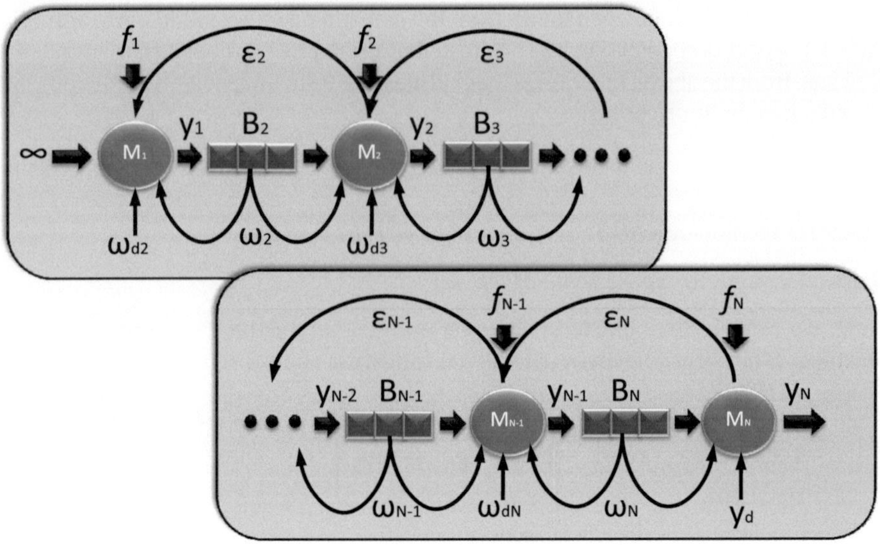

Fig. 21.1 Flow model diagram for a line of N manufacturing machines

The flow model of the manufacturing line is defined as

$$\begin{aligned}
\Delta y_1(k) &= \beta_1(k)\text{sign}_(w_2(k) - \gamma_2),\\
\Delta y_j(k) &= \beta_j(k)\text{sign}_{\text{Buff}}(w_j(k) - \beta_j(k))\text{sign}_(w_{j+1}(k) - \gamma_{j+1}),\\
&\qquad j = 2, \ldots, N-1,\\
\Delta y_N(k) &= \beta_N(k)\text{sign}_{\text{Buff}}(w_N(k) - \beta_N(k)),
\end{aligned}$$

where $\Delta y_j(k) = y_j(k+1) - y_j(k)$, $y_j(k)$ is the cumulative output of machine M_j in time k, $w_j(k) = y_{j-1}(k) - y_j(k)$ is the buffer content of buffer B_j, $\beta_j(k) = (\mu_j + f_j(k))u_j(k), \forall j = 1, \ldots, N$, f_j is the external disturbance affecting machine M_j (e.g., production speed variations, undesired delay, or setup time), μ_j is the nominal processing speed of machine j, u_j is the control input of machine M_j, $\text{sign}_{\text{Buff}}(x) = (1, \text{if } x \geq 0|0, \text{otherwise})$, $\text{sign}\_(x) = (1, \text{if } x \leq 0|0, \text{otherwise})$, and γ_{j+1} is the threshold value of the buffer content w_{j+1}.

The control inputs are defined as follows:

$$\begin{aligned}
u_j(k) &= \text{sign}_+(\varepsilon_{j+1}(k) + (w_{d_{j+1}} - w_{j+1}(k))),\\
&\qquad \forall j = 1, \ldots, N-1\\
u_N(k) &= \text{sign}_+(y_d(k) - y_N(k)),
\end{aligned}$$

where $w_{d_{j+1}}$ is the desired buffer level (base stock) of buffer B_{j+1} and the partial tracking errors are given by

$$\begin{aligned}
\varepsilon_j(k) &= \varepsilon_{j+1}(k) + (w_{d_{j+1}} - w_{j+1}(k)),\\
&\quad \forall j = 1, \ldots, N-2,\\
\varepsilon_{N-1}(k) &= \varepsilon_N(k) + (w_{d_N} - w_N(k)),\\
\varepsilon_N(k) &= y_d(k) - y_N(k).
\end{aligned}$$

Basically, for each machine, we introduce an extra restriction on production which is based on the buffer content of its upstream and downstream buffer. In this case, any machine M_j, with $j = 2, \ldots, N-1$, is activated only if three authorizations are given. The first authorization comes from control input $u_j(k)$ of M_j. The second authorization comes from the restriction on the upstream buffer content ($\text{sign}_{\text{Buff}}(\cdot)$), which is granted if the buffer contains at least the minimal number of products required ($\beta_j(k)$) in order for the machine M_j to start its work. The third authorization ($\text{sign}\_(\cdot)$) comes from the downstream buffer of given machine. This authorization is possible only if the downstream buffer has sufficient storage in order to accept incoming production, i.e., $w_{j+1}(k)-\gamma_{j+1} \leq 0$. The control actions are decentralized throughout the network. In other words, the control action of each machine in the line depends only on the production error of its neighboring downstream machine (except for machine M_N, which depends directly on cumulative demand input) and the current buffer content of its upstream and downstream buffer. The decentralized nature of the controller simplifies the implementation of the algorithm and gives our flow model an extra robustness with respect to the undesired events such as temporal machine setup or breakdown and flexibility in case of a modification to a configuration of the network.

E. Results on performance (Production Line with unbounded buffers)

Assuming that each machine in a manufacturing line is capable to process the arriving production demand and the base stock w_d of every intermediate buffer j satisfies the desired level of

$$w_{d_j} \geq \mu_j + \mu_{j-1} + \alpha_3 + \alpha_2 - \alpha_1,$$

and each buffer content is limited by $\gamma_j = \mu_j + \alpha_2 - \alpha_1 + w_{d_j}$ (for details, see [7, 8]), then the following results on performance of a line of N manufacturing machines hold true:

- Each tracking error (ε_j) of the system satisfies the following bounds

$$\begin{aligned}
&\limsup_{k\to\infty} \varepsilon_j(k) \leq v_d + \alpha_2,\\
&\liminf_{k\to\infty} \varepsilon_j(k) \geq v_d + \alpha_1 - \mu_j.
\end{aligned}$$

Further details on a manufacturing line with bounded buffers and the complete relation between the demand tracking accuracy and the base stock levels can be found in [9].

Now, in order to support the presented theoretical results, let us extend our analysis to a simulation example.

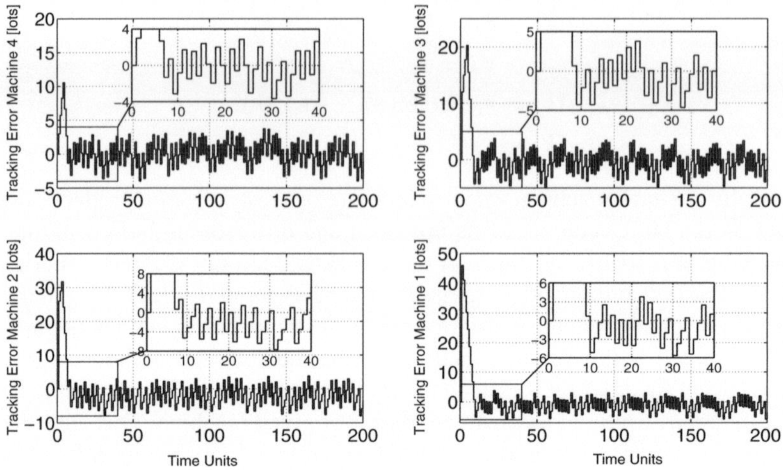

Fig. 21.2 Tracking errors $\varepsilon_j(k)$, with $v_d = 3$ and $\Delta\varphi(k) = \sin(50k)$

21.4 Examples

Consider the following example of a production line that consists of 4 manufacturing machines operating under variable structure regulators. The processing speed for each machine was set to $\mu_j = < 8, 10, 7, 6 >$ (lots per time unit), with $j = 1, \ldots, 4$, and the base stock level of each buffer was selected as $w_{d_j} = < 20, 18, 14 >$ (lots), with $j = 2, \ldots, 4$. The initial conditions of experiment $(y_{d0}, y_1(0), y_2(0), y_3(0), y_4(0))$ were set to the zero value. Further, the tracking error of each machine is depicted in Fig. 21.2. Here, it is noticeable that after the first 7 time steps, the output of machine M_4 reaches its steady-state level. Tracking errors are maintained inside [−6,4] lots for machine M_1, [−8,4] lots for machine M_2, [−5,4] lots for machine M_3, and [−4,4] lots for machine M_4, which satisfy the obtained theoretical bounds. The inventory level of each buffer is depicted in Fig. 21.3. The figure shows that the content of each buffer satisfies its upper bound.

21.5 Overview of Research Contributions

The main contributions of this research are the following. An important industrial problem, which consists in finding the balance between customer satisfaction and inventory levels in a manufacturing network, is tackled in our research. The industrial problem is interpreted in the form of a trajectory (demand) tracking control problem. Flow models for commonly used network topologies such as a single machine and a line are derived. The surplus-based production policy is proven to be optimal for a single manufacturing machine. Furthermore, the worst case scenario relation

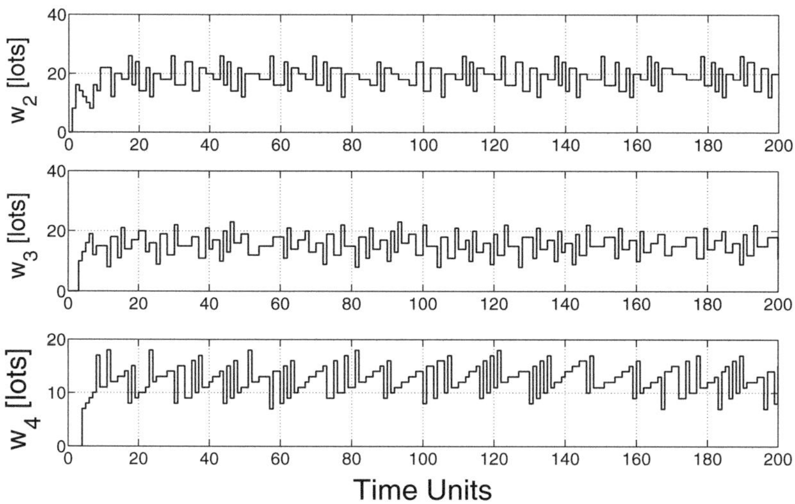

Fig. 21.3 Buffer contents $w_j(k)$, with $v_d = 3$ and $\Delta\varphi(k) = \sin(50k)$

between the demand tracking accuracy and base stock levels is obtained for a line of N machines subject to perturbations and market fluctuations.

21.6 Further Reading

For a reader new to the subject, the following references are recommended, [1, 2, 8, 9].

References

1. Bonvik A, Couch C, Gershwin S (1997) A comparison of production-line control mechanisms. Int J Prod Res 35(3):789–804
2. Dallery Y, Gershwin S (1992) Manufacturing flow line systems: a review of models and analytical results. Queueing Syst 12:3–94
3. Bielecki T, Kumar PR (1988) Optimality of zero-inventory policies for unreliable manufacturing systems. Oper Res 36(4):532–541
4. Gershwin SB (2000) Design and operation of manufacturing systems: the control-point policy. IIE Trans 32:891–906
5. Nilakantan K (2010) Enhancing supply chain performance with improved order-control policies. Int J Syst Sci 41(9):1099–1113
6. Starkov KK, Feoktistova V, Pogromsky AY, Matveev A, Rooda JE (2011) Optimal production control method for tandem manufacturing line. In: Submitted to Physcon, Leon, Spain
7. Starkov KK, Pogromsky AY, Rooda JE (2010) Towards a sustainable control of complex manufacturing networks. In: XV Summer School F. Turco, Monopoli, Italy

8. Starkov K, Pogromsky A, Rooda J (2012) Performance analysis for tandem manufacturing lines under variable structure production control method. Int J Prod Res 50(8):2363–2375
9. Starkov K, Feoktistova V, Pogromsky A, Matveev A, Rooda J (2012) Performance analysis of a manufacturing line operated under optimal surplus-based production control. Math Probl Eng doi:10.1155/2012/602094
10. Starkov KK (2012) Performance analysis of manufacturing networks: surplus-based control. Eindhoven University of Technology ISBN: 978-90-386-3266-7

Part V
Distributed Control with Communication

Chpater 22 provides an introduction to distributed control with direct communication between controllers. This control architecture was defined in Chap. 11. The next two chapters define information structures and show how control theory depends on the information structure used. Chapter 26 defines concepts of common and private information for control of distributed systems. The next chapter shows how this can be used for state estimation. The last chapter of this part deals with communication constraints in distributed systems.

Chapter 22
What Is Distributed Control with Direct Communication Between Controllers?

Jan H. van Schuppen

22.1 Introduction

The purpose of this chapter is to introduce to the reader the topic of distributed control with communication between controllers. The control synthesis problem for the control architecture of distributed control with communication between controllers is to synthesize control laws of this type which result in closed-loop systems meeting the control objectives as well as possible.

The motivation for the control architecture of distributed control with direct communication between controllers is that without such communication, the control objectives cannot be met satisfactorily. Then, direct communication between controllers seems the next best option. Direct communication means that these controllers sent each other messages at every time step, or when perceived as useful. At every time step, each controller may send zero, one, or more messages to other controllers and also receive zero, one, or more messages from other controllers. Several decades ago, the communication equipment was costly. Nowadays, the costs of the online use of the communication equipment are comparable to the costs of the control performance; hence, this form of communication for control may be used to lower the overall performance.

The choice for the use of the control architecture of distributed control with direct communication between controllers depends therefore on the answer to the question: Is the decrease in the overall performance costs of the closed-loop distributed control system, due to direct communication between controllers, higher or lower than the costs of online communication and of the associated computations? In the opinion of the author, these costs need to combined so as to achieve an integrated design. This question is both a complexity issue and an economic issue.

J.H. van Schuppen (✉)
Van Schuppen Control Research, Gouden Leeuw 143,
1103 KB Amsterdam, The Netherlands
e-mail: jan.h.van.schuppen@xs4all.nl

J.H. van Schuppen and T. Villa (eds.), *Coordination Control of Distributed Systems*,
Lecture Notes in Control and Information Sciences 456,
DOI 10.1007/978-3-319-10407-2_22

In several examples, the control architecture of distributed control with direct communication between controllers has resulted in closed-loop systems which meet the control objectives well. Examples include control with nearest-neighbor information structures.

Control theory for this class of systems is underdeveloped and difficult, while there is a need in control engineering for control synthesis with this control architecture. It is a fact that there are not so many recent references on this research topic, while the approach can be quite effective for engineering.

22.2 Examples

Control of an urban road network as studied in the C4C Project, see Chap. 5, is an example where this control architecture of control with communication between controllers is used. The road network is in the state of Flanders of Belgium. The control architecture is such that the controller of a particular intersection acts as a leader and, as such, exchanges information with the controllers of the nearest-neighbor intersections. This is a form of nearest-neighbor control that is applied in many other control engineering problems.

Mobile sensors such as a set of underwater vehicles or a set of aerial vehicles, as studied in the C4C Project, is another example. The control architecture is such that the underwater vehicles exchange information with other vehicles so as to avoid collision, see Chap. 3 for references. In aerial vehicles with a search mission, the control architecture is such that any vehicle communicates with its neighboring vehicles and they may exchange information on their findings and their travel plans.

In platoons of vehicles on motorways, the objective is to stabilize the platoon, so as to guarantee that the vehicles will follow the trajectory set by the lead vehicle and to guarantee that the vehicles never collide. Two control architectures are distinguished: (1) nearest-neighbor control architecture in which any car only receives information from its nearest neighbors, thus from the car directly in front or from the car directly following it, and (2) the coordinated-control architecture in which the lead vehicles communicates directly with all other vehicles of the platoon. A research issue is whether the control objective of stability can be met with the nearest-neighbor control architecture. The reader is referred to [14–16] for further information.

The control architecture of direct communication between controllers is used in various forms of communication networks. The most well-known form is the back-pressure algorithm of communication networks in which partial state information of the nearest neighbors is used in each local controller. Another one is the alternating bit protocol. These control laws can be found in most recent books on communication networks.

22.3 Problem

Problem 22.1 The *problem of distributed control with direct communication between controllers* is to synthesize and to design control laws in this class for both the control and the communication tasks so as to achieve the control objectives as well as possible.

22.4 Research Issues

Specific research issues of control theory for the control architecture of distributed control with direct communication between controllers include the following:

1. *When*? When should a controller communicate with other controllers? Forms discussed in the literature include the following: (a) *Communication after receipt of an observation.* A controller sends after receipt a subset of its observations. Which subset? (b) *Communication when useful.* A controller sends its latest observation when it has evaluated that this information is useful to other controllers at that time. (c) *Communication when requested.* A controller requests information of another controller if it presumes that that controller has information which is useful for its own control task.
2. *Whom*? To which other controllers should the information be sent? Answers could be to the nearest neighbors, to particular subsystems depending on the contents, etc. The answer depends much on the structure of the interaction of the distributed system.
3. *What*? Which information about the system is useful to the control task of a controller which that controller does not observe but other controllers observe? This requires an analysis of the distributed system, of the observations, and of the control objectives. A principle could be that the stronger the interaction of two or more subsystems of the distributed system, the more information the respective controllers should exchange. In case of almost no interaction, then communication is likely not to be useful.
4. *How to communicate*? How is the encoding and the decoding to be formulated in a protocol for the communication from a controller to another? The theory of real-time communication needs to be used in this context. The form and the structure of the communication network are factors of this investigation.
5. *Processing of the received information for the control task*? How to formulate a framework for processing the information received by a controller? The information received is not only that received directly from the control system but also that received from other controllers. Hans Witsenhausen has formulated the concept of an information structure for this problem issue and this concept deserves to be better known and used for theory development. See Chap. 24. A related question is how to integrate the two streams of observations received by a controller into a control law. The difficulty here is the fact that the events received of two or more

streams have to be ordered by the distributed system. For this, the concept of state of a controller of a decentralized control system is needed.

6. *Control synthesis.* How to carry out control synthesis for distributed control with communication of a distributed system? It seems that even after the information of other controllers has been received, one still is faced with the distributed control problem discussed in Chap. 11. Determination of an equilibrium like the controller-by-controller equilibrium is a way to proceed, see Chap. 18 where this is a discussed for a team-by-team equilibrium.

22.5 Distributed Control with Communication for Linear Systems

The problem is briefly made more concrete by describing it for linear systems structured by the geometric structure of a line piece, thus in dimension 1. In this structure, every subsystem is connected to a left-hand and right-hand neighbor, except for the left and right terminal subsystems which are only connected to one neighbor.

Definition 22.1 Consider a line-structured deterministic linear system of the form,

$$x(t+1) = \begin{pmatrix} A_{11} & A_{12} & 0 & 0 & \dots & & 0 \\ A_{21} & A_{22} & A_{23} & 0 & \dots & & 0 \\ 0 & A_{32} & A_{33} & A_{34} & \dots & & 0 \\ \vdots & & \ddots & \ddots & \ddots & & \vdots \\ 0 & 0 & & A_{k-1,k-2} & A_{k-1,k-1} & & A_{k-1,k} \\ 0 & 0 & \dots & 0 & A_{k,k-1} & & A_{kk} \end{pmatrix} \begin{pmatrix} x_1(t) \\ x_2(t) \\ x_3(t) \\ \vdots \\ x_{k-1}(t) \\ x_k(t) \end{pmatrix}$$
$$+ \begin{pmatrix} B_{11} & 0 & 0 & 0 & \dots & 0 \\ 0 & B_{22} & 0 & 0 & \dots & 0 \\ 0 & 0 & B_{33} & 0 & \dots & 0 \\ \vdots & & \ddots & \ddots & \ddots & \\ 0 & 0 & & 0 & B_{k-1,k-1} & 0 \\ 0 & 0 & \dots & 0 & 0 & B_{kk} \end{pmatrix} u(t), \tag{22.1}$$

$$y(t) = \begin{pmatrix} C_{11} & 0 & 0 & 0 & \dots & 0 \\ 0 & C_{22} & 0 & 0 & \dots & 0 \\ 0 & 0 & C_{33} & 0 & \dots & 0 \\ \vdots & & \ddots & \ddots & \ddots & \\ 0 & 0 & & 0 & C_{k-1,k-1} & 0 \\ 0 & 0 & \dots & 0 & 0 & C_{kk} \end{pmatrix} x(t), \tag{22.2}$$

$$z(t) = \begin{pmatrix} z_1(t) \\ z_2(t) \\ z_3(t) \\ \vdots \\ z_{k-1}(t) \\ z_k(t) \end{pmatrix} = \begin{pmatrix} H_{11} & 0 & 0 & 0 & \ldots & 0 \\ 0 & H_{22} & 0 & 0 & \ldots & 0 \\ 0 & 0 & H_{33} & 0 & \ldots & 0 \\ \vdots & & \ddots & \ddots & \ddots & \\ 0 & 0 & & 0 & H_{k-1,k-1} & 0 \\ 0 & 0 & \ldots & 0 & 0 & H_{kk} \end{pmatrix} y(t). \quad (22.3)$$

In the linear system representation described above, there are $k \in \mathbb{N}_+$ local subsystems with states, inputs, and outputs denoted by (x_i, u_i, y_i) for $i = 1, 2, \ldots, k$. The communication signals from controller j to Controller i are denoted by z_j. A more refined model would distinguish the signal also on the basis of its destination i.

Definition 22.2 Define for the line-structured linear system, the following control laws each of which respects the structure of the linear system. Define the *nearest-neighbor state-feedback control law* by the formulas,

$$g_{sf}(x) = Fx, \quad (22.4)$$

$$u(t) = Fx(t), \quad (22.5)$$

$$F = \begin{pmatrix} F_{11} & F_{12} & 0 & 0 & \ldots & 0 \\ F_{21} & F_{22} & F_{23} & 0 & \ldots & 0 \\ 0 & F_{32} & F_{33} & F_{34} & \ldots & 0 \\ \vdots & & \ddots & \ddots & \ddots & \\ 0 & 0 & & F_{k-1,k-2} & F_{k-1,k-1} & F_{k-1,k} \\ 0 & 0 & \ldots & 0 & F_{k,k-1} & F_{kk} \end{pmatrix} = F_{lc} + F_{nn}, \quad (22.6)$$

where F_{lc} is a block-diagonal matrix and F_{nn} is a matrix with as nonzero elements only the upper and lower block-diagonal matrices.

The *static-output nearest-neighbor feedback control law* is defined by the structure

$$u(t) = G_{lc} y(t) + G_{nn} z(t), \quad (22.7)$$

where the dimensions of the matrices G_{lc} and of G_{nn} are different than in the nearest-neighbor state-feedback case, but the block structure of the matrices is the same.

Note that the backpressure algorithm for communication networks is a form of static-output feedback described above though it applies to a stochastic system.

There follows now a specialization of Problem 22.1 for the line-structured linear system described above.

Problem 22.2 Consider the line-structured linear system defined above. Consider control objectives, say of stability, performance, and robustness.

1. Does there exist a control law in the class of nearest-neighbor state-feedback laws such that the closed-loop system is asymptotically stable? What are necessary and sufficient conditions for the existence of such control laws?

2. In case a stable control law exists, what are the smallest dimensions of the z_i vectors for $i = 1, 2, \ldots, k$ such that there exists a control law for which the closed-loop system is asympotically stable? More specifically, which information components of the communication signals are really needed for stability?
3. Which conditions on the system and on the performance criterion do imply that the optimal control law has one of the structures described in the previous definition?
4. If particular robustness properties of stability or of performance of the closed-loop system are needed, which conditions on the system matrices imply those robustness properties? One expects conditions on the dimensions of the system.

The example of a string of road vehicles in a platoon has been investigated, see [9, 11, 16].

The author has not investigated the above problems deeply. Apparently, a new concept of controllability for these structured systems is required. The second and third questions seem to be most the most interesting.

22.6 Overview of Theoretical Contributions in Supervisory Control

Distributed control with direct communication between controllers has been investigated for about 15 years in the research area of supervisory control. The problem refers to control of discrete-event systems as described by automata which are subject to control, see Chap. 16 for an introduction to this research topic.

The problem was formulated in the paper [20] where a sufficient condition for its usefulness appears. Thus, the control objective of attaining the specification language can be met if the combined observations allow that controller to meet the control objective. However, the control synthesis was incomplete partly due to the many synthesis choices that are available. The corresponding results for nonlinear systems are likely to be somewhat different.

Major research contributions to supervisory control for the control architecture of control with communication between controllers were provided by Barrett and Lafortune [1, 3, 4], and by Ricker and Rudie [12, 13]. Both provide algorithms on how to request communication and how to integrate the received communications into the control for the system.

22.7 Further Research

The main research issues need to be disentangled because the complexity of the problem is too high. The research issues stated in Sect. 22.4 all need attention.

Various forms of nearest neighbor communication have been explored in parts of engineering. It seems useful to explore the boundary of the subset of all cases for which nearest-neighbor control architecture can meet the control objectives either optimally or at least in a satisfactory way.

22.8 Further Reading

The control architecture of this chapter is defined in Chap. 5.

A reader who is new to the topic may want to first read the following introductory papers [18, 19].

The main theoretical framework for control with communication between controllers is the paper by Witsenhausen on separation properties, see [19]. See also the following chapters of this book, Chaps. 22, 24, and 25.

The problem of distributed control with direct communication between controllers for Gaussian stochastic control systems was briefly treated in [5, 18].

Control of distributed systems with a nearest-neighbor control law for problems structured in two dimensions is discussed in [9]. Control of structured linear systems is investigated in [7, 8]. Control of discretized partial differential equations also lead to problems with this control architecture, see [6]. See also the papers on control of platoons of vehicles [14, 15].

In communication networks, the backpressure algorithms are used which are a form of nearest-neighbor control laws. Any node of the system also receives partial state information from its nearest neighbors but not from those further away from it, see Chap. 30 and the paper [17]. There has been a large research effort to explore this approach for other communication networks. Currently, researchers active in cloud computing are exploring this control architecture.

For discrete-event systems also, control with communication between controllers has been investigated. The problem of supervisory control of discrete-event systems was first formulated in [20]. Research was carried out on this subject by Barrett and Lafortune [2–4] and Ricker and Rudie [12, 13].

The engineering, computer engineering, and computer science have publications which cover this control architecture, in particular multi-agent systems, see, for example [10].

References

1. Barrett G, Lafortune S (1998) A novel framework for decentralized supervisory control with communication. In: Proceedings of 1998 IEEE systems, man, and cybernetics conference, New York. IEEE Press
2. Barrett G, Lafortune S (1998) On the synthesis of communicating controllers with decentralized information structures for discrete-event systems. In: Proceedings of IEEE conference on decision and control, New York. IEEE Press, pp 3281–3286
3. Barrett G, Lafortune S (1999) Some issues concerning decentralized supervisory control with communication. In: Proceedings of 38th IEEE conference on decision and control, New York. IEEE Press, pp 2230–2236
4. Barrett G, Lafortune S (2000) Decentralized supervisory control with communicating controllers. IEEE Trans Autom Control 45:1620–1638
5. Boel RK, van Schuppen JH (2010) Control of the observation matrix for control purposes. In: Edelmayer A (ed) Proceedings of the international symposium on the mathematical theory of networks and systems (MTNS.2010), Budapest. University of Budapest, pp 1261–1268

6. Brockett RW, Willems JL (1974) Discretized partial differential equations: examples of control systems defined on modules. Automatica 10:507–515
7. Hamilton SC, Broucke ME (2012) Geometric control of patterned linear systems. Number 428 in LNCIS. Springer, Berlin
8. Hamilton Sarah C, Broucke Mireille E (2012) Patterned linear systems. Automatica 48:263–272
9. Jadbabaie A, Lin J, Morse AS (2003) Coordination of groups of mobile agents using nearest neighbor rules. IEEE Trans Autom Control 48:988–1001
10. Kopetz H (1997) Real-time systems—Design principles for distributed embedded applications. Kluwer Academic Publishers, Dordrecht
11. Melzer SM, Kuo BC (1971) A closed form solution for the optimal error regulation of a string of moving vehicles. IEEE Trans Autom Control 16:50–52
12. Ricker S, Rudie K (1997) Know means no: incorporating knowledge into decentralized discrete-event control. In: Proceedings of 1997 American control conference
13. Ricker SL, Rudie K (1999) Incorporating communication and knowledge into decentralized discrete-event systems. In: Proceedings of 38th IEEE conference on decision and control, New York. IEEE Press, pp 1326–1332
14. Sheikholeslam S, Desoer CA (1992) Control of interconnected nonlinear dynamical systems: the platoon problem. IEEE Trans Autom Control 37:806–810
15. Sheikholeslam S, Desoer CA (1993) Longitudinal control of a platoon of vehicles with no communication of lead vehicle information: a system level study. IEEE Trans Veh Technol 42:546–554
16. Swaarop D, Hedrick JK (1996) String stability of interconnected systems. IEEE Trans Autom Control 41:349–357
17. Tassiulas L, Ephremides A (1992) Stability properties of constrained queueing systems and scheduling policies for maximal throughput in multi hop radio networks. IEEE Trans Autom Control 37:1936–1948
18. van Schuppen Jan H (2011) Control of distributed stochastic systems—Introduction, problems, and approaches. Proc IFAC World Congr 2011:6029–6035
19. Witsenhausen Hans S (1971) Separation of estimation and control for discrete time systems. Proc IEEE 59:1557–1566
20. Wong KC, van Schuppen JH (1996) Decentralized supervisory control of discrete-event systems with communication. In: Proceedings of international workshop on discrete event systems 1996 (WODES96), London. IEE, pp 284–289

Chapter 23
Communication Constraints in Control and Observation of Distributed Systems

A. Yu. Pogromsky and A.S. Matveev

23.1 Motivation

Recent advances in communication technology have created the possibility of large-scale control systems, where the control tasks are distributed among numerous processors via a communication network. As the size of those systems grows, limitations caused by the network finite capacity become a bottleneck that cannot be neglected in designs of control algorithms. This motivated development of a new chapter of control theory, where control and communication issues are integrated, and gave rise to an intensive recent research on control under communication constraints—see e.g., the surveys in [1–3].

In this area, one of the basic questions is about the smallest communication data rate required to achieve the given control objective. This is similar to the Shannon source coding theory, which is interested in the smallest data rate above which a given stationary stochastic process can be reliably communicated. However, the critical difference is that in control systems, the processes are not necessarily stationary and the requirement of the real-time processing, typical for such systems, challenges the block-coding approach, typical for the classic information theory.

A.Yu. Pogromsky (✉)
Eindhoven University of Technology, Eindhoven, The Netherlands
e-mail: a.pogromsky@tue.nl

A.S. Matveev
St. Petersburg State University, St. Petersburg, Russia
e-mail: almat1712@yahoo.com

J.H. van Schuppen and T. Villa (eds.), *Coordination Control of Distributed Systems*,
Lecture Notes in Control and Information Sciences 456,
DOI 10.1007/978-3-319-10407-2_23

23.2 Problem

The answer to the above question is given, in various settings, by the fundamental data rate theorem, see e.g., [1–3] for a recent survey. Basically, this theorem states that the rate at which the channel is capable of reliable data communication should exceed the topological entropy of the open-loop system.

It is not difficult to show that for a smooth system of n differential equations with global Lipschitz constant L, its topological entropy is bounded from above by $Ln/\ln 2$. This estimate grows linearly with respect to n and for large-scale systems can impose unnecessarily severe constraints on communication in controller design. Hence, keeping in mind control and observation problems for large-scale distributed systems, it is interesting to find less conservative estimates of the topological entropy.

Inspired by engineering practice, which is often concerned with non-autonomous plants, e.g., driven by reference signals and/or exposed to external disturbances, a particular attention should be paid to time-varying systems.

The majority of the results dealing with the topological entropy evaluates this entropy by means of the Lyapunov exponents of the linearized system (the first Lyapunov method). This approach is not computationally efficient and not suitable for *design* problems.

The focus of our approach is on the estimates via the second Lyapunov method. This method is widely accepted in control practice.

23.3 Theory and Concepts

We consider a system of differential equations

$$\dot{x} = f(x, t), \quad x \in \mathbb{R}^n \tag{23.1}$$

that satisfies some minor regularity assumptions. For such a system, following [4] we can define the topological entropy on a bounded tube K. For time-invariant systems, this definition coincides with the standard definition of the topological entropy on an invariant compact set.

Now, we introduce the Jacobi matrix

$$A(t, \tau, \xi) = \frac{\partial f}{\partial x}[x(t, \tau, \xi), t]$$

and the following key assumption.

Assumption 23.1 There exist continuous and continuously differentiable with respect to t scalar $v_i(t, \tau, x_0)$, $i \in [1 : n]$ and $n \times n$-matrix valued $P(t, \tau, x_0) = P(t, \tau, x_0)'$ functions of $t \geq \tau$, $(x_0, \tau) \in K$ such that the following statements hold:

1. For all $t \geq s \geq t_0, x_0 \in K_{t_0}, i \in [1 : n]$,

$$P(t, t_0, x_0) = P[t, s, x(s, t_0, x_0)], \qquad v_i(t, t_0, x_0) = v_i[t, s, x(s, t_0, x_0)];$$

2. $P(t, s, x_0)$ is uniformly positive definite and bounded: $\exists \mu_1, \mu_2 > 0$:

$$\mu_1^2 I_n \leq P(t, s, x_0) \leq \mu_2^2 I_n \quad \forall t \geq s, (x_0, s) \in K; \tag{23.2}$$

3. The functions $v_i(t, s, x_0)$ are uniformly bounded:

$$\exists V > 0 \; : \; |v_i(t, s, x_0)| \leq V \quad \forall t \geq s, (x_0, s) \in K, i \in [1 : n]; \tag{23.3}$$

4. Let $\lambda_1(t, s, x_0) \geq \lambda_2(t, s, x_0) \geq \cdots \geq \lambda_n(t, s, x_0)$ be the roots of the following algebraic equation

$$\det[A'(t, s, x_0)P(t, s, x_0) + P(t, s, x_0)A(t, s, x_0) + \dot{P}(t, s, x_0) - \lambda P(t, s, x_0)] = 0,$$

repeated in accordance with their algebraic multiplicities. There exist constants $\Lambda_d \geq 0, d \in [1 : n]$ such that

$$\sum_{i=1}^{d} \lambda_i(t, s, x_0) + \dot{v}_d(t, s, x_0) \leq \Lambda_d \qquad \forall d \in [1 : n], t \geq s, (x_0, s) \in K. \tag{23.4}$$

Theorem 23.1 *The topological entropy of the dynamic flow Φ generated by this system on the bounded tube at hand obeys the estimate:*

$$H(\Phi, K) \leq \frac{\max_d \Lambda_d}{2 \ln 2}.$$

A similar statement can be formulated for discrete-time systems, see [5].

At this moment, we would like to clarify the assumptions imposed. Apart from the technical assumptions 1–3, the main condition involves solutions of the algebraic equation

$$\det(A'P + PA + \dot{P} - \lambda P) = 3.$$

This equation is of common use (perhaps reformulated in a different form) $i = n$ studies related to dynamical systems. For example, the condition $\lambda_1 + \dot{v}_1 < 0$ implies uniform asymptotic stability of forced oscillations. The condition $\lambda_1 + \lambda_2 + \dot{v}_2 < 0$ or $\lambda_1 + \lambda_2 + \dot{v}_2 > 0$) being verified inside a simply connected domain D for time-invariant systems guarantees that D contains no whole periodic orbits and hence serves as a generalization of the Bendikson criterion. Inequalities involving partial sums of λ_i's are common in upper estimations of various dimensions (i.e., Hausdorff, fractal, and Lyapunov) of invariant sets. For details, see [6].

23.4 Examples

For the Lorenz system

$$\dot{x} = -\sigma x + \sigma y,$$
$$\dot{y} = rx - y - xz,$$
$$\dot{z} = -bz + xy$$

with positive parameters σ, b, r we prove that if all the equilibria are unstable, then the topological entropy of any invariant compact set K satisfies

$$H(\Phi, K) \le \frac{1}{2\ln 2}\left(\sqrt{(\sigma-1)^2 + 4r\sigma} - (\sigma+1)\right).$$

For the Duffing oscillator

$$\ddot{x} + \delta\dot{x} + x^3 = u(t), \quad \delta > 0$$

with bounded input $u(t)$ and with bounded derivative $\dot{u}(t)$, we prove that the topological entropy of the corresponding dynamic tube K satisfies

$$H(\Phi, K) \le \frac{1}{2\ln 2}\left(-\delta + \sqrt{\min_{a>0}\frac{1}{a}\left[\left(a + \frac{\delta^2}{4}\right)^2 + \frac{27w^2}{8\left(a+\frac{\delta^2}{4}\right)}\right]}\right)$$

where

$$w := \limsup_{T\to\infty}\sup_{\tau\ge t_0}\left[\left(\frac{1}{T}\int_\tau^{\tau+T} u^2(t)\mathrm{d}t\right)^{\frac{1}{2}} + \delta^{-1}\left(\frac{1}{T}\int_\tau^{\tau+T}\dot{u}^2(t)\mathrm{d}t\right)^{\frac{1}{2}}\right].$$

For the Hénon system

$$x(k+1) = a + by(k) - x^2(k),$$
$$y(k+1) = x(k),$$

where $x \in \mathbb{R}$, $y \in \mathbb{R}$, and $a > 0$, $0 < b < 1$, we prove that the topological entropy for any invariant compact set K satisfies

$$H(\phi, K) \le \log_2\left(\sqrt{x_-^2 + b} - x_-\right), \quad x_- := \frac{b - 1 - \sqrt{(b-1)^2 + 4a}}{2}.$$

For the bouncing-ball system generated by the map on the cylinder $x_1 \in S^1$, $x_2 \in \mathbb{R}$

$$\phi(x) = \begin{pmatrix} x_1 + x_2 \\ \alpha x_2 - \beta \cos(x_1 + x_2) \end{pmatrix}, \quad 0 < \alpha < 1, \quad \beta > 0$$

we prove that the topological entropy for any invariant compact set K satisfies

$$H(\phi, K) \le \log_2 \left(1 + \alpha + \beta + \sqrt{(1 + \alpha + \beta)^2 - 4\alpha}\right) - 1.$$

Based on the theory presented above, we designed coder/decoder pairs for Hénon and bouncing-ball systems and showed that the corresponding observers are optimal in the sense that the observation problem is solved with minimal possible maximal data rate of the digital channel between coder an decoder, see [5].

23.5 Overview of Research Contributions

The main contribution of the approach is that it allows to find *constructive* estimates of the topological entropy for nonlinear systems. Once those estimates are found for discrete-time systems, one can design a coder/decoder pair to solve the observation problem via channels under data rate constraints. In particular situations for some examples, it is possible to show that such an observation scheme is optimal with respect to communication speed [5].

23.6 Most Relevant Literature

- A.S. Matveev and A.V. Savkin. Estimation and Control over Communication Networks. Birkhäuser, Boston, 2009.
- G.N. Nair, F. Fagnani, S. Zampieri, and R.J. Evans. Feedback control under data rate constraints: an overview. *Proceedings of the IEEE*, 95(1):108137, 2007.
- G.N. Nair, R.J. Evans, I.M.Y. Mareels, and W. Moran. Topological feedback entropy and nonlinear stabilization. *IEEE Transactions on Automatic Control*, 49(9):15851597, 2004.
- V.A. Boichenko, G.A. Leonov, and V. Reitman. *Dimension Theory for Ordinary Differential Equations*. Teubner Verlag, Wiesbaden, Germany, 2005.
- A. Katok. Fifty years of entropy in dynamics: 19582007. *Journal of Modern Dynamics*, 1(4):545596, 2007.

References

1. Antsaklis P, Baillieul J (2007) Special issue on the technology of networked control systems. Proc IEEE, 95(1)
2. Matveev AS, Savkin AV (2009) Estimation and control over communication networks. Birkhäuser, Boston
3. Nair GN, Fagnani F, Zampieri S, Evans RJ (2007) Feedback control under data rate constraints: an overview. Proc IEEE 95(1):108–137
4. Pogromsky A, Matveev A (2011) Estimation of topological entropy via direct Lyapunov method. Nonlinearity 27(7):1937–1959
5. Pogromsky A, Matveev A (2011) A topological entropy approach for observation via channels with limited data rate. IFAC World Congress, Milano
6. Boichenko VA, Leonov GA, Reitman V (2005) Dimension theory for ordinary differential equations. Teubner Verlag, Wiesbaden, Germany

Chapter 24
Information Structures

Jan H. van Schuppen

24.1 Motivation

In a problem of decentralized or distributed control with direct communication between controllers, each controller is not only provided observations directly from the control system but is also provided signals directly from one or more other controllers. There is also indirect communication between the controllers via the plant but that is not addressed in this chapter. Any model for control of such a system should include a specification of what is received from whom and for this the concept of an information structure is useful.

Control synthesis for an information structures can be simplified based on theoretical analysis. For example, the subset of control laws based on the outputs and inputs specified by the information structure can mostly be reduced to a smaller subset of control laws based on a conditional distribution or on a subset of the outputs and inputs without loss of performance.

The concept of information structure was formulated, and the associated theory was initiated by H.S. Witsenhausen. An example of Witsenhausen [11] shows that for a simple stochastic control problem with a private information structure, linear equations, Gaussian disturbances, and quadratic costs, a particular nonlinear control law achieves a strictly lower cost than the best linear control law. This example indicates that control synthesis for control with private information structures is not so simple and needs further investigation.

24.2 Problem

The problem of this chapter is formally stated.

J.H. van Schuppen (✉)
Van Schuppen Control Research, Gouden Leeuw 143, 1103 KB Amsterdam, The Netherlands
e-mail: jan.h.van.schuppen@xs4all.nl

J.H. van Schuppen and T. Villa (eds.), *Coordination Control of Distributed Systems*,
Lecture Notes in Control and Information Sciences 456,
DOI 10.1007/978-3-319-10407-2_24

Problem 24.1 Consider a distributed control system. Formulate the concept of an information structure, classify information structures, and develop control synthesis for control of distributed systems with any of the defined information structures.

24.3 Concepts

To be definite, information structures are defined for discrete-time nonlinear stochastic systems. The continuous-time version is similar though technically more demanding and hence not suitable for a short chapter.

Definition 24.1 Consider a discrete-time nonlinear stochastic system with the representation,

$$
\begin{aligned}
x(t+1) &= f(t, x(t), u(t), v(t)),\ x(t_0) = x_0,\\
y(t) &= h(t, x(t), u(t), v(t)),\\
&k \in \mathbb{Z}_+ \text{ number of controllers or stations}, n, m, m_v \in \mathbb{Z}_+,\\
&\{m_1, m_2, \ldots, m_k \in \mathbb{Z}_+\}, m_1 + m_2 + \ldots + m_k = m,\\
&\mathbb{Z}_k = \{1, 2, \ldots, k\}, \text{ the index set of the controllers},\\
&U_i = \mathbb{R}^{m_i}, U = \prod_{i=1}^{k} U_i, Y_i = \mathbb{R}^{m_i}, Y = \prod_{i=1}^{k} Y_i, X = \mathbb{R}^n, V = \mathbb{R}^{m_v},\\
T &= \{t_0, t_0+1, \ldots, t_1\} \subset \mathbb{Z}, \text{ the } \textit{time index set},\\
&x_0 \colon \Omega \to \mathbb{R}^n, \text{ a random variable, the } \textit{initial state},\\
&v \colon \Omega \times T \to V, \text{ a sequence of independent identically-distributed}\\
&\text{random variables},\\
&f \colon T \times X \times U \times V \to X, h \colon T \times X \times U \times V \to Y,\\
&\text{Borel measurable functions},\\
&u \colon \Omega \times T \to \mathbb{R}^m, u_i \colon \Omega \times T \to \mathbb{R}^{m_i}, x \colon \Omega \times T \to \mathbb{R}^n,\\
&y \colon \Omega \times T \to \mathbb{R}^m, y_i \colon \Omega \times T \to \mathbb{R}^{m_i}, \text{ stochastic processes},\\
y(t) &= \left(y_1(t)\ y_2(t)\ \ldots\ y_k(t)\right)^T, u(t) = \left(u_1(t)\ u_2(t)\ \ldots\ u_k(t)\right)^T,\\
y_i(t_0{:}\,t) &= \left(y_i(t_0),\ y_i(t_0+1),\ \ldots,\ y_i(t)\right) \in Y_i^{t-t_0+1}, = \emptyset, \quad \text{if } t < t_0,\\
y(t_0{:}\,t) &= \left(y_1(t_0{:}\,t),\ y_2(t_0{:}\,t),\ \ldots,\ y_k(t_0{:}\,t)\right) \in Y^{t-t_0+1},\\
u_i(t_0{:}\,t) &= \left(u_i(t_0),\ u_i(t_0+1),\ \ldots,\ u_i(t)\right) \in U_i^{t-t_0+1}, = \emptyset, \quad \text{if } t < t_0,\\
u(t_0{:}\,t) &= \left(u_1(t_0{:}\,t),\ u_2(t_0{:}\,t),\ \ldots,\ u_k(t_0{:}\,t),\right) \in U^{t-t_0+1}.
\end{aligned}
$$

At time $t \in T$, controller $i \in \mathbb{Z}_k$ receives directly from the system the observation process $y_i(t-1)$ and provides to the system the input $u_i(t)$.

Definition 24.2 Consider the distributed control system of Definition 24.1. Define the *information structure* of this system as the sets,

$$\begin{aligned}
\text{IS}(i,t) &\subseteq \prod_{j=1}^{k}(y_j(t_0{:}\,t-1), u_j(t_0{:}\,t-1)) \in \prod_{j=1}^{k}(Y_j^{t-t_0} \times U_j^{t-t_0}), \forall i \in \mathbb{Z}_k, \\
&\quad \text{the } \textit{information structure of Controller i at time t} \in T, \\
\text{IS}(t) &= \{\text{IS}(1,t), \ldots, \text{IS}(k,t)\}, \\
&\quad \text{the } \textit{information structure of the system at time t} \in T, \\
\text{IS} &= \{\text{IS}(t_0), \text{IS}(t_0+1), \ldots, \text{IS}(t_1)\}, \\
&\quad \text{the } \textit{information structure of the system}.
\end{aligned}$$

The information structure $\text{IS}(i, t)$ at time $t \in T$ includes only outputs and inputs up to time $t - 1$ due to the delay in the stochastic control system. Witsenhausen in [10, 13] uses for the information structure not the actual observations but index sets for the labels of the controllers and for the time in the index set. In this paper, the formulation provided above is preferred.

There follow properties and a classification of information structures. The classification is more refined and with different terms than that of [10]. Another expository paper on information structures is [6].

Definition 24.3 Define the following properties of information structures.

(a) An information structure is said to possess *perfect recall* if

$$\text{IS}(i,t) \subseteq \text{IS}(i,t+1), \quad \forall t \in T\backslash\{t_1\}, \quad \forall i \in \mathbb{Z}_k.$$

(b) An information structure is said to be *partially nested* if

$$\begin{aligned}
&\forall i \in \mathbb{Z}_k, \quad \exists I_i \subseteq \mathbb{Z}_k \text{ possibly empty, such that,} \\
&\text{IS}(j,t) \subseteq \text{IS}(i,t), \quad \forall t \in T, \quad \forall j \in I_i.
\end{aligned}$$

There is yet another distinction of the class of information structures: *sequential information structures* and *nonsequential information structures*. In this paper, the sequentiality is related to an information structure, it is not so emphasized by Witsenhausen in [14, 16].

An information structure is called *sequential* if the order of arrival of the inputs is fixed in advance before the system starts to progress in time. In a *nonsequential information structure*, the order of arrival of inputs is itself subject to the evolution of the system. For a nonsequential information structure, the reader may think of the controllers being connected by a communication network. Due to the use of different routes and to varying delays, a message sent first from a sending controller may arrive at the receiving controller later than a second message sent later from the sender controller due to them having followed different routes through the network. With nonsequential information structures, the state of the controller has to take care of the different orders in which the inputs may arrive, see [16].

If a control system has a sequential information structure, then Witsenhausen has proven that various results of control theory hold like determination of the reachable subsets and dynamic programming, see [14]. Control with nonsequential information structures has been investigated by Andersland and Teneketzis, see [1, 2].

Definition 24.4 Define the following special information structures.

(a) The *classical information structure* is that defined by the conditions that (1) all controllers receive the same information and (2) have perfect recall. In terms of mathematical notation,

$$\begin{aligned}
&\mathrm{IS}(i,t) = \mathrm{IS}(j,t), \quad \forall i,j \in \mathbb{Z}_k, \quad \forall t \in T; \\
&\mathrm{IS}(i,t) \subset \mathrm{IS}(i,t+1), \quad \forall i \in \mathbb{Z}_k, \quad \forall t \in T\backslash\{t_1\}.
\end{aligned}$$

An information structure is called *strictly classical* if it is classical and if there is only one controller ($k = 1$). It is usually assumed that in this information structure the structure at any time contains past outputs and past inputs of all controllers though this is not so specified above.

(b) The *overlapping information structure* is the information structure defined by the condition that it is not a classical information structure and there exists at least a distinct tuple of controllers which have a nonempty intersection of their information structures while the information structure has perfect recall; thus

$$\begin{aligned}
&\exists i,j \in \mathbb{Z}_k, \quad i \neq j, \quad \exists t \in T\backslash\{t_1\}, \text{ such that,} \\
&\mathrm{IS}(i,t) \neq \mathrm{IS}(j,t), \quad \mathrm{IS}(i,t) \cap \mathrm{IS}(j,t) \neq \emptyset; \quad \text{and,} \\
&\mathrm{IS}(i,t) \subset \mathrm{IS}(i,t+1), \quad \forall i \in \{1,\ldots,k\}, \quad \forall t \in T\backslash\{t_1\}.
\end{aligned}$$

There follows two special cases of overlapping information structures.

(c) The *r-step delayed-sharing information structure* is defined by

$$\begin{aligned}
\mathrm{ISDSrs}(i,t) = {} & ((y_i(t_0{:}\,t-1), u_i(t_0{:}\,t-1)), \\
& \prod_{j=1, j\neq i}^{k} (y_j(t_0{:}\,t-r-1), u_j(t_0{:}\,t-r-1))), \\
& \forall i \in \{1,2,\ldots,k\}, \quad \forall t \in T.
\end{aligned}$$

A *multi-step delayed-sharing information structure* is then defined as those of the r-step delayed sharing for any $r \in \mathbb{Z}_+$.

The one- or multi-step delayed-sharing information structures are appropriate models for control of communication networks in which all network nodes share their outputs and inputs but where the communication between network nodes of the information takes one or more time steps.

The r-step delayed-sharing information structure may be partitioned into the *common information structure* of all controllers and the *private information structure* of each controller. Define the common and the private information structure of the i-th controller by

$$\text{ISDSrsCommon}(i,t) = \left(\prod_{j=1}^{k} (y_j(t_0\colon t-r-1), u_j(t_0\colon t-r-1)) \right),$$
$$\forall i \in \mathbb{Z}_k, \quad \forall t \in T;$$
$$\text{ISDSrsPrivate}(i,t) = \prod_{s=t-r}^{t-1} ((y_i(s), u_i(s))), \forall i \in \mathbb{Z}_k, \quad \forall t \in T.$$

(d) The *nearest-neighbor-sharing information structure* is the information structure in which any controller receives its local outputs and local inputs and, in addition, outputs and inputs of its nearest neighbors, and it has perfect recall. In terms of mathematical notation,

$\forall i \in \{1, 2, \ldots, k\}, \quad \exists \mathrm{I}_{\mathrm{nn}}(i) \subset \mathbb{Z}_k,$
the *index set of the nearest neighbors of Controller i*,
such that, $i \in \mathrm{I}_{\mathrm{nn}}(j) \Leftrightarrow j \in \mathrm{I}_{\mathrm{nn}}(i), \quad \forall i, j \in \mathbb{Z}_k,$

$$\text{IS}(i,t) = \left((y_i(t_0\colon t-1), u_i(t_0\colon t-1)), \prod_{j \in \mathrm{I}_{\mathrm{nn}}(i)} (y_j(t_0\colon t-1), u_j(t_0\colon t-1)) \right)$$
$$\in Y_i^{t-t_0} \times U_i^{t-t_0} \times \prod_{j \in \mathrm{I}_{\mathrm{nn}}(i)} (Y_j^{t-t_0} \times U_j^{t-t_0}), \quad \forall i \in \mathbb{Z}_k;$$
$$\text{IS}(i,t) \subseteq \text{IS}(i,t+1), \quad \forall i \in \mathbb{Z}_k, \quad \forall t \in T \backslash \{t_1\}.$$

The nearest-neighbor information structure is a special case of the overlapping information structure. One can of course also define the nearest-neighbor information structure with delayed sharing. A further refinement of the nearest-neighbor information structure is to specify that not the outputs of nearest neighbors are observed but a specially generated signal is sent from that neighbor to the considered controller.

The r-step delayed-sharing information structure and the nearest-neighbor information structure are complementary in that the first one applies to the time axis while the second one applies to a spatial axis of network nodes.

The nearest-neighbors information structures have been used and investigated, see for example [5]. In communication networks, the class of *backpressure algorithms* and its generalizations are nearest-neighbor information structures, see [8].

(e) The *nonclassical information structure* is the negation of the concept of a classical information structure. Thus, (1) either there is no perfect recall or (2) the equality relation of the information structures of all controllers does not hold.

(f) The *private information structure* is defined by the conditions that the information structures of different controllers are pairwise disjoint. In terms of mathematical notation,

$$\text{IS}(i,t) \cap \text{IS}(j,t) = \emptyset, \forall i,j \in \mathbb{Z}_k, \text{ such that } i \neq j, \forall t \in T. \tag{24.1}$$

It is called a *local–private information structure* if each controller receives all its local outputs and local inputs, and if it has perfect recall. If the outputs and the inputs of the system of the different controllers are disjoint, then a local–private information structure is a private information structure. In terms of mathematical notation, a local–private information structure is defined as,

$$\text{IS}(i,t) = (y_i(t_0{:}\,t-1), u_i(t_0{:}\,t-1)) \in Y_i^{t-t_0} \times U_i^{t-t_0}, \ \forall i \in \mathbb{Z}_k, \forall t \in T.$$

Note that Controller i only receives the observations y_i from the plant and it recalls its own inputs u_i. This information structure corresponds to distributed control or to distributed filtering/state estimation both without sharing between controllers. It is a special case of the nonclassical information structure and the private information structure.

The relation of information structures with control laws is best illustrated by the Witsenhausen counterexample, see [11]. The counterexample presents a simple decentralized control problem with a nonclassical information structure for a system with linear equations and with Gaussian random variables. Consider the conjecture that the optimal control laws for the considered stochastic system and for a quadratic cost function are affine control laws. The counterexample of Witsenhausen shows that a nonlinear control law can achieve a strictly lower cost than the best affine control law. Details of the example follow.

Denote a Gaussian random variable taking values in $\mathbb{R}^n$, $x_0 : \Omega \to \mathbb{R}^n$, with mean value $m_0 \in \mathbb{R}^n$ and variance $Q_0 \in \mathbb{R}^{n\times n}$ by $x_0 \in G(m_0, Q_0)$.

Definition 24.5 Consider a decentralized stochastic system with Gaussian random variables, linear dynamics, and linear observation equations. There are two controllers: labeled Controller 1 and Controller 2. The notation of the system is,

$$\begin{aligned}
T &= \{0,1,2\},\\
&\quad x_0{:}\,\Omega \to \mathbb{R},\ x_0 \in G(0,Q_0), \text{ the initial state of the system},\\
y_0 &= x_0, \text{ the observation of Controller 1},\\
&\quad g_0{:}\,\mathbb{R} \to \mathbb{R}, \text{ a Borel measurable control law},\\
u_0 &= g_0(y_0), \text{ the input at time 0},\\
x_1 &= x_0 + u_0, \text{ the system dynamics},\\
y_1 &= x_1 + v, \text{ the observation of Controller 2},\\
&\quad v{:}\,\Omega \to \mathbb{R},\ v \in G(0,Q_v), \text{ the observation noise},\\
&\quad F^{x_0}, F^{v}, \text{ are independent } \sigma\text{-algebras},\\
&\quad g_1{:}\,\mathbb{R} \to \mathbb{R}, \text{ a Borel measurable control law},\\
u_2 &= g_1(y_1), \text{ the input at time1},\\
x_2 &= x_1 - u_2, \text{ the system dynamics.}
\end{aligned}$$

The information structure of the system at the two time steps is then,

$$\begin{aligned}\text{IS}(1,0) &= \{y_0\} \subset Y(0) \times U(0),\\ \text{IS}(1,1) &= \{y_1\} \subset Y(1) \times U(1);\\ &\text{hence, IS}\,(1,0) \not\subset \text{IS}(1,1) \text{ and IS}(1,1) \not\subset \text{IS}(1,0).\end{aligned}$$

This information structure does not have perfect recall. Hence, it is nonclassical.

In another interpretation, one can consider the time axis not as time but as an index of the controllers. Thus, Controller 1 sends information of the state x_1 to Controller 2 who receives, via the channel with noise v, the observation y_1. Then, the information structure is also nonclassical.

The cost function is,

$$J(g_0, g_1) = E[x_2^2 + k^2 u_0^2],\ k \in \mathbb{R}_+, \text{ a cost function parameter.}$$

The computations are simplified by the following expressions,

$$\begin{aligned}x_1 &= x_0 + u_0 = x_0 + g_0(y_0) = x_0 + g_0(x_0),\\ y_1 &= x_1 + v = x_0 + g_0(x_0) + v,\\ x_2 &= x_1 - u_2 = x_0 + g_0(x_0) - g_1(y_1) = x_0 + g_0(x_0) - g_1(x_0 + g_0(x_0) + v).\end{aligned}$$

The conclusion for this example is, as stated above, that there exists a nonlinear control law which has a strictly lower cost then the optimal affine control law.

24.4 Further Research

There are several open-research issues for the formulation and use of information structures for control of decentralized systems.

Research issues are as follows:

1. Explore the relation of information structures and of common and private information in decentralized control. See Chap. 26. One expects a further refinement of the concepts of common and of private information structure.
2. Explore the approach of realization theory for factorizing the map from the information structure to the input space, including a concept of a minimal realization.

24.5 Further Reading

The reader is advised to continue the reading of this chapter with that of its companion chapter, Chap. 25. In the latter chapter, control theory is developed based on information structures. This chapter is also linked to the following other chapters of the book, 20, 22, 26 and 30.

For the reader new to the subject, the author recommends the paper of Witsenhausen, [10], and the tutorial papers by Ho, [3, 4].

Related papers by Witsenhausen are [9–17]. Applications of information structures are discussed in [7].

References

1. Andersland MS, Teneketzis D (1992) Information structures, causality, and nonsequential stochastic control I: design-independent properties. SIAM J Control & Opt 30:1447–1475
2. Andersland MS, Teneketzis D (1994) Information structures, causality, and nonsequential stochastic control: II design-dependent policies. SIAM J Control & Opt 32:1726–1751
3. Ho YC (1980) Team decision theory and information structures. Proc IEEE 68:644–654
4. HoYC (2008) Review of the Witsenhausen problem. In Proceedings 47th IEEE conference on decision and control, IEEE, IEEE Press, New York
5. Jadbabaie A, Lin J, Morse AS (2003) Coordination of groups of mobile agents using nearest neighbor rules. IEEE Trans Automatic Control 48:988–1001
6. Mahajan A, Martins NC, Rotkowicz M, Yüksel S (2012) Information structures in optimal decentralized control. In: Proceedings 51st IEEE conference on decision and control, IEEE, IEEE Press, New York, pp 1291–1306
7. Stabile DA, Levis AH (1984) The design of information structures: basic allocation strategies for organizations. Large Scale Syst. 6:123–132
8. Tassiulas L, Ephremides A (1992) Stability properties of constrained queueing systems and scheduling policies for maximal throughput in multi hop radio networks. IEEE Trans Automatic Control 37:1936–1948
9. H. Witsenhausen (1980) Informational aspects of stochastic control. In: Jacobs OLR et al. (eds.) Analysis and optimization of stochastic systems, Academic Press, London, pp 273–284
10. Witsenhausen HS (1971) Separation of estimation and control for discrete time systems. Proc IEEE 59:1557–1566
11. Witsenhausen HS (1968) A counter example in stochastic optimal control. SIAM J Control Opt 6:131–147
12. Witsenhausen HS (1970) On performance bounds for uncertain systems. SIAM J Control 8:55–89
13. Witsenhausen HS (1971) On information structures, feedback and causality. SIAM J Control 9:149–160
14. Witsenhausen HS (1973) A standard form for sequential stochastic control. Math Syst Theory 7:5–11
15. Witsenhausen HS (1975) The intrinsic model for discrete stochastic control: some open problems. Lecture Notes in Economic and Mathematical Systems, vol 107. Springer, Berlin, pp 322–335
16. Witsenhausen HS (1976) Some remarks on the concept of state. In: Directions in large-scale systems, many-person optimization, and decentralized control (Proc. symposium, Wakefield, MA), Plenum, New York, pp 69–75
17. Witsenhausen HS (1988) Equivalent stochastic control problems. Math Control Sig Syst 1:3–11

Chapter 25
Control Theory with Information Structures

Jan H. van Schuppen

25.1 Introduction

Control theory of decentralized control with communicating controllers is well motivated by control engineering. Many control problems of communication networks and with vehicles are of this class. Control synthesis of such decentralized control problems is best distinguished based on the information structure used.

The reader is expected to have read Chap. 24 in which information structures are defined. The concepts defined in that chapter are used in this chapter on control theory with information structures. The main concepts are those of an information structure, and a classical, overlapping, and private information structure. A control law based on an information structure is such that the control law at any time depends only on the online information provided by the information structure.

Control theory with classical information structures can basically be deduced from control theory with only one controller and is well developed.

Control theory with overlapping information structures has so far produced few results. Well known is the case of a one-step delayed-sharing information structure for which a theorem of no loss of performance can be proven when restricting attention to a set of control laws depending on the conditional distribution of the state conditioned on the common information and depending on the latest local information. The case of a nearest-neighbor information structures is less well investigated. It is used in control of communication networks where the backpressure control law may use partial state information of nearest neighbors.

Control theory with private information structures is an open-research problem. The current focus of research includes the following: real-time communication, common and private information, signaling, and control synthesis. Control theory with private information structures is well motivated and useful for engineering.

J.H. van Schuppen (✉)
Van Schuppen Control Research, Gouden Leeuw 143, 1103 KB Amsterdam, The Netherlands
e-mail: jan.h.van.schuppen@xs4all.nl

J.H. van Schuppen and T. Villa (eds.), *Coordination Control of Distributed Systems*,
Lecture Notes in Control and Information Sciences 456,
DOI 10.1007/978-3-319-10407-2_25

25.2 Problem

Problem 25.1 Consider a decentralized/distributed control system either with the control architecture of distributed control or of distributed control with direct communication between controllers. Develop control synthesis for control of such a decentralized/distributed system for each of the defined information structures, in particular for private information structures.

25.3 Control Theory with Classical and Overlapping Information Structures

Control synthesis for control of decentralized or distributed systems is best distinguished per subclass of the information structures as described below. The structure of control laws and the character of the associated control theory differs by case.

Control synthesis for the classical information structure was the first case treated and is regarded as almost completed. The reader is referred to the references below. This case will not be discussed further in this chapter.

Control synthesis for overlapping information structures has been considered but is far from solved. Below control synthesis for the case of a one-step delayed-sharing information structure is summarized even though that case cannot be extended to multi-step delayed-sharing information structures with two or more steps.

Definition 25.1 *Set of control laws depending on common conditional distribution and private information.* Consider the stochastic control system of Definition 24.1 and the information structure of one-step delayed sharing, ISDS1s. Define the set of control laws,

$$G_1 = \left\{ \begin{array}{l} (g_1, \ldots, g_k) | g_i = (g(i, t_0), \ldots, g(i, t_1 - 1)), \\ g(i, t) : \text{ISDS1s}(i, t) \rightarrow U_i, \ \forall\, i \in \{1, 2 \ldots, k\}, \ \forall\, t \in T \backslash \{t_1\}. \end{array} \right\}.$$

Define the subset $G_2 \subset G_1$ of control laws,

$$G_2 = \left\{ \begin{array}{l} (g_1, \ldots, g_k) | g_i = (g(i, t_0), \ldots, g(i, t_1 - 1)), \\ g(i, t) = g_{ds.cpdf} \left([cpdf(.,.; x(t) | \text{ISDS1sCommon}(i, t))], y_i(t-1) \right), \\ \forall\, i \in \{1, 2 \ldots, k\}, \ \forall\, t \in T \backslash \{t_1\} \end{array} \right\},$$

$cpdf(.,.; x(t)|\text{ISDS1sCommon}(i, t))$,
denotes the conditional probability distribution
of the state $x(t)$ conditioned on the common information structure at time t.

Definition 25.2 Consider the stochastic control system of Definition 24.1 and the information structure of one-step delayed sharing, $ISDS1s$. One says that there is *no loss in performance when restricting attention from the set of control laws* G_1 *to the subset* $G_2 \subset G_1$ if

$$\inf_{g_1 \in G_1} J(g_1) = \inf_{g_2 \in G_2} J(g_2). \tag{25.1}$$

Note that $G_2 \subseteq G_1$ implies that $\inf_{g_1 \in G_1} J(g_1) \leq \inf_{g_2 \in G_2} J(g_2)$. The above definition therefore requires equality of the infima.

Theorem 25.1 No loss in performance. *Consider the stochastic control system of* Definition 24.1, *the information structure of one-step delayed-sharing* $ISDS1s$, *and the subsets of control laws* G_1, G_2 *of* Definition 25.1. *There is no loss in performance when restricting attention from the set of control laws* G_1 *to the subset* $G_2 \subset G_1$.

For a proof of the above theorem, see the papers, [28, 30, 43]. Control design for control with this information structure is because of the above theorem simplified because attention may be restricted to the subclass G_2 of control laws.

The control law of Controller $i \in \mathbb{Z}_k$ at time $t \in T$ depends only on (1) the conditional distribution of the state provided the common part of the information structure of one-step delayed sharing and (2) on the latest local observation $y_i(t-1)$. The above theorem does not hold in case the time step of the delayed sharing is two or more as has been shown by an example in [43].

There exist two methods to prove theorems or conjectures as formulated above for the no loss in performance by restricting attention to a subset of control laws. The first method is to solve the optimal control problem for the set G_1 of control laws and, if there exists an optimal control law g_2^*, to take any subset G_2 such that $g^* \in G_2 \subset G_1$. Then, it follows directly that there is no loss in restricting attention to the class of G_2. But this method requires a solution to the problem.

The second method does not use the solution of the problem. It is based on formulating an information state and the associated information system. As discussed in Chap. 24, the information state needs to be: (1) a state of a system; (2) the information state of the information system at any time needs to be measurable with respect to the information structure at that time; and (3) the expected future cost condition on the information structure at a time moment has to be a function of the information state at that time. The problem of determining the information state is likely to be equivalent to solving a stochastic realization problem in which the expected cost is an output of the information system; this approach is to be developed further. This approach for the centralized control problem was formulated by Striebel in [36] and was used in [11]. But that approach has to be extended to decentralized control and this will involve the signaling with the other controllers via the plant as discussed in more detail below.

25.4 Control Theory with Private Information Structures

Control synthesis with *private and nonclassical information structures* has very few substantial contributions and many open problems. From a viewpoint of engineering, particularly from the viewpoint of control of networks, control synthesis with private information structures is a well-motivated problem of control theory.

The problem is to determine the set of control laws with the private information structure which achieves the minimal cost.

Example 25.1 Consider a distributed linear system structured by a line. The information structure is that of nearest-neighbor type. The distributed control system is defined by the equations,

$$x(t+1) = \begin{pmatrix} A_{11} & A_{12} & 0 & 0 \\ A_{21} & A_{22} & A_{23} & 0 \\ 0 & A_{32} & A_{33} & A_{34} \\ 0 & 0 & A_{43} & A_{44} \end{pmatrix} x(t) + \begin{pmatrix} B_{11} & 0 & 0 & 0 \\ 0 & B_{22} & & 0 \\ 0 & 0 & B_{33} & 0 \\ 0 & 0 & 0 & B_{44} \end{pmatrix} u(t), \quad (25.2)$$

$$y(t) = \begin{pmatrix} C_{11} & 0 & 0 & 0 \\ 0 & C_{22} & & 0 \\ 0 & 0 & C_{33} & 0 \\ 0 & 0 & 0 & C_{44} \end{pmatrix} x(t), \quad (25.3)$$

$$y_i(t_0 : t) = \{y_i(t_0),\ y_i(t_0+1),\ \ldots,\ y_i(t)\},\ = \emptyset,\ \text{if } t < t_0, \quad (25.4)$$

$$\text{ISNN}_i(t) = \{y_i(t_0 : t-1),\ y_{i-1}(t_0 : t-1),\ y_{i+1}(t_o : t-1)\},\ i = 2,\ 3; \quad (25.5)$$

$$\text{ISNN}_1(t) = \{y_1(t_0 : t-1),\ y_2(t_0 : t-1)\}, \quad (25.6)$$

$$\text{ISNN}_4(t) = \{y_4(t_0 : t-1),\ y_3(t_0 : t-1)\}. \quad (25.7)$$

Note that the distributed system consists of four (deterministic) linear systems with the structure of a line. Thus, Subsystem 2 is connected to its nearest left and right neighbors, Subsystem 3 likewise, while the Subsystems 1 and 4 are only connected to one nearest neighbor. The information structures of the Subsystems 2 and 3 are such that they contain the observations of their locals subsystems and those of their nearest neighbors. For the Subsystems 1 and 4, the information structures are also those of the local subsystems and their nearest neighbor though there is only one such neighbor in each case.

The control problem is to determine a control law based on the defined information structure which minimizes an optimization criterion not stated here. A centralized control synthesis based on complete observations would yield a control law such that the control law of any subsystem depends on the state of all subsystems. Of interest is therefore the case in which the control law of any subsystem depends only on the local state and that of its nearest neighbors. For example, when is the optimal control law linear and does it not depend on state x_4, thus such that $u_2(t) = F_{2,2}x_2(t) + F_{2,1}x_1(t) + F_{2,3}x_3(t)$? When the optimal control law is not of this form, how much is the performance loss by restricting attention to control laws which depend on the information structure of nearest neighbors only?

The approach of restricting attention to a *controller-by-controller equilibrium* for control of decentralized systems with the private information structurecan always be applied. This approach was described for a team decision problem in which every team takes only one decision, see Chap. 18. The dynamic team problem is discussed in Chap. 19 and in the report [13]. This approach was used to obtain properties of the control laws, see for example [44, 53]. References for this approach to control problems of dynamic games include [9, 27].

Current research in control theory with private information structures focusses on the investigation of properties of optimal control laws, in particular of the signaling of information between controllers via the control system.

For the control problem with private information structures, the following subproblems can be distinguished. The solution of these subproblems is expected to contribute the solution of the control problem with private information structures. But there is still much uncertainty as to what can be achieved for this problem.

1. *Real-time communication*. In decentralized control with private information structure, any controller can exchange information with other controllers via the decentralized control system. This has been called *signaling* of information between controllers via the plant. This communication exchange is called in information theory and in communication theory, *real-time communication*. Note that in this case the communication channel is the control system, hence the communication channel has dynamics. Moreover, there is feedback from the channel to both the sender and the receiver. Yet, the feedback is not of the type classically considered in communication theory from the receiver to the sender but from the channel to the sender. The reader is referred to the recent paper [42]. Control with real-time communication has received limited attention from the communication theory research community but recently much from control theory. The communication is to be carried out without delay, though communication takes usually a time step. The communication is further nonanticipatory. The reader is referred to [41] and other references mentioned below. References on real-time communication include [44, 53].
2. *Common and private information*. Which information should a controller send to another controller? The information available to any controller can be distinguished into common, private, and correlated information. The associated decomposition can then be used to analyze the questions: What to send? and To whom to send? This research issue is discussed further in Chap. 26.
3. *Signaling*. How to synthesize and to design a communication law for the information to be sent from one controller to one or more other controllers? This research issue requires an integration of communication with the points (1) and (2) described above. See for further information Chap. 20.
4. *Control synthesis*. How to synthesize control laws such that each controller uses the information provided by other controllers via signaling?

Decentralized control with private information structures is a research issue in which only the first steps have been taken.

25.5 Further Research

There are many open-research issues of control with overlapping and with private information structures.

Research issues of control theory with information structure include the following:

1. Develop methods to determine information structures for which the no loss of performance holds in particular for control with private information structures.
2. Develop methods to prove that control laws need to depend only on particular information structures, possibly with the concept of a sufficient statistic of the cost function, see [36].
3. Investigate further control problems with special cases of the nearest-neighbor information structures.
4. Develop control theory and communication theory for private information structures in particular for the subproblems of real-time communication, common and private information, signaling, and control synthesis.
5. Develop further stochastic control theory with nonsequential information structures.
6. Explore the approach of realization theory for factorizing the map from the information structure to the input space, including a concept of minimal realization.

Solutions for control problems with private information structures are much needed in control of vehicles, in control of communication networks, and in control of road networks.

25.6 Further Reading

This chapter is linked to the following other chapters of the book, 20, 22, and 26.

For the reader new to the subject, the author recommends the paper of Witsenhausen, [45], and the tutorial papers by Ho, [19, 22]. The recently published book may also be of interest, [57].

The reader is advised to read the papers by Witsenhausen, [45–54].

Stochastic control for a centralized system, thus with one controller station, in which a subclass of control laws depending only in the state of the Kalman filter has no loss of performance, was developed by Striebel, [36] and Wonham, [55]. Control theory with classical information structures is discussed in [25].

Control theory with the overlapping information structure has addressed the partitioning of information structures into common and private information, see [15, 16, 18, 21]. In addition, it has addressed control synthesis of stochastic systems with the one-step or multi-step delayed-sharing information structure see [7, 8, 23, 24, 28–30, 34, 35, 43, 45, 56]. The conclusion is that the one-step delay case is solved satisfactorily but that the case of multi-step delays larger than one time step, needs further research.

Control synthesis of stochastic systems with the common past information structure is treated in [1–3, 12].

Control theory with the nearest-neighbor information structure is discussed in the papers, [10, 26, 33, 37, 38]. Yet, control and communication engineering require further development for this case.

Expository papers on control with nonclassical information structures include [31, 39, 40].

Stochastic control for nonclassical information structures was developed by Teneketzis and co-workers, see [4–6, 32, 41]. Other references include [20]. Papers on signaling in control of decentralized systems with nonclassical information structures include [14, 17].

References

1. Aicardi M, Casalino G, Minciardi R, Zoppoli R (1992) On the existence of stationary optimal receding-horizon strategies for dynamic teams with common past information structures. IEEE Trans Autom Control 37:1767–1771
2. Aicardi M, Davoli F, Mincardi R (1987) Decentralized optimal control of Markov chains with a common past information set. IEEE Trans Autom Control 32:1028–1031
3. Aicardi M, Davoli F, Minciardi R (1988) Correction to 'Decentralized optimal control of Markov chains with a common past information set'. IEEE Trans Autom Control 33:891–892
4. Andersland MS, Teneketzis D (1991) Solvable systems are usually measurable. Stoch Anal Appl 9:233–244
5. Andersland MS, Teneketzis D (1992) Information structures, causality, and nonsequential stochastic control I: design-independent properties. SIAM J Control Opt 30:1447–1475
6. Andersland MS, Teneketzis D (1994) Information structures, causality, and nonsequential stochastic control: II design-dependent policies. SIAM J Control Opt 32:1726–1751
7. Bagchi A, Basar T (1980) Team decision theory for linear continuous-time systems. IEEE Trans Autom Control 26:1154–1161
8. Basar T (1978) Two-criteria LQG decision problems with one-step delay observation sharing pattern. Info Control 38:21–50
9. Basar T, Olsder GJ (1999) Dynamic noncooperative game theory (2nd edn). Number 23 in Classics in Applied Mathematics. SIAM, Philadelphia
10. Bamieh B, Paganini F, Dahleh MA (2002) Distributed control of spatially invariant systems. IEEE Trans Autom Control 47:1091–1107
11. Benes VE (1976) Full "bang" to reduce predicted miss is optimal. SIAM J Control Opt 14:62–84
12. Casalino G, Davoli F, Mincardi R, Puliafita PP, Zappoli R (1984) Partially nested information structures with a common past. IEEE Trans Autom Control 29:846–850
13. Charalambous CD, Ahmed NU (2013) Dynamic team theory of stochastic differential decision systems with decentralized noisy information structures via Girsanovs measure transformation. Technical report arXiv:1309.1913v2, ArXiv, 10 Oct 2013.
14. Charlambous CD, Farhadi A (2008) LQG optimality and separation principle for general discrete time partially observed stochastic systems over finite capacity communication channels. Automatica 44:3181–3188

15. Cremers AB, Hibbard TN (1978) Orthogonality of information structures. Acta Informatica 9:243–261
16. Dick R, Hexner G, Ho Y (1977) On common information in decision problems. IEEE Trans Inf Theor 23:393–396
17. Grover P (2011) Actions can speak more clearly than words. PhD thesis, Department of Electrical Engineering and Computer Science, University of California at Berkeley, Berkeley, 5 Jan 2011
18. Hexner G, Ho YC (1977) Information structure: common and private. IEEE Trans Inf Theor 23:390–393
19. Ho YC (1980) Team decision theory and information structures. Proc IEEE 68:644–654
20. Ho YC, Chu KC (1973) On the equivalence of information structures in static and dynamic teams. IEEE Trans Autom Control 18:187–188
21. Ho YC, Hexner G, Martin C (1976) Two person information structure. Part I: information structure: common and private. Part II: private information structures of A with respect to B. Report 663, Division of Engineering and Applied Physics, Harvard University, Cambridge, MA, USA, July 1976
22. Ho YC (2008) Review of the Witsenhausen problem. In: Proceedings of 47th IEEE conference on decision and control, IEEE Press, New York
23. Hsu K, Marcus SI (1980) Decentralized control of finite state markov processes. In: (ed) Proceedings of the 19th IEEE conference on decision and control, IEEE Press, New York, pp 143–148
24. Hsu K, Marcus SI (1982) Decentralized control of finite-state Markov chains. IEEE Trans Autom Control 27:426–431
25. Ikeda M, Siljak DD, White DE (1981) Decentralized control with overlapping information sets. J Optim Theor Appl 37:279–310
26. Jadbabaie A, Lin J, Morse AS (2002) Coordination of groups of mobile autonomous agents using nearest neighbor rules. Report, Department of Electrical Engineering, Yale University, New Haven
27. Kumar PR, van Schuppen JH (1980) On Nash equilibrium solutions in stochastic dynamic games. IEEE Trans Autom Control 25:1146–1149
28. Kurtaran B (1976) Decentralized stochastic control with delayed sharing information patterns. IEEE Trans Autom Control 21:576–581
29. Kurtaran B (1979) Corrections and extensions to 'Decentralized stochastic control with delayed sharing information patterns'. IEEE Trans Autom Control 24:656–657
30. Kurtaran BZ, Sivan R (1974) Linear quadratic Gaussian control with one step delay sharing. IEEE Trans Autom Control 19:571–574
31. Mahajan A, Martins NC, Rotkowicz M, Yüksel S (2012) Information structures in optimal decentralized control. In: Proceedings of conference on decision and control (CDC 2012), IEEE Press, New York, pp 1291–1306
32. Mahajan A, Teneketzis D (2009) Optimal performance of networked control systems with nonclassical information structures. SIAM J Control Opt 48:1377–1404
33. Motee N, Jadbabaie A (2008) Optimal control of spatially distributed systems. IEEE Trans Autom Control 53:1616–1629
34. Nayyar A, Mahajan A, Teneketzis D (2011) Optimal control strategies in delayed sharing information structures. IEEE Trans Autom Control 56:1606–1620
35. Sandell NR, Athans M (1974) Solution of some nonclassical LQG stochastic decision problems. IEEE Trans Autom Control 19:108–116
36. Striebel C (1965) Sufficient statistics in the optimum control of stochastic systems. J Math Anal Appl 12:576–592
37. Swaarop D, Hedrick JK (1996) String stability of interconnected systems. IEEE Trans Autom Control 41:349–357
38. Tassiulas L, Ephremides A (1992) Stability properties of constrained queueing systems and scheduling policies for maximal throughput in multi hop radio networks. IEEE Trans Autom Control 37:1936–1948

39. Tatikonda S, Sahai A, Mitter S (2004) Stochastic linear control over a communication channel. IEEE Trans Inf Theor 49:1549–1561
40. Teneketzis D (1997) On information structures and nonsequential stochastic control. CWI Q 10:179–199
41. Teneketzis D (2006) On the structure of optimal real-time encoders and decoders in noisy communication. IEEE Trans Inf Theor 52:4017–4035
42. Uribe CA, van Schuppen JH (2013) Analysis of signaling in a finite stochastic system motivated by decentralized control. In: Proceedings of 52nd IEEE conference on decision and control (CDC 2013), IEEE Press, New York, pp 5884–5889
43. Varaiya P, Walrand J (1978) On delayed sharing patterns. IEEE Trans Autom Control 23:443–445
44. Walrand JC, Varaiya P (1983) Optimal causal coding–decoding problems. IEEE Trans Inf Theor 29:814–820
45. Witsenhausen HS (1971) Separation of estimation and control for discrete time systems. Proc IEEE 59:1557–1566
46. Witsenhausen HS (1980) Some aspects of convexity useful in information theory. IEEE Trans Inf Theor 26:265–271
47. Witsenhausen HS (1968) A counter example in stochastic optimal control. SIAM J Control Opt 6:131–147
48. Witsenhausen HS (1970) On performance bounds for uncertain systems. SIAM J Control 8:55–89
49. Witsenhausen HS (1971) On information structures, feedback and causality. SIAM J Control 9:149–160
50. Witsenhausen HS (1973) A standard form for sequential stochastic control. Math Syst Theor 7:5–11
51. Witsenhausen HS (1975) The intrinsic model for discrete stochastic control: some open problems, vol 107 of lecture notes in economic and mathematical systems. Springer, Berlin, pp 322–335
52. Witsenhausen HS (1976) Some remarks on the concept of state. In: Directions in large-scale systems, many-person optimization, and decentralized control (Proceedings, Symposium, Wakefield, MA, 1975), pp 69–75, New York
53. Witsenhausen HS (1979) On the structure of real-time source coders. Bell Syst Tech J 58:1437–1451
54. Witsenhausen HS (1988) Equivalent stochastic control problems. Math Control Signals Syst 1:3–11
55. Wonham WM (1968) On the separation theorem of stochastic control. SIAM J Control 6:312–326
56. Yoshikawa T (1975) Dynamic programming approach to decentralized stochastic control problems. IEEE Trans Autom Control 20:796–797
57. Yüksel S, Basar T (2013) Stochastic networked control systems—Stabilization and optimization under information constraints. Birkhäuser, Boston

Chapter 26
Common, Correlated, and Private Information in Control of Decentralized Systems

Jan H. van Schuppen

26.1 Introduction

Decentralized control involves the exchange of information of the two or more controllers influencing the decentralized control system. The communication between controllers may be carried out via the plant, then called signaling, or directly in the case of control with direct communication between controllers. Therefore, there is a need for understanding the communication exchange in particular: What information of a controller is of interest to which other controllers?

The problem of common and private information in decentralized control is then to formulate concepts, to develop theory and algorithms, and to use this theory for control of distributed or of decentralized systems.

In game theory, the same concepts and theory are useful. In case of zero-sum games, each of the players wants to prevent the communication of private information so as to obtain an advantage over other players.

In this short chapter, definitions of common, of private, and of correlated information are proposed. The case of Gaussian random variables is analyzed using the canonical variable decomposition. For a particular Gaussian stochastic control system, a decomposition is presented. Further research is mentioned.

The research issue of common and of private information in decentralized control was initiated by Ho, see [1, 2]. There is a clear link of the concepts of common and private information with those of information theory. The common information of two random variables was formulated and explored by Wyner, see [3, 4]. But that theory goes less far than is proposed in this paper.

J.H. van Schuppen (✉)
Van Schuppen Control Research, 1103 KB Amsterdam, The Netherlands
e-mail: jan.h.van.schuppen@xs4all.nl

J.H. van Schuppen and T. Villa (eds.), *Coordination Control of Distributed Systems*,
Lecture Notes in Control and Information Sciences 456,
DOI 10.1007/978-3-319-10407-2_26

26.2 Motivation

Why is decomposition of the information available to a controller of a decentralized control system into common and private information useful for decentralized and for distributed control?

Suppose that there exist appropriate definitions of common, private, and correlated information of any controller with respect to the other controllers. Suppose also that a decomposition of the decentralized control system has been made such that the structure of the system matrix and that of the observation matrix of any controller are decomposed into the common, private, and correlated parts. To simplify the discussion, first assume that the decentralized control system has only two controllers.

In a decentralized control system, which information components should a controller send to the other controller? Recall that in decentralized control all controllers have the same control objective. There is no need to send the common information because both controllers observe the same output process. The private information of a controller may be useful to the other controller but only if it would help the other controller to better achieve the common control objectives. The case of information that is neither common nor private is more complicated; it depends on the amount of correlation involved. This aspect has to be treated in the planned theory. There are additional aspects in case there are many controllers; this case requires slightly adjusted concepts.

The communication of the information requires concepts and theory of coding theory and of real-time communication not discussed here at length.

26.3 Problem

Problem 26.1 The *common and private information problem*. Formulate concepts, develop theory, and formulate algorithms for the use of common, private, and correlated information in control of decentralized control systems. Formulate decompositions of decentralized control systems in which the different concepts are distinguished.

26.4 Concepts

As to the formulation of the concepts of common and of private information, it is required to be as general as possible. From this point of view, a formulation in terms of stochastic systems with σ-algebras as spaces is preferred. In this chapter, attention is first focused on a σ-algebraic formulation and subsequently, attention is restricted to Gaussian random variables and to decentralized systems with Gaussian distributions for which more explicit concepts can be formulated.

Thus, for the observation vectors (y_1, y_2) with $y_i : \Omega \rightarrow \mathbb{R}^{p_i}$ for $i = 1, 2$, consider the σ-algebras generated by these variables (F^{y_1}, F^{y_2}). Instead, one works with abstract spaces, the tuple of σ-algebras (F_1, F_2), where one may associate with each F_i an observation space. Call these objects the *observation* σ-algebras. Denote the positive real line by $\mathbb{R}_+ = [0, \infty) \subset \mathbb{R}$ and the associated Borel σ-algebra on that set by $B(\mathbb{R}_+)$.

Call the tuple of σ-algebras F_1, F_2 of a probability space (Ω, F, P) *conditionally independent* given a third σ-algebra $G \subseteq F$ if the following factorization property holds,

$$E[x_1 x_2 | G] = E[x_1|G]E[x_2|G], \ \forall \, x_i \in L(\Omega, F_i; \mathbb{R}_+, B(\mathbb{R}_+)); \tag{26.1}$$

$$L(\Omega, F_i; \mathbb{R}_+, B(\mathbb{R}_+)) = \{x : \Omega \rightarrow \mathbb{R}_+ | \text{a random variable, } F_i\text{-measurable}\}.$$

Denote by $(F_1, F_2|G) \in \mathrm{CI}$ that this triple of σ-algebras is *conditional independent*.

The problem is to define concepts on the relation of the two observation spaces. This will be done by the concept of common and privated information defined next.

Definition 26.1 Consider two observation σ-algebras (F_1, F_2).

(a) Define the *common information* of the two observation σ-algebras as their intersection,

$$G_{\text{com}} = F_1 \cap F_2. \tag{26.2}$$

It follows immediately that the common information is a σ-algebra.

(b) Call a σ-algebra G a *sufficient σ-algebra* for the tuple of σ-algebras (F_1, F_2) if it makes F_1 and F_2 conditionally independent, $(F_1, F_2|G) \in \mathrm{CI}$.

(c) Define the *private information* of the observation σ-algebra F_1 with respect to the σ-algebra G as a σ-algebra $F_{1,p} \subseteq F_1$ which is a complement of the σ-algebra G. Thus, $F_{1,p}$ is characterized by the three conditions,
(1) $F_{1,p}$ is a σ-algebra; (2) $F_{1,p} \subseteq F_1$; and (3) $F_1 = G \vee F_{1,p}$. A σ-algebra which is the private information of F_1 with respect to G always exists; $F_{1,p} = F_1$ will do, but that is also a trivial choice. More interesting is the next concept. The *independent private information* is defined to be a private information σ-algebra $F_{1,p}$ if in addition the condition holds
(4) $F_{1,p}$ and G are independent σ-algebras.

A private information $F_{2,p} \subseteq F_2$ is then defined by symmetry.

A sufficient σ-algebra which makes the two σ-algebras F_1, F_2 conditionally independent always exists, for example, $F_1 \vee F_2$ will do. Such a σ-algebra is not unique; in general, there are many in continuous probability spaces even uncountably many. Of interest is therefore a minimal σ-algebra which makes a tuple of σ-algebras conditionally independent. Minimality refers to the ordering defined by set inclusion. A minimal sufficient σ-algebra is not unique in general. Of interest is therefore a classification of all such σ-algebras. See [5, 6] for further information.

The private σ-algebra $F_{1,p}$ always exists. But it is not unique, there exist many such objects. Of interest is therefore a minimal private observation space. How this is to be formulated is not yet clear. An independent private observation space may not exist; there exists an example where the observation spaces are generated by finite-valued random variables. These concepts are subject of further investigation.

26.5 Common and Private Information of Gaussian Random Variables

Based on the concept of the canonical variable decomposition, one can define the concepts of common, correlated, and private information for a tuple of Gaussian random variables.

Definition 26.2 Consider a tuple of Gaussian random variables $y_i : \Omega \to \mathbb{R}^{p_i}$, $i = 1, 2$. Assume that these random variables have been transformed by a linear transformation, $(y_1, y_2) \mapsto (S_1 y_1, S_2 y_2)$ for matrices S_1, S_2, to the *canonical variable decomposition* defined by the representation,

$$
\begin{aligned}
& (y_1, y_2) \in G(0, Q), \\
Q = & \left(\begin{array}{ccc|ccc} I_{p_{11}} & 0 & 0 & I_{p_{21}} & 0 & 0 \\ 0 & I_{p_{12}} & 0 & 0 & D & 0 \\ 0 & 0 & I_{p_{13}} & 0 & 0 & 0 \\ \hline I_{p_{21}} & 0 & 0 & I_{p_{21}} & 0 & 0 \\ 0 & D & 0 & 0 & I_{p_{22}} & 0 \\ 0 & 0 & 0 & 0 & 0 & I_{p_{23}} \end{array} \right) \in \mathbb{R}^{p \times p}, \qquad (26.3)
\end{aligned}
$$

$$
\begin{aligned}
& p,\ p_1,\ p_2,\ p_{11},\ p_{12},\ p_{13},\ p_{21},\ p_{22},\ p_{23} \in \mathbb{N}, \\
& p = p_1 + p_2,\ p_1 = p_{11} + p_{12} + p_{13},\ p_2 = p_{21} + p_{22} + p_{23}, \\
& p_{11} = p_{21},\ p_{12} = p_{22}, \\
D = & \operatorname{Diag}(d_1, \ldots, d_{p_{12}}),\ 1 > d_1 \geq d_2 \geq \cdots \geq d_{p_{12}} > 0, \qquad (26.4) \\
y = & \begin{pmatrix} y_1 \\ y_2 \end{pmatrix} = \begin{pmatrix} y_{11} \\ y_{12} \\ y_{13} \\ y_{21} \\ y_{22} \\ y_{23} \end{pmatrix}, \quad y_{ij} : \Omega \to \mathbb{R}^{p_{ij}},\ i = 1, 2,\ j = 1, 2, 3.
\end{aligned}
$$

Call then Q the *covariance matrix* of the random variables (y_1, y_2) in the canonical variable decomposition. This is a slight abuse of terminology because only its (1,2)-block is usually called the covariance matrix of (y_1, y_2).

Define then

$y_{11} = y_{21}$	*Common information* of y_1 and y_2
y_{12}	*Correlated information* of y_1 with respect to y_2
y_{13}	*Independent private information* of y_1 with respect to y_2
(y_{11}, y_{12})	*Sufficient information* of y_1 for the tuple y_1, y_2
$y_{21} = y_{11}$	*Common information* of y_1 and y_2
y_{22}	*Correlated information* of y_2 with respect to y_1
y_{23}	*Independent private information* of y_2 with respect to y_1
(y_{21}, y_{22})	*Sufficient information* of y_2 for the tuple y_1, y_2

Proposition 26.1 *The following properties hold for the common, correlated, and private information of a pair of Gaussian random variables.*

(a) *The three components y_{11}, y_{12}, y_{13} of y_1 are independent random variables.*
(b) *The three components y_{21}, y_{22}, y_{23} of y_2 are independent random variables.*
(c) *The equality $y_{11} = y_{21}$ of these random variables holds almost surely; hence, the term common information is appropriate.*
(d) *The tuple of random variables (y_{12}, y_{22}) is correlated as shown by the formula*

$$E[y_{12} y_{22}^T] = D = \text{Diag}(d_1, \ldots, d_{p_{12}}). \tag{26.5}$$

Note that the different components of y_{12} and of y_{22} are independent random variables; thus, $y_{12,i}$ and $y_{12,j}$ are independent, and $y_{22,i}$ and $y_{22,j}$ are independent, and $y_{12,i}$ and $y_{22,j}$ are independent, for all $i \neq j$, and that $y_{1,j}$ and $y_{2,j}$ for $j = 1, \ldots, p_{12} = p_{22}$ are correlated.
(e) *The random variable y_{13} is independent of y_2, hence justifying the term of independent private information of y_1 with respect to y_2. Similarly, the random variable y_{23} is independent of y_1*

Proof The results are immediately obvious from the fact that the random variables are all jointly Gaussian and from the covariance matrix (26.3) of the tuple of random variables (y_1, y_2) in the canonical variable decomposition.

Proposition 26.2 *Consider the concepts of common, correlated, and private information for a tuple of Gaussian random variables of* Definition 26.2. *Then, the following conditional independence properties hold:*

$$(F^{y_1}, F^{y_2} | F^{y_{11}, y_{12}}) \in \text{CI}; \quad (F^{y_1}, F^{y_2} | F^{y_{21}, y_{22}}) \in \text{CI}; \tag{26.6}$$

$$(F^{y_{11}, y_{12}}, F^{y_{21}, y_{22}} | F^{y_{11}, y_{12}}) \in \text{CI}; \quad (F^{y_{11}, y_{12}}, F^{y_{21}, y_{22}} | F^{y_{21}, y_{22}}) \in \text{CI}. \tag{26.7}$$

These formulas display the sufficient information of the tuple of random variables (y_1, y_2).

The above decomposition of a tuple of Gaussian random variables allows conjectures about the use of common, private, and sufficient information for information exchange and control in decentralized or distributed systems. Consider the setting of the decomposed Gaussian random variables of the previous section, see Definition 26.2. The independent private information of Controller 1 with respect to Controller 2, y_{13}, is useful to Controller 2 for its tasks of estimation and control. Therefore, that independent private information y_{13} is best sent from Controller 1 to Controller 2. The information of Controller 1 which is correlated with the information of Controller 2, above denoted by y_{12}, may be of interest to Controller 2 only if the correlation coefficient d_i is sufficiently high. If the correlation coefficient is low, then it does not seem of interest to send the information. See Chap. 27 for an application.

26.6 Common and Private Information of Gaussian Systems

There follows a first definition of a Gaussian system in which the common and private information is made explicit in the decomposition of the system. At the time this chapter is completed, there does not yet exist a procedure to decompose an arbitrary Gaussian system into its common and private components.

Definition 26.3 A *Gaussian system with a common–private information decomposition*. Consider a Gaussian system with the following rather specific decomposition.

$$x(t+1) = \begin{pmatrix} A_{11} & 0 & 0 & \\ 0 & A_{22} & & 0 \\ 0 & 0 & A_{33} & 0 \\ 0 & 0 & 0 & A_{44} \end{pmatrix} x(t) + \begin{pmatrix} M_{11} & 0 & 0 & \\ 0 & M_{22} & & 0 \\ 0 & 0 & M_{33} & 0 \\ 0 & 0 & 0 & M_{44} \end{pmatrix} v(t), \quad (26.8)$$

$$x(t_0) = x_0,$$

$$y(t) = \begin{pmatrix} y_{11}(t) \\ y_{12}(t) \\ y_{13}(t) \\ y_{21}(t) \\ y_{22}(t) \\ y_{23}(t) \end{pmatrix} = \begin{pmatrix} C_{11} & 0 & 0 & 0 \\ 0 & C_{12} & 0 & 0 \\ 0 & 0 & C_{13} & 0 \\ C_{11} & 0 & 0 & 0 \\ 0 & C_{22} & 0 & 0 \\ 0 & 0 & 0 & C_{24} \end{pmatrix} x(t) + \begin{pmatrix} N_{11} & 0 & 0 & 0 \\ 0 & N_{22} & 0 & 0 \\ 0 & 0 & N_{33} & 0 \\ N_{11} & 0 & 0 & 0 \\ 0 & N_{52} & 0 & 0 \\ 0 & 0 & 0 & N_{44} \end{pmatrix} w(t).$$

In this decomposition, y_{11} and y_{21} are called the *common output processes* of Observer 1 and Observer 2. The tuple (y_{12}, y_{22}) is called the *correlated output processes* of the two output processes. The process y_{13} is called the *private output process of Observer* 1 relative to Observer 2, while y_{23} is called the *private output process of Observer* 2 relative to Observer 1. The stochastic processes v and w are discrete time-independent Gaussian white noise processes.

The usefulness of the decomposition of the outputs and states of the above-defined common–private Gaussian system is then directly obvious from the terms and from the decomposition of the system matrices.

Further research is needed into the general decomposition of a Gaussian system by linear transformations of state and of output spaces. A further problem is to relate the informational decomposition also to the cost function or the control objectives.

26.7 Further Research

The following research issues seem useful for control of decentralized and of distributed control systems.

1. Investigate further the σ-algebraic concepts of common, private, and sufficient information. Investigate decompositions of finite-valued random variables in regard to common and private information.
2. Explore decompositions of a triple and of a higher number of tuples of random variables in regard to common, private, and correlated information. Particular cases include Gaussian random variables and finite-valued random variables.
3. Investigate decompositions of decentralized stochastic control systems such that the decomposition displays the common, private, and correlated information.
4. Investigate the use of the concepts of common and private information for state estimation or filtering by two or more controllers. See Chap. 27 for an initial approach to this problem.
5. Investigate the interaction of information and control, which requires new concepts of information decomposition.

26.8 Further Reading

A reader novel to the topic is advised to read the tutorial paper by Ho, [7], and a paper relating information decompositions to information theory, see [8]. This chapter is related to the following other chapters of the book: Chaps. 18, 20, and 24.

Early papers on the concepts of common and of private information are [1, 9]. The σ-algebraic concept of common information was published by Aumann, see [10].

The canonical variable decomposition of Gaussian random variables was proposed by Hoteling, see [11]. A book about the canonical variable decomposition and its applications is [12]. The use of this decomposition for stochastic realization of Gaussian random variables is described in [13].

The decomposition of Gaussian systems in regard to common and private information sketched in this paper is under development by the author. The geometric theory of linear systems used may be found in the book by Wonham, [14]. Decompositions of coordinated linear systems are derived in [15]. The σ-algebraic stochastic realization problem is treated in [5, 6].

Books on information theory include [16–18]. The concept of mutual information of two variables has been investigated from the information theoretic viewpoint in [3, 4, 19]. Related to the topic of this chapter is the *agreement problem*. The reader is referred to the papers of Teneketzis and Varaiya for agreement in the context of control theory [20, 21], and to the paper by Aumann [10].

References

1. Hexner G, Ho YC (1977) Information structure: common and private. IEEE Trans Inf Theor 23:390–393
2. Ho YC, Hexner G, Martin C (1976) Two person information structure. Part I: information structure: common and private. Part II: private information structures of A with respect to B. Report 663, Division of Engineering and Applied Physics, Harvard University, Cambridge, MA, USA, July 1976
3. Wyner AD (1975) The common information of two dependent random variables. IEEE Trans Inf Theor 21:163–179
4. Wyner AD (1978) A definition of conditional mutual information for arbitrary ensembles. Inf Control 38:51–59
5. van Schuppen JH (1982) The strong finite stochastic realization problem—Preliminary results. In: Bensoussan A, Lions JL (eds) Analysis and optimization of systems, vol 44., Lecture Notes in Control and Information SciencesSpringer, Berlin, pp 179–190
6. van Schuppen JH (2000) Stochastic realization of σ-algebras. In: Proceedings of 14th international symposium MTNS (published on CD-ROM only), Université de Perpignan, Perpignan
7. Ho YC (1980) Team decision theory and information structures. Proc IEEE 68:644–654
8. Ho YC, Karstner P, Wong E (1978) Teams, signaling and information theory. IEEE Trans Autom Control 23:305–311
9. Dick R, Hexner G, Ho Y (1977) On common information in decision problems. IEEE Trans Inf Theor 23:393–396
10. Aumann RJ (1976) Agreeing to disagree. Ann Statist 4:1236–1239
11. Hotelling H (1936) Relation between two sets of variates. Biometrika 28:321–377
12. Gittens RD (1985) Canonical analysis—A review with applications in ecology. Springer, Berlin
13. van Putten C, van Schuppen JH (1983) The weak and strong Gaussian probabilistic realization problem. J Multivariate Anal 13:118–137
14. Wonham WM (1979) Linear multivariable control: a geometric approach. Springer, Berlin
15. Kempker PL, Ran A, van Schuppen JH (2012) Controllability and observability of coordinated linear systems. Linear Algebra its Appl 437:121–167
16. Cover TM, Thomas JA (1991) Elements of information theory. Wiley, New York
17. Gallager RG (1968) Information theory and reliable communication. Wiley, New York
18. Ihara S (1993) Information theory for continuous systems. World Scientific, Singapore
19. Witsenhausen HS (1976) Some remarks on the concept of state. In: Directions in large-scale systems, many-person optimization, and decentralized control (Proceedings Symposium, Wakefield, MA, 1975), Plenum, New York, pp 69–75
20. Teneketzis D, Varaiya P (1984) Consensus in distributed estimation with inconsistent beliefs. Syst Control Lett 4:217–221
21. Washburn RB, Teneketzis D (1984) Asymptotic agreement among communicating decisionmakers. Stochastics 13:103–129

Chapter 27
Distributed State Estimation with Communication of Observations

André C.M. Ran and Jan H. van Schuppen

27.1 Introduction

The purpose of the chapter is to formulate algorithms and theorems on distributed state estimation with communication between observers of a decentralized Gaussian system.

The problem is motivated by the problem of observer synthesis for decentralized systems. In case of decentralized deterministic systems, one works with observers, while for decentralized stochastic systems one works with filters. The research topic is also described as decentralized filtering and decentralized observer synthesis. Examples are in state estimation of power systems or on large-scale communication systems.

Often, the decentralized systems are an interconnection of two or more subsystems in which case the term of *distributed system* is used. Most networks of systems are of this type. The main problem is then also called filtering of distributed systems.

The ultimate aim is to develop distributed control with communication. Therefore, one has to first investigate the problem of distributed observer synthesis with communication of observations. The distributed control with communication is much more complicated due to the feedback of information via the control system.

For discrete-event systems, distributed control with communication has been treated for many years, see [2, 7, 10]. The case of communication of a subset of the observations after every observation of a discrete-event systems is treated in [8].

A.C.M. Ran (✉)
Department of Mathematics, VU University Amsterdam, De Boelelaan 1081,
1081 HV Amsterdam, The Netherlands
e-mail: ACM.Ran@few.vu.nl

A.C.M. Ran
Unit for BMI, North-West University, Potchefstroom, South Africa

J.H. Van Schuppen
Van Schuppen Control Research, Gouden Leeuw 143, 1103 KB Amsterdam, The Netherlands
e-mail: jan.h.van.schuppen@xs4all.nl

J.H. van Schuppen and T. Villa (eds.), *Coordination Control of Distributed Systems*,
Lecture Notes in Control and Information Sciences 456,
DOI 10.1007/978-3-319-10407-2_27

For selection of a component of the output of a Gaussian system, there is the general approach of Athans [1] and for the case of when to communicate an observation, see [3].

27.2 The Problem

Consider then a decentralized systems.

In distributed control with communication, the main questions are as follows: What? When? and to Whom? to communicate one's observations. There are various forms of communication between subsystems; the two major ones are (1) send part of the observations after receipt of any observation and (2) send part of the observations only at those time moments when they seem to be needed by the other subsystems. In this paper, attention is restricted to the first form.

To make progress on the problem, several restrictions are imposed. As a first step, attention is restricted to distributed state estimation. The second step is in the restriction to Gaussian stochastic systems consisting of linear systems with Gaussian disturbances. The situation is further restricted to two subsystems and to one-directional communication only; Subsystem 2 sends at any time a linear combination of its observations to Subsystem 1. The subsystem that receives the observations of the subsystem then has to estimate the state of the decentralized system as best as possible.

The problem treated in this paper is thus the selection of which linear combination of observations to communicate. Consider a distributed Gaussian system with two observation streams,

$$\begin{aligned} x(t+1) &= Ax(t) + Mv(t), x(0) = x_0, \\ y_1(t) &= C_1 x(t) + N_1 v(t), \\ y_2(t) &= C_2 x(t) + N_2 v(t). \end{aligned} \tag{27.1}$$

Here, $x(t) \in \mathbb{R}^n$, and for $i = 1, 2$, $y_i(t) \in \mathbb{R}^{p_i}$, and finally, $v(t) \in \mathbb{R}^{m_v}$. We assume that the stochastic process $v(t)$ is Gaussian white noise, and, as a first approach, that the noise on state and outputs are uncorrelated, that is $MN_1^T = 0$, $MN_2^T = 0$, $N_2N_1^T = 0$. Moreover, we shall assume that N_1 and N_2 are full rank, that is, $N_1N_1^T$ is invertible and $N_2N_2^T$ is invertible. In particular, $m_v \geq p_i$ for $i = 1, 2$.

The communications from Observer 2 to Observer 1 can be any time-invariant linear combination $Ly_2(t)$ of the observations of Observer 2, where $L \in \mathbb{R}^{p \times p_2}$. Hence, for the purpose of state estimation, Observer 1 has available the vector

$$y(t) = \begin{bmatrix} C_1 \\ LC_2 \end{bmatrix} x(t) + \begin{bmatrix} N_1 \\ LN_2 \end{bmatrix} v(t).$$

We shall assume that $p \leq p_2$ and that L is chosen to be full rank.

The task of Observer 1 is to estimate the state of the Gaussian system based on the output which is the combination of the two observation streams. This can be done with a Kalman filter. It is well known that the error covariance matrix for the state estimation is the solution of the filter Riccati equation. For the steady-state case, we may as well consider the algebraic Riccati equation. Denote by $P(L) \in \mathbb{R}^{n \times n}$ the covariance of the estimation error of the state x of the system based on the observation vector y. Below the ordering on the set of variance matrices is the usual partial ordering on the set of positive semidefinite matrices.

Variants of the problem include to send from Observer 2 to Observer 1 only the state estimate of Observer 2. In that case, the complexity of the communication should be considered.

Problem 27.1 Consider the decentralized Gaussian system described above and the linear combination of observations of Observer 2 to be sent to Observer 1. Determine the optimal *communication matrix* L^* by solving the optimization problem

$$\inf_{L \in \mathbb{R}^{p \times p_2}} P(L) = P(L^*), \quad L^* \in \mathbb{R}^{p \times p_2} \tag{27.2}$$

Consider the following two subproblems:

(a) The size of the observation communication $p \in \mathbb{Z}_+$ is free to be chosen, and
(b) the size $p \in \mathbb{Z}_+$ is fixed a priori.

In particular, we shall be interested in the question whether there exists an L for which we have the smallest possible error covariance matrix for the state estimation obtained from the Kalman filter based on the output $y(t)$. We consider both the case where we do not fix p in advance, as well as the case when we do fix p in advance. It turns out that an answer to the first of these question is more or less straightforward, whereas the second question is far more involved.

27.3 Communication Law of Variable Rank

We shall first focus on the following set of admissible matrices: let $r = \text{rank}\, C_2$, and define $\mathcal{M}(r) = \{L \in \mathbb{R}^{p \times p_2} \mid \text{rank}\,(L) = p \le r\}$, where $p \le p_2$ is some natural number, which is allowed to vary with L. Note that every L in $\mathcal{M}(r)$ is of full rank, but that this rank varies. Thus, the problem we consider is to minimize $P(L)$ over $L \in \mathcal{M}(r)$, provided a minimum exists, and if so to find a minimizing L. The problem was described in [9].

After several rephrasings (see [6]) of the problem, using results in [5], one sees that the algorithm to compute an optimal choice of $L \in \mathcal{M}(r)$ is as follows (for computational details on the steps of the algorithm see [4]):

Algorithm 3 Computation of the optimal communication law in case of a range of ranks.

1. Construct the positive definite square root N_0 of $N_2 N_2^T$ (from e.g., the polar decomposition of N_2^T).
2. Compute $C_0 = N_0^{-1} C_2$, and let $r = \text{rank}\, C_2$.
3. Compute the SVD of C_0: $C_0 = UDV^T$.
4. Take $L_{\text{opt}} = \begin{bmatrix} I_r & 0 \end{bmatrix} U^T N_0^{-1}$. □

Theorem 27.1 *For the system* (27.1)*, let L_{opt} be given by the construction above. Then for any L with* rank $L \leq r$ *we have that the solution $P(L)$ of the Kalman filter Riccati equation satisfies $P_{\text{opt}} \leq P(L) \leq P_1$, where $P_1 = P(0)$, and $P_{\text{opt}} = P(L_{\text{opt}})$.*

Let us discuss this result. Introduce $w(t) = U^T N_0^{-1} v(t)$, then $w(t)$ has covariance matrix I_{p_2}, and

$$\begin{aligned} U^T N_0^{-1} y_2(t) &= U^T N_0^{-1} C_2 x(t) + U^T N_0^{-1} v(t) \\ &= U^T C_0 x(t) + w(t) \\ &= DV^T x(t) + w(t) = \begin{bmatrix} \sigma_1 & 0 & \cdots & 0 \\ 0 & \ddots & & \vdots \\ \vdots & & \sigma_r & 0 \\ 0 & \cdots & \cdots & 0 \end{bmatrix} V^T x(t) + w(t). \end{aligned} \tag{27.3}$$

It is obvious from this computation why the algorithm for producing L gives the desired effect: $Ly_2(t)$ transforms the noise component to $Lv(t)$ but keeps its variance identity, and then disregards those components of y_2 that are only affected by the noise and which do not carry any information concerning the state of the system.

27.4 Communication Law of Fixed Rank

Next, we change the point of view slightly and consider matrices L of fixed rank only. Let p be a fixed number with $p \leq r$, where $r = \text{rank}\, C_2$. Let $\mathscr{L}(p) = \{L \in \mathbb{R}^{p \times p_2} \mid \text{rank}\, L = p, \|L\| = 1\}$. When rank $L = p < \text{rank}\, C_2 = r$ is fixed, we have to investigate whether or not there is a minimal $P(L)$ over $L \in \mathscr{L}(p)$. It turns out that a minimum in this case does not always exist. Examples of this are given in [6].

One may try to circumvent this in several ways: For instance, one may try to minimize the trace of $P(L)$, i.e., one may consider $\min_{L \in \mathscr{L}(p)} \text{trace}\, P(L)$ and to look for a minimizing $L \in \mathscr{L}(p)$. Although this is appealing, since the interpretation is natural, it is not clear how to minimize the trace of $P(L)$ over the set $\mathscr{L}(p)$, and it is not clear whether or not a minimum does exist.

However, we argue that the choice described in the following algorithm is more natural.

Algorithm 4 Computation of the communication law in case of a fixed rank.

1. Construct the positive definite square root N_0 of $N_2 N_2^T$ (from e.g., the polar decomposition of N_2^T).
2. Compute $C_0 = N_0^{-1} C_2$, and let $r = \text{rank } C_2$.
3. Compute the SVD of C_0: $C_0 = U D V^T$.
4. Check whether $\sigma_p > \sigma_{p+1}$. If not, stop.
5. Take $L_{\text{trace-opt}} = \begin{bmatrix} I_p & 0 \end{bmatrix} U^T N_0^{-1}$. □

It remains then to investigate whether or not for given fixed p, the corresponding solution $P_{\text{trace-opt}}$ of the algebraic Riccati equation also minimizes $P(L)$, with rank $L = p$ in the partial order, or perhaps whether it minimizes the trace of $P(L)$. Numerical evidence shows that at least in special cases this may be true, but that in general it will not be true.

Continuing the discussion at the end of the previous section, from formula (27.3), we see that the effect of taking $L y_2(t)$ is to transform the noise component, keeping its variance identity, and then to select the p components of the output which have the highest singular values in the matrix that carries the state to the output. One might conjecture that under the condition in step 4 of the algorithm, one always has an optimal L; however, examples show that this may or may not be the case (see [6]).

27.5 Concluding Remarks

We have shown that the problem of communication of a linear combination of output observations so as to minimize the state estimate error covariance is solvable when the only restriction on the rank of the communication matrix L is that it is of rank less than or equal to the rank of C_2. However, when we fix the rank of L, the problem does not have a solution when we consider minimization of $P(L)$. An upper and lower bound for $P(L)$ are provided, and we propose an alternative for the minimization of $P(L)$ which is computationally feasible.

27.6 Further Research and Further Reading

Communication of observations from a controller to an other controllers is a problem area requiring further attention. Research issues include the investigation of the decomposition of the system so that it becomes clear what to send and what not to send; when should the observations be sent if the costs of doing so are nonneglible; and to whom should observations be sent if there are three or more observers? The case of a decentralized finite stochastic system is of particular interest in communication theory.

The problem addressed in this contribution finds its source in [9], see also [1]. The essay is based on [6].

References

1. Athans M (1972) On the determination of optimal costly measurement strategies for linear stochastic systems. Automatica 8:397–412
2. Barrett G, Lafortune S (2000) Decentralized supervisory control with communicating controllers. IEEE Trans Autom Control 45:1620–1638
3. Boel RK, van Schuppen JH (2010) Control of the observation matrix for control purposes. In: Edelmayer A (ed) Proceedings of the international symposium on the mathematical theory of networks and systems (MTNS.2010), University of Budapest, Budapest, pp 1261–1268
4. Golub G, VanLoan C (1996) Matrix computations, 3rd edn. Johns Hopkins University Press, Baltimore
5. Lancaster P, Rodman L (1995) Algebraic Riccati equations. Oxford Science Publications, Oxford
6. Ran ACM, van Schuppen JH (2013) Distributed state estimation with communication of observations. Linear Algebra Appl 439:600–612
7. Ricker S, Rudie K (1997) Know means no: incorporating knowledge into decentralized discrete-event control. In: Proceedings of 1997 American control conference
8. Rohloff KR, van Schuppen JH (2005) Approximating minimal communicated event sets for decentralized supervisory control. In: Proceedings of IFAC World Congress. Elsevier, London
9. van Schuppen JH (2011) Control of distributed stochastic systems—introduction, problems, and approaches. In: Proceedings of IFAC World Congress 2011
10. Wong K, van Schuppen J (1996) Decentralized supervisory control of discrete-event systems with communication. In: Proceedings of international workshop on discrete event systems 1996 (WODES96). IEE, London, pp 284–289

Part VI
Multilevel Control

Both chapters describe research into control of hierarchical systems with the second chapter focusing on control of road networks.

Chapter 28
Complexity and Scalability of Hierarchical Control for Networked Infrastructures

Jos Vrancken

28.1 Introduction

Hierarchical control (HC) is a form of multilevel control. Hierarchical means that a set of control instances are organized in a number of levels, where a control instance at one level controls a target system that itself consists of a number of control systems at the next lower level. This recursive way of organizing a control system is a very old and well-known way of managing large systems: the government of countries, and the governance of large organizations is always done with a degree of hierarchy: A country has a government for the whole country but this government cannot possibly handle all the details that have to be dealt with. Therefore, a country is split up into smaller units (states or provinces), each with its own form of government. And this is repeated recursively, down to the level of municipalities and even single households. At each level, there is not only a geographic subdivision, but also a thematic subdivision (the different departments of a government). Both kinds of divisions serve the purpose of reducing complexity to the level that single human beings can handle.

There are examples, such as the former communist countries, which also show the limitations of the hierarchical approach. As we will see below, this can better be considered as improper balancing of the different levels, giving too much power and too many responsibilities to the top, making the task too complex for humans at the top to handle adequately. The validity of the hierarchical concept itself is not necessarily impaired by this example, which above all shows the importance of proper balancing of responsibilities. The same applies to large organizations [1, 2].

The key-driving factor behind hierarchy is people's limited capacity for handling complexity. The systems that people build, however, are obviously hugely complex. Society is an example, but even subsystems of society, such as the networked in-

J. Vrancken (✉)
Delft University of Technology, Delft, The Netherlands
e-mail: j.l.m.vrancken@tudelft.nl

J.H. van Schuppen and T. Villa (eds.), *Coordination Control of Distributed Systems*,
Lecture Notes in Control and Information Sciences 456,
DOI 10.1007/978-3-319-10407-2_28

frastructures, already exceed by far the level of complexity at which some form of hierarchy becomes needed.

However well-known the hierarchical approach may be in politics and company governance, its application to other systems lags way behind demand. For instance, many of the problems we have with road traffic, most notably congestion, are strongly network-related, which implies a level of complexity that points toward hierarchy, but most of today's traffic management measures are still purely local and not network aware at all. The exceptions are still limited in number, geographic extent, and functionality.

An important difference between politics and, for instance, the control of networked infrastructures is the involvement of automated systems. The lower levels are completely automated (for example, traffic signals in road traffic control) and the higher levels, with operators behind operator consoles, strongly depend on automated systems. The limited understanding of HC is an important hurdle for the construction of the (semi-)formal models needed to build automated systems. Moreover, we should not forget that the models used in politics took centuries of trial and error to obtain their present form.

There is a strong societal demand to apply control to huge systems, such as the networked infrastructures. It is a matter of making control scalable. This is hampered by the huge complexity of large systems. HC is an attractive candidate to achieve this scalability, following the examples of politics and company governance. HC offers an overall system organization, which forms the context for the more localized applications of mathematical, control, engineering, and computer science methods. The hierarchical approach is well developed in, and more or less synonymous with, the field of systems engineering (SE) [3], which in turn is based on systems theory and systems thinking. Key notions from these fields are *systems*, the *recursive structure* of systems, and system *boundaries* that function as interfaces between systems. HC is in essence the application of SE to control engineering.

Chapter 29 contains a description of the way HC can be applied to road traffic management.

28.2 Problem: Complexity

Complexity is a key notion in hierarchical control as it is the main problem in large systems: The huge complexity that usually grows exponentially with system size. The hierarchical approach is first of all a way to manage complexity, by making sure that complexity is limited for each control instance involved.

The key notion is that of *perceived* complexity. What matters is only the complexity that we have to deal with for a given purpose, the complexity that we perceive when pursuing this purpose. In the sequel, complexity is synonymous with this perceived complexity. This kind of complexity can be defined as the number of details one perceives and the number of relationships or dependencies between these details. Objects can be extremely complex for one purpose and simple for another purpose.

An important property of complexity (i.e., the perceived complexity for a given purpose) is whether it is *reductionistic* or not. Reductionistic means that when zooming in on part of an object, the complexity (expressed in number of details and relationships) becomes less. Looking at a Mandelbrot picture with the purpose of copying it, is an example where zooming in does not make it simpler. Real-life systems usually are hugely complex, but for many purposes, we only need to deal with very little of that complexity. On the other hand, the more complexity one can handle, the more one can do with a system. In practice, one often has to search for a way of modeling a system with the right level of complexity. Too much makes the model incomprehensible; too little limits the model in its capabilities.

Pursuing a purpose that confronts people with non-reductionistic complexity is always a no-go. But different ways of looking at a system and getting to know a system better can help to find a way of looking at a system that turns out as reductionistic. The simple mathematical description of a Mandelbrot picture and limiting the level of detail offers a reductionistic approximation that helps in, for instance, drawing such a picture on a computer screen.

28.3 Splitting up a System into Parts

Splitting a system up into parts, aka subsystems, is a key means to obtain complexity reduction. But it is essential to do this recursively. Just splitting up a huge system into many smaller ones can make each individual subsystem manageable, but then there are too many of them. Splitting it up into a small number of still large subsystems limits the number of subsystems, but the subsystems are still to big to be managed directly. Then, recursion comes into play: The subsystems can also be split up into parts, but these parts stay within their subsystem and are not visible from the level of the system we started with. This process can be repeated until the remaining systems are small enough to be managed directly.

This recursive process of splitting up a system results in a tree structure, aka hierarchical, model of a system, which is common in the field of SE. Crucial in the complexity reduction offered by SE is the notion of the boundary of a system. A system has first of all a boundary which determines what is inside the system and belongs to it, and what is outside of it. In addition, the boundary is where the interactions between a system and its environment take place: It is the interface between a system and its environment. The boundary helps greatly in limiting the complexity of a system: For the interactions with the environment, the internal behavior of a system can be abstracted to effects on the boundary. And the other way around: What happens outside of the system is relevant for the internals only if it has an effect on the boundary.

In the case of hierarchically organized control systems, a system S in the tree (the recursive split up) has to control a number of subsystems at the next lower level. The boundary of these subsystems shields their internal complexity from S. It only needs to manage the boundary behavior of its subsystems (which is the behavior visible from outside of the subsystem). A system's boundary is not always easy to identify, but when looking at a networked infrastructure, it is usually geographically determined. The split up of such a system can be done geographically, making the boundary a finite set of points. Indeed, for a networked infrastructure, what happens outside of it, is only relevant if it has an effect on this geographic boundary. The reason for choosing a geographical split up is indeed this strong reduction of complexity offered by such a boundary. But the geographical way does not need to be the only way. In the case of the government of countries, we saw the example of a thematic split up (economy, education, defense, etc.). Such a split up can be combined with a geographic one, as is done in governments all over the world. It certainly helps to reduce complexity, although for themes it is harder to define boundaries and the complexity reduction is not as obvious as in the case of the geographic split up.

28.4 Hierarchical Control

Given the recursive split up of a system and the complexity shielding by the boundaries of all these systems in the tree, we will now describe in more detail how control is organized in such a tree. Consider a system S controlled by a controller C. If C is a human being, the complexity of S as seen by C must be reductionist and limited as people can only deal with a limited amount of complexity. Replacing the human C by a machine does not change this fundamentally, as the machine has to be programmed by a human being who is hampered by the same limitation. Systems that are too complex have to be split up into smaller ones recursively, as described above. If reductionism applies, subsystems are simpler. Each smaller part P has its own controller D, dealing with part of the complexity of S, such that this is largely shielded from C. Parts of S, such as P, should be chosen such that most of P's complexity can be handled inside P and only little of it is visible from outside (P's boundary behavior). In this way, the controller C only needs to deal with the boundary behavior of the subsystems of S.

HC is a form of multilevel control: In the tree usually a number of levels can be distinguished, determined by the distance from the root. This applies when parts of a system are largely similar, such as the parts of an infrastructure in a geographic split up. Then, different levels in different branches are comparable and a level extends over the full breadth of a tree. Hierarchy makes that complexity is spread over a number of levels, and within each level, over a number of components, such that each control instance has a limited amount of complexity to deal with. The fact that new levels can be introduced makes that the number of parts of a system (at one level) can be kept limited. This works on condition that the boundary behavior of a component is considerably less complex than the component's full complexity, which typically holds for networked infrastructures, using a geographic split up. Finding the right size of subsystems is an example of the balancing between feasibility and utility mentioned above.

Some terminology is needed to talk about tree-structured systems. When a system is divided into parts, we call these parts the *direct children*. The grandchildren (parts of parts of a system), great grandchildren, etc., are all called children. Likewise, a system is part of a larger system, unless it is the root of the tree. This containing system is called the *direct parent*. Grandparents and higher are all called parents of a system. The lowest systems in the tree, the ones that are not divided into parts, are called the *leaves*.

Apart from their relative position in the tree, systems may have other relationships. An important one is *adjacency*, which, for instance, applies in case of networked infrastructures in a geographic split up. Then, systems may be *neighbors*, sharing part of their boundaries. Moreover, part of a boundary may be shared by parts at different levels, i.e., a system and several of its children and/or parents (same side of the boundary), and by systems not related by parenthood (at the other side). Such a border is called a *multi-border*.

For full scalability, we must consider the number of levels, the size of boundaries and the number of different control processes within each level. In principle, the number of levels can grow indefinitely (there is no instance dealing with the set of levels), but this may reduce scalability if the control processes at the different levels show great variety. Full scalability can only be obtained if the number of different control processes at different levels is limited, which, practically speaking, means that it is uniform for the majority of levels, with only the bottom and the root being different.

Boundaries of large systems are themselves large and complex objects, whose complexity can reduce scalability. For full scalability, the abstraction of a system's behavior to its boundary must be such that this stays limited even for arbitrarily large systems. This can be handled by splitting up boundaries into parts too, but this is beyond the scope of this chapter.

Finally, the number of different control processes within one level must be kept limited. Often the split up of a system into subsystems is such that most subsystems are similar and the number of different kinds of subsystems is small, which means that also the number of different control processes is small.

If not all of these conditions are fulfilled, a partial scalability remains, which may still be useful. The hierarchical approach then still offers a way to control systems of much larger size than is possible by direct control using a single control instance.

Once a system is organized as a hierarchical system, there is considerable freedom within each subsystem. Any level of local complexity is allowed, as long as the boundary behavior of the subsystem is not made more complex. This means that the hierarchical approach can be combined with other control approaches, adding scalability to methods that are not scalable by themselves. Moreover, it shields off developments within subsystems. Typically, each system in the tree is subject to a learning process of continuous improvement of its internal control strategy. This is practically feasible if the results of this learning process can be shared by many similar systems.

The task of a system's controller can be further alleviated by using peer-to-peer (p2p) interaction between its direct children in addition to control instructions from the controller: If one has a problem with a neighbor, it is best to first talk to this neighbor, before resorting to authorities. It works even better in the case of a shared interest with a neighbor. Whatever can be handled by p2p interaction, does not need to bother the controller. Both mechanisms have their pro's and con's. Decision making is more efficiently done by a single instance, the controller. But responding to a sudden problem between two subsystems, a problem that requires a speedy answer, is likely to be faster when done by p2p interaction than by asking the controller. Typically, control systems have a control objective and therefore can be seen as having interests (anything that contributes to achieving their objective). Conflict of interests in general has to be settled by the controller. But matters that serve the interests of all subsystems involved, may be handled by p2p interaction. P2p interaction is strongly related to agent-based control and plays an important role in the research on infrastructure management [4, 5].

At a multi-border, the boundary behavior of the systems sharing this border must be such that it is consistent for all levels involved. We call this *recursively consistent* boundary behavior. The controller C of a system S essentially manages the boundary behavior of its direct children, thereby fulfilling the boundary behavior requirements imposed on S. It helps if the required boundary behavior of children (at least the aspect that has to be controlled by C) changes on a time scale, substantially longer than the internal behavior of the children. If this is the case, the internal control of each child is effectively decoupled from its external behavior. Then, the control actions of C can be considered as setting the system configuration for the behavior of the children. This configuration, in its turn, then guides the p2p interaction between the children of S. Setting this configuration is a matter of negotiation between the controller C of a system S and the direct children of S. C itself got a request from its parent for such a configuration. It proposes a configuration to its children. They respond with whether they can realize this. Or, if they have a certain autonomy of their own, whether they are willing to comply with this proposal in addition to whether they are physically capable. This is a repetitive and recursive process, going up and down the tree structure of systems, in order to obtain a configuration that is recursively consistent over the whole tree.

28.5 Summary

In summary, the hierarchical approach is increasingly effective and scalable as more of the following properties hold:

1. There is a recursive way of subdividing systems into subsystems.
2. The subsystems shield off much of their internal complexity from the parent system. In other words: A system's boundary behavior is much simpler than the system's internal behavior.
3. There is a small number of different control strategies at different levels.

4. There is a small number of different control strategies at different subsystems within one level.
5. Complexity of boundary behavior can be kept limited, even for bigger systems.
6. The control configuration for the boundary behavior of the subsystems of a system changes at a longer time scale than the internal control of subsystems.
7. The control configuration at boundaries is recursively consistent at multi-borders.
8. HC applies both top-down parent-to-child control and p2p interaction between children.

References

1. Sutton RI , Rao H (2014) Scaling up excellence. Crown Business, Penguin Random House Company, New York, USA
2. Williamson OE (1967) Hierarchical control and optimum firm size. J Polit Econ 75(2):123–138
3. Sage AP, Rouse WB (2011) Handbook of systems engineering and management. Wiley, New York
4. Bazzan ALC , Klügl F (2013) A review on agent-based technology for traffic and transportation. The Knowl Eng Rev 1–29
5. Dou C, Liu B (2014) Hierarchical management and control based on MAS for distribution grid via intelligent mode switching. Int J Electr Power Energy Syst 54:352–366

Chapter 29
Hierarchical Control for Road Traffic Management

Jos Vrancken

29.1 Introduction

Road traffic management (RTM) at the network level, also known as network management (NM), is a key challenge in traffic engineering, to which hierarchical control can be applied. As we will see below, the hierarchical approach actually fits very well with NM.

The starting point is a network N of roads, a connected part of the total network of roads on the world. One would like to manage traffic inside N according to some traffic management policy that may vary in details, but that usually includes congestion reduction, improved safety, reduction of environmental damage, and improved liveability in the area that contains N. This chapter offers a coherent framework for both operational traffic management and for the development and deployment of the monitoring and control systems involved. The framework was introduced in [17] under the name QHM: Quantitative Hierarchical Model. The two main components of the model are:

- A recursive split-up of the network N.
- A multilevel control synthesis.

29.2 Literature Review

Splitting up a network into parts is part of virtually every approach to NM. There is an increasing number of articles addressing network partitioning together with a control approach based on this partitioning, for instance: [2, 4, 6, 8, 10, 14], but most are not recursive approaches, using only one or a few levels. The control approaches

J. Vrancken (✉)
Delft University of Technology, Delft, The Netherlands
e-mail: j.l.m.vrancken@tudelft.nl

J.H. van Schuppen and T. Villa (eds.), *Coordination Control of Distributed Systems*,
Lecture Notes in Control and Information Sciences 456,
DOI 10.1007/978-3-319-10407-2_29

are mostly based on MFD (the Macroscopic Fundamental Diagram). The role of recursion is usually underestimated or absent in these publications. The scalability of their control approaches is most often not mentioned or argued, and certainly not evident. Many publications (for instance [3, 5, 11, 12]) hint at hierarchical approaches, but none pursue this concept systematically for the purpose of traffic management.

29.3 The Recursive Split-up of Networks

The recursive split-up of a network results in a tree structure of networks, down to the level of individual segments and nodes. The boundary of each network S in the tree consists of a number of entry points and exit points. A point is a cross section through the carriageway of a road, in one driving direction. It is recommended to choose the boundary points at quiet places (to be determined on the basis of historical data), such that the shared control problem at such a point is easy. A second important requirement is about shortest paths: it makes control of a network easier if it is closed under 'shortest path': the shortest path between any two nodes within S should fall within S. There are more requirements for the recursive split-up of N. They are listed in [16].

This way of splitting-up networks obeys several of the requirements mentioned in the summary of Chap. 28. The surroundings of a network S only see the traffic behavior on the boundary of S. Anything that happens within S without an effect on the boundary of S is irrelevant to the network outside S. In addition, it is clear that traffic behavior on the boundary of S, which consists of a finite number of points, is far simpler than traffic behavior within S. This is really a strong reduction of complexity. The effect increases for a bigger S. On the other hand, this boundary behavior can still be very complex, but it is also clear that only part of this complexity is relevant to the surrounding network.

One of the ways by which boundaries can be made simpler is by grouping entry points (same for exit points). A group of boundary points in many ways behaves the same as a single point: it has a flow (the sum of the flows of the member points) and there are shortest routes to the group (the shortest of the routes to any member point) and there are routes via such a group (via any of the member points). Unimportant, low capacity entries (exits) that are close to each other can be grouped into such a composed entry (exit) point and then be treated like one entry (exit) point, thereby strongly reducing the boundary complexity of a large network.

29.4 The Multilevel Control Synthesis

Traffic management policies may pursue a variety of different goals and may stress different goals at different times, but some goals, such as congestion reduction and improving safety, are always present and they contribute positively to the other goals mentioned earlier. Among safety improvement and congestion reduction, the latter is most strongly network-related. This is the reason we focus on congestion reduction, which is, according to [7, 9] already an effective traffic management approach.

Considering a network N, the occurrence of congestion is strongly related to the amount of traffic within the network. The fact that the distribution of traffic inside N also plays a role, is handled by recursion, therefore does not need to be addressed explicitly. The future amount of traffic in a network is a function of current amount of traffic, inflow, and outflow. For the outflow, a network depends on its neighbors. But the inflow can be controlled. In principle, the inflow should be such that, on average, the inflow equals outflow. In this way, a fluid network will stay fluid. In reality, it is a bit more complicated of course, but this is the general principle. On the other hand, a network should try to be as productive as possible, which means allowing in as much as possible without becoming congested, and it should maintain at least a minimal flow (specified in traffic policies) on each of its entries; otherwise, some vehicles would be blocked for an indefinite amount of time. The traffic that is generated inside a network, or that has its final destination within the network, would complicate the description below somewhat and is therefore left out of consideration. A second assumption is that all actuators and all sensors needed are always present. The reasons for this are given in [17].

Modeling Traffic

Traffic can be considered as the sum of regular, predictable *patterns*, such as the morning rush hour, or the traffic generated by a large sports event, and an (often strong) *unpredictable* component. These patterns can be expressed as demand patterns (demand on boundaries). The unpredictable deviations from this regular pattern are mostly caused by events outside of the traffic system. External events have an obvious influence on traffic, so there will always be such an unpredictable component, unless we would be able to predict every relevant event outside of traffic. In addition, there are exceptional events, such as accidents, which in a sense change the network itself. Each of these three components is treated in a different way in the multilevel control synthesis.

Regular Patterns: Network Configuration

We consider a network N, subnetwork of the total network roads. N is partitioned recursively into a tree structure of subnetworks. We associate a network with the control instance controlling that network. The regular traffic patterns are addressed proactively by a configuration of all the networks in the tree. N does this by setting boundary conditions on all internal boundaries, between direct children of N, and on the outer boundary of N. This is done recursively. N is in this case the top of the tree, so does not receive any configuration requests from above, but below the top, boundary conditions are determined by means of an iterative process of negotiation between parent and children and between parent and the next higher parent (for the outer boundary). The agreements can be expressed in all the relevant traffic quantities that apply to points: speed, flow, etc. The agreements have to be recursively consistent at multiborders. For most quantities, this is not a problem (a speed for the parent is the same speed for the children), but for the destination-specific partial flows at an entry point, this needs a bit of math (details in [17]). If demand is higher than outflow, one has to decide how much of each destination-specific flow at each entry point is allowed. This can be expressed in a flow-priorities matrix.

Network configuration is closely related to the phenomenon of scenarios, which, at least in the Netherlands, are heavily used in current practice of traffic management [18].

Unpredictable Component: Peer-to-Peer Interaction

The unpredictable component can only be handled reactively. We consider a network A with a neighbor B. Given the boundary agreements expressed in flow-priorities, two neighboring networks can agree on actual maximal flows by peer-to-peer (p2p) interaction among networks. As this can consist of many interactions, this can best be handled by p2p, in order not to overload the parent. It is based on a shared interest as neighbors have a shared interest that they both stay fluid and highly productive.

The typical p2p request concerns a point e on the shared boundary of A and B, entry to A. According to its current outflow, A sets the maximal inflow on e by a request to B. Most of the time, B will fulfill this request, knowing that A tries to be as productive as possible (this maximal flow will be as high as possible without causing congestion in A) and knowing that a congested A could make outflow even worse. If necessary, B will adapt its own inflow settings, on its entry points accordingly. This is a recursive chain reaction of maximal flow requests in the network. In order to avoid generating too many requests, this kind of adaptations is done with a maximum frequency. Network A keeps some buffer space free, in order to keep internal density at a safe distance from congestion. The free buffer space is also used to dampen unpredictable variations in inflow and outflow at the border. Not every change in traffic flow leads to p2p request. Density must not be too low, however, as that would reduce productivity. A sudden increase in supply (i.e., what is allowed to flow out of A) would not be benefitted from. At a multiborder, each boundary point e has a lowest network to which it is entry point and a lowest network to which it is exit point. In principle, a p2p request concerning maximal inflow is sent from lowest network to lowest neighbor, with the higher networks that share the same boundary point being updated on the change.

Child Priorities

Cyclic p2p request chains are to be avoided. One way to do this is by introducing *child priorities*. A network N may assign a priority value to each of its children on the basis of how important a child is for the overall network. This can be done for instance by looking at how many shortest routes from entry points of N to exit points of N go through the child. This may be weighed by the amount of traffic or by the priority level of the entry points. The optimal way to do this is still under investigation. For two networks neighboring at a multiborder, not being children of the same parent, both networks look at the lowest pair of parents that have a common parent and are therefore comparable in terms of child priorities. The child priority graph, with directed edges from higher to lower priority, should always be acyclic. If this does not result automatically, it can be made acyclic artificially. In principle, p2p requests go from high to low area priority. In this way, cyclic request chains are avoided. The p2p requests have the effect that traffic is held back from high-priority areas to low-priority areas were congestion is the least damaging. In this way, the core of the network is always kept fluid (unless an exceptional event occurs), and congestion occurs only in residential areas, parking lots, and other low-priority areas.

From Robust to High Performance Control

The p2p process is admittedly rather complex. But to a large extent, the complexity can be kept local to each network involved and can grow from relatively simple, robust control, focussing on congestion prevention, to more optimized (but also more complex) forms of control. The way the TCP process in data communications [1], constantly optimizes parameter values amidst unpredictable behavior of its surroundings, is an important source of inspiration here. In this process of gradual optimization, other traffic policy goals besides congestion reduction may also be addressed.

Exceptional Events: Network Reconfiguration

Finally, exceptional events can be handled by changing the network configuration. Because of their exceptional character, this kind of events cannot be fully modeled, so we limit ourselves here to one category of events that temporarily block part of the network. A traffic accident is an example of this category. Such an event can be modeled as a sudden change of the network itself. The traffic demand pattern does not change, it is just the network that changes. Therefore, such an event is handled by a new network configuration which has to be determined and deployed as soon as possible after the event.

Summarizing, the multilevel control considers traffic to consist of three parts: the regular, predictive part; the unpredictable part; and the part that consists of exceptional, network changing events. The regular part and the exceptional events can be handled by network configuration and rapid re-configuration, respectively, implemented by parent to child instructions. The unpredictable part can best be handled by p2p interaction between peers. In this way, the high-priority parts of the road network can be kept fluid, by pushing back imminent congestion toward low-priority areas. When the core network is kept fluid, it is not likely that the low-priority areas will suffer from serious congestion either. The p2p process is rather complex, but there are many ways in which this complexity can be kept local to each network involved and can be built up and improved gradually. This learning process can even be automated to some extent, as shown by the example of the TCP protocol in data communications.

Traffic engineering offers a number of optimized algorithms, such as model predictive control, for network management in limited areas. These are rarely scalable, but they can always be fitted into this scheme, as long as they can maintain the required boundary behavior. One might use the QHM framework just to join a number of areas with such advanced algorithms, in this way giving scalability to algorithms that do not have this property themselves.

29.5 The Implementation of QHM: Interface Standardization

For the implementation and deployment of network management systems, an interface standard for connections between traffic control systems is needed. A NM system needs connections with virtually all the existing control systems within its scope. Practice showed [15] that making and maintaining connections with all these legacy systems, that were not designed to function under the coordination of an NM system, is prohibitively expensive. The semantics part of such an interface standard refers to the operational traffic control process. For this reason, such a standard has to be a well-integrated part of an overall NM system architecture. QHM is the traffic management part of such an architecture. It offers a model for the functionality of an arbitrary traffic control system, for both local and network-level control systems. Such a standard, called DVM-Exchange, is currently under development in the Netherlands [15] and has reached a basic level of practical usability. It responds to a clearly and broadly felt need, especially from the side of the traffic management authorities. If this standard achieves broad acceptance from both customers and suppliers in the traffic management systems market, it is likely to have a transformative influence on the market, transforming it from offering primarily closed and vertically integrated proprietary products, to offering open and horizontally integrated components, allowing combinations of components from different manufacturers.

29.6 Evaluation of QHM

The QHM framework is currently in the process of being evaluated. There is already an operational implementation, the Scenario Coordination Module in the Amsterdam area [18], that uses part of the concepts in the framework: it has a network split-up in three levels, applies to motorways and main urban arteries, and uses priorities and network configurations (called *scenarios*).

Further evaluations of QHM that are currently taking place or will take place in the near future, are as follows:

- Coordinated ramp metering in the QHM style, discussed as an example in [17], is currently being evaluated in a traffic simulator. First results show that this approach does indeed give a much better control of a series of motorway on-ramps and allows a greater variety of traffic policies to be implemented than existing approaches such as the HERO algorithm [13].
- Effectiveness of congestion prevention by recursive holdback of traffic. This is about testing the effectiveness of the p2p process, which consists of a recursive chain of flow reduction requests, pushing back traffic from high-to-low-priority areas.
- Operation in isolation. Early implementations will cover only small areas within networks that are not controlled at all or only with traditional local means. There

are good reasons to assume that the QHM approach can still function in this setting, but this has to be tested thoroughly, as early adoptions depend on this property.
- Effectiveness of network (re-)configuration after (unpredictable) accidents or (predictable) sports events and road works, as described above.
- The usefulness of the approach in case of emergency evacuations of whole areas, due to imminent flooding or other calamities.
- The prioritization of public transport in cities. This will be an important addition to the framework for urban applications.

29.7 Summary and Conclusions

The problem of road traffic fits very well with the requirements of hierarchical control. Although many details in the application of hierarchical control to network management of road traffic still need further research, this approach offers a clear prospect of more or less solving the congestion problem in the coming decade, as it offers many ways to deploy and improve it gradually, from simple and robust to highly optimized. Moreover, the hierarchical approach can be combined with virtually any existing control algorithm from the field of traffic engineering, on condition that such an algorithm can maintain/approximate the required boundary condition. To the best of our knowledge, none of these algorithms is scalable by itself. So for real scalability and for managing really large networks, the hierarchical approach seems to be a very promising option.

29.8 Future Research

Important items for further research are as follows:

- The development of traffic models and simulation tools for traffic in hierarchically controlled networks. This will be an important tool for the deployment of hierarchical traffic management. Traditional models and tools have serious scalability issues for larger networks, which makes novel approaches here necessary.
- Efficient algorithms for a number of items:
  - network partitioning
  - negotiating network configurations
  - effective p2p interaction
  - TCP-like continuous adaptation of the parameters that govern p2p interaction
  - Child priorities and acyclic priority graphs.

- Stochastic treatment of the relationship between variability of demand and the optimal free buffer size to maintain in networks.
- The relationship between hierarchical control and the problem of traffic assignment and network equilibrium.

References

1. Allman M, Paxson V, Blanton E (2009) TCP congestion control. Technical report, RFC 5681, Sept 2009
2. Chomkatek L, Poniszewska-Marańda A (2012) Multi-agent system for parallel road network hierarchization. In: Artificial intelligence and soft computing, Springer, Berlin, pp 424–432
3. Dupont E, Papadimitriou E, Martensen H, Yannis G (2013) Multilevel analysis in road safety research. Accident analysis and prevention
4. Etemadnia H, Abdelghany K, Hassan A (2013) A network partitioning methodology for distributed traffic management applications. Transportmetrica A: Transp Sci, 1–15
5. Frederix R, Viti F, Himpe WWE, Tampère CMJ (2014) Dynamic origin-destination matrix estimation on large-scale congested networks using a hierarchical decomposition scheme. J Intell Transp Syst 18(1):51–66
6. Geroliminis N, Haddad J, Ramezani M (2013) Optimal perimeter control for two urban regions with macroscopic fundamental diagrams: a model predictive approach. IEEE Trans Intell Transp Syst 14(1):348–359
7. Hoogendoorn S, Hoogendoorn-Lanser S, van Kooten J, Polderdijk S (2011) Integrated network management: towards an operational control method. In: TRB 90th annual meeting compendium of papers
8. Ji Y, Geroliminis N (2012) On the spatial partitioning of urban transportation networks. Transp Res B: Methodol 46(10):1639–1656
9. Kerner BS (2011) Optimum principle for a vehicular traffic network: minimum probability of congestion. J Phys A: Math Theor 44(9)
10. Keyvan-Ekbatani M, Kouvelas A, Papamichail I, Papageorgiou M (2012) Exploiting the fundamental diagram of urban networks for feedback-based gating. Transp Res B: Methodol 46(10):1393–1403
11. Kim J, Yu H (2013) Processing time-dependent shortest path queries without pre-computed speed information on road networks. Inf Sci
12. Ma Y, Kuik R, van Zuylen HJ (2013) Day-to-day origin-destination tuple estimation and prediction with hierarchical bayesian networks using multiple data sources. In: Transportation Research Board 92nd annual meeting, number 13–2861
13. Papamichail I, Papageorgiou M, Wang Y (2007) Motorway traffic surveillance and control. Eur J Control, 13(2):297–319
14. Potuzak T (2012) Methods for division of road traffic networks focused on load-balancing. Adv Comput 2(4):42–53
15. Vrancken JLM, Brokx A, Olsthoorn R, Schreuders A, Valé M (2012) DVM-exchange, the interoperability standard for network management. In: Proceedings of the 19th ITS world congress, Vienna, Austria, pp 1–11
16. Vrancken JLM, Wang Y, van Schuppen JH (2013) QHM: the quantitative hierarchical model, a scalable framework for road traffic network management. In: Proceedings of the 20th ITS World Congress, Tokyo, Japan, pp 1–7

17. Vrancken JLM , Wang Y, van Schuppen JH (2013) QHM: the quantitative hierarchical model for network-level traffic management. In: Proceedings of the IEEE ITSC 2013, The Hague, The Netherlands
18. Wang Y, Vrancken JLM, Ma Y (2012) Evaluation of the scenario coordination module, a traffic network management control system. In: Proceedings of the 19th ITS World Congress, Vienna, Austria, pp 1–10

Part VII
Communication and Control of Distributed Systems

Control of communication networks including layered backpressure control, cache control, and modeling may be found in the Chaps. 30–32. Chapter 33 provides an approach for performance evaluation of model predictive controllers for services over packet networks. The next chapter presents results for co-simulation of communication and control.

Chapter 35 introduces the problem of communication in case of feedback using the concept of directed information. Feedback in communication systems requires researchers to work with the concept of directed information rather than with the usual concept of information. The next three chapters provide research results on how to compute with directed information, how to determine rate distortion, and how to design coders with directed information.

Chapter 30
Layered Backpressure Scheduling for Delay-Aware Routing in Ad Hoc Networks

Dimitrios Katsaros

30.1 Introduction

Packet transmission scheduling in wireless ad hoc (multi-hop) networks is a fundamental issue since it is directly related to the achievement of a prescribed quality of service (QoS). QoS is usually measured in terms of the average packet delay. Additionally, any packet scheduling/routing algorithm for ad hoc networks must be resilient to topology changes (link/node failures, node mobility) and strive for throughput optimality. The development of a throughput-optimal routing algorithm for packet radio networks which is also robust to topology changes was first presented in [8], and it is known as the *backpressure* (***BP***) packet scheduling algorithm. Nevertheless, this algorithm and many of its variations suffer from delay problems. Certain features of the backpressure algorithm suggest that long delays will be common. Firstly, backpressure routing uses queue buildup at nodes to create a "gradient" within the network that guides routing. However, this may come at the cost of increased queuing delay. Secondly, backpressure routing tends to explore all paths in a network, including paths with loops and "dead-end" paths that cannot lead to the desired destination. Hence, packets generally may not take the shortest path to their destination, thereby leading to additional delay.

30.2 Motivation

The performance of backpressure deteriorates in conditions of low, and even of moderate, load in the network, since the packets "circulate" in the network increasing the packet delay. To circumvent the delay problems of backpressure, the *shadow queue*

D. Katsaros (✉)
University of Thessaly, Volos, Greece
e-mail: dkatsar@inf.uth.gr

J.H. van Schuppen and T. Villa (eds.), *Coordination Control of Distributed Systems*,
Lecture Notes in Control and Information Sciences 456,
DOI 10.1007/978-3-319-10407-2_30

algorithm [3] (***SQ-BP***) forces the links to stay inactive in order to lead the network to work in a burst mode, since for periods where the load of the network is low or moderate, link activation is prevented by a parameter M. On the other hand, the relatively high computational cost incurred at each node by backpressure (maintenance of a queue for each possible destination and update of these queues at each new arrival) inspired the *cluster-based backpressure* , which is based on node grouping in order to reduce the number of these queues [10] (***CB-BP***), and thus, the computational overhead, and as a side effect, reduces the delay. To alleviate the delay problems of backpressure, scheduling based on the combination of backpressure and shortest paths has been proposed [9], i.e., the *shortest paths backpressure* (***SP-BP***) policy. Finally, some ideas [6] could be incorporated in an orthogonal way to improve analogously the delay performance of all policies at the expense of throughput optimality, but this is a radically different problem.

30.3 Examples

We shall first present the basic backpressure mechanism with a small example assuming slotted time. Let us suppose that we have a 4-node ad hoc network with two flows from node A to D depicted in Fig. 30.1. Each node maintains a separate queue for each flow. For each queue, the number of backlogged packets is shown. Assume that we have two link sets, $\{(A, B), (C, D)\}$ and $\{(A, C), (B, D)\}$. The links in each set do not interfere and can transmit in the same time slot. The scheduler executes the following three steps at each slot. First, for each link, it finds the flow with the maximum differential queue backlog. For example, for link (A, B), the blue flow has a difference of 2 packets and the black flow has a difference of 7 packets. The maximum value is then assigned as the weight of the link (see Fig. 30.1). Second, the controller selects the set of noninterfering links with the maximum sum of weights for transmission. This requires to compute the sum of link weights for each possible set. In the example, set $\{(A, B), (C, D)\}$ sums to $7 + 4 = 11$ and set $\{(A, C), (B, D)\}$ sums to $6 + 6 = 12$. The scheduler then selects the set with the maximum sum of weights, i.e., $\{(A, C), (B, D)\}$, to transmit at this slot. Finally, packets from the selected flows are transmitted on the selected links, i.e., blue flow on link (A, C) and black flow on link (B, D).

A simulation-based evaluation of the aforementioned backpressure variations revealed their shortcomings. The shadow queues method forces the packets stay in queues longer, which leads to higher delays (see Fig. 30.2). The ***CB-BP*** policy requires maintaining one queue per *gateway* at each relay node which leads to an excessive number of gateways, which in turn alleviates any performance gains (i.e., increases delays) when the number of clusters becomes, say, more than ten (see Fig. 30.3). The ***SP-BP*** method assumes the precomputation of all pairwise-node distances. Apart from this computational-type problem, frequent topology changes would lead the method to break down, since many shortest paths would not exist anymore (see Fig. 30.4).

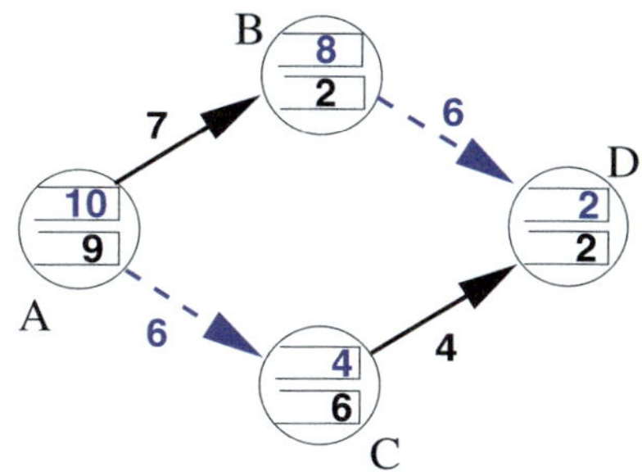

Fig. 30.1 Backpressure scheduling in a network with two flows, black and blue from A to D. Links in sets (A, B), (C, D) (*continuous*) and (A, C), (B, D) (*dashed*) can be scheduled in the same slot

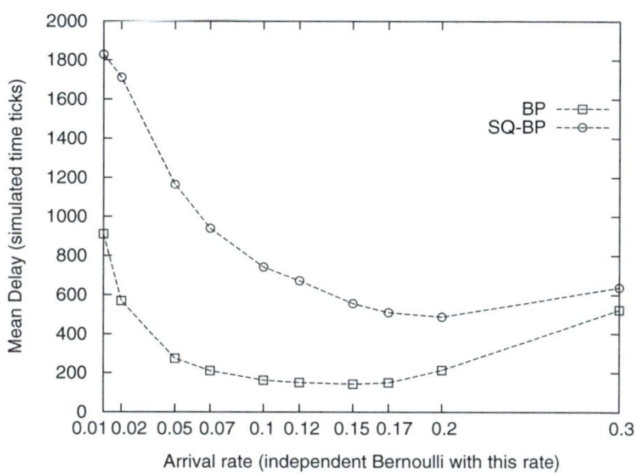

Fig. 30.2 Performance of the ***SQ-BP*** policy in a 4×4 grid network ($M = 2$)

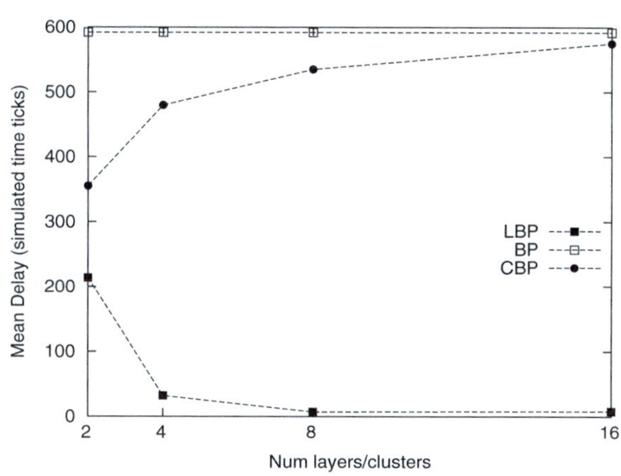

Fig. 30.3 The impact of the number of clusters

Despite the aforementioned efforts, the delay problem of backpressure has not been adequately solved. The challenge is to take a *holistic* approach in designing an efficient delay-aware backpressure policy, that is also practical—one that will have low computational overhead and that will be robust to topology changes.

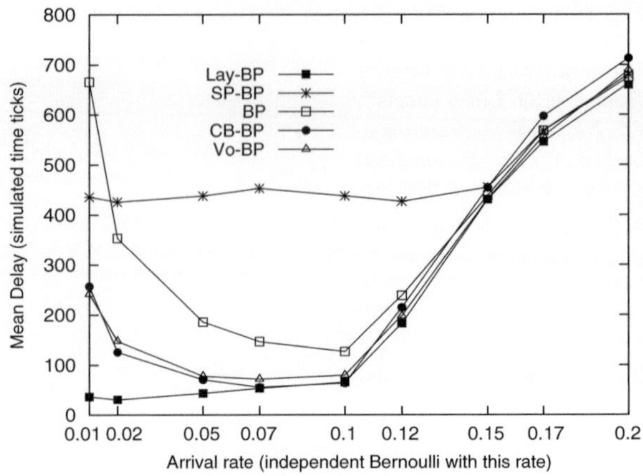

Fig. 30.4 Impact of topology changes on the delay performance of ***SP-BP***. The network consists of two clusters connected to each other with two links only. We inactivate one of the two links without recalculating the shortest paths among all node pairs

30.4 Theory and Concepts

Formally, the backpressure algorithm performs the following actions for routing and scheduling decisions at every time slot t.

- *Resource allocation*
  For each link (n, m), assign a temporary weight according to the differential backlog of every destination in the network:
  $$wt_{nmd}(t) = \max(Q_n^d - Q_m^d, 0).$$
  Then, define the maximum difference of queue backlogs according to
  $$w_{nm}(t) = \max_{d \varepsilon D} wt_{nmd}(t).$$
  Let $d^*_{mn}[t]$ be the destination with maximum backpressure for link (n, m) at time slot t.
- *Scheduling*
  The network controller chooses the control action that solves the following optimization problem:
  $$\mu^*(t) = \text{argmax} \sum_{n,m} \mu_{nm} w_{nm}(t),$$
  subject to the one-hop interference model. In our model, where the capacity of every link μ_{nm} equals to one, the chosen schedule maximizes the sum of weights. Ties are broken arbitrarily.
- *Routing*
  At time slot t, each link (n, m) that belongs to the selected scheduling policy forwards one packet of the destination $d^*_{mn}[t]$ from node n to node m. The routes

are determined on the basis of differential queue backlog providing adaptivity of the method to congestion.

30.5 Overview of Research Contributions

To provide a holistic solution, we designed the *Layered Backpressure* (***LayBP***), which divides the network into "layers" according to the connectivity of nodes. This method maintains the same number of queues with the original ***BP*** and one order of magnitude less number of queues compared to ***SP-BP***. It does not require the existence of *gateways* and aggregated queues as does ***CB-BP***. In addition, it can be seen as a "relaxed" version of the ***SP-BP*** between layers where packets are not forced to travel the shortest path among nodes, but the packets are "suggested" to follow the shortest path from the source to the destination layer. Therefore, the ***LayBP*** is a hybrid among ***CB-BP*** and ***SP-BP***, compromising a little delay for robustness, low computational complexity, and simplicity.

After the completion of the grouping and the assignment of IDs to the layers, the actual packet scheduling is performed as follows. Each node n maintains a separate queue of packets for each destination. The length of such queue is denoted as $Q_n^d[t]$. For every queue $Q_n^d[t]$, the node computes the parameter $D\text{level}_n^d$ which represents the absolute difference between current and destination node's layer numbers: $D\text{level}_n^d = |\text{Layer}(n) - \text{Layer}(d)|$. At each time slot t, the network controller observes the queue backlog matrix $Q(t) = (Q_n^d(t))$ and performs the following actions for routing:

Layered Backpressure at time slot t:

- Each link (n, m) is assigned a temporary weight according to the differential backlog $wt_{nmd}(t) = \left(Q_n^d - Q_m^d\right)$ and parameter A_{nmd} according to

$$A_{nmd} = \begin{cases} 2, & \text{if } D\text{level}_n^d > D\text{level}_m^d \\ 1/2, & \text{if } D\text{level}_n^d < D\text{level}_m^d \\ 1 & \text{otherwise.} \end{cases}$$

- Each link is assigned a final weight according to w_{nm} and parameter A_{mnd} $w_{nm}(t) = \max_{d \varepsilon D}(wt_{nmd}(t) * A_{nmd})$.
- The network controller chooses the control action that solves the following optimization: $\mu^*(t) = \text{argmax} \sum_{n,m} \mu_{nm} w_{nm}(t)$ subject to the interference model where adjacent links are not allowed to be active simultaneously.

The ***LayBP*** algorithm is throughput optimal, in the sense that it can stabilize queues for any stabilizable arrival rate.

In order to prove that the ***LayBP*** is throughput optimal, the Lyapunov stability criterion is used. The idea behind the Lyapunov drift technique is to define a nonnegative function, called the Lyapunov function, which represents the aggregate congestion of all queues (Q_n^d) of the network. The drift of the function at two successive time slots is then taken, and in order for the policy to be throughput optimal, this drift must be negative, when the sum of queue backlogs is sufficiently large. For both strategies, we use

$$L(Q) = \sum_{nd} \theta_n^d (Q_n^d)^2$$

as the Lyapunov function.

Recall that each link is assigned a final weight according to w_{nm} and parameter A_{mnd}:

$$w_{nm}(t) = \max_{d\varepsilon D}(wt_{nmd}(t) * A_{nmd}).$$

This equation can be rewritten in the following form:

$$w_{nm}(t) = \max_{d\varepsilon D}\left(A_{nmd} * Q_n^d - A_{nmd} * Q_m^d\right).$$

which is equivalent to

$$w_{nm}(t) = \max_{d\varepsilon D}\left(\theta_n^d * Q_n^d - \theta_m^d * Q_m^d\right),$$

where weights θ_i^d are used to offer priority service.

Queue dynamics at each time slot satisfy

$$Q_n^d(t+1) \leq \max[Q_n^d(t) - \sum_b \mu_{nb}^d(t), 0] + A_n^d(t) - \sum_a \mu_{an}^d(t),$$

where $\mu_{nm}^d(t)$ are routing control variables, representing the amount of commodity d data delivered over link (n, m) during slot t, and $A_n^d(t)$ represents the process of exogenous commodity d data arriving at source node n.

$$(Q_n^d(t+1))^2 \leq (Q_n^d(t))^2 + (\sum_b \mu_{nb}^d(t))^2 + (A_n^d(t) + \sum_a \mu_{an}^d(t))^2 -$$
$$2[Q_n^d(t)(\sum_b \mu_{nb}^d(t) - A_n^d(t) - \mu_{an}^d(t))]$$

Multiplying both sides with θ_n^d, summing over all valid entries (n, d), and using the fact that the sum of squares of nonnegative variables is less than or equal to the square of the sum, we take

$$\sum_{nd} \theta_n^d (Q_n^d(t+1))^2 \leq \sum\nolimits_{nd} \theta_n^d (Q_n^d(t))^2 + \sum\nolimits_{nd} \theta_n^d (\sum\nolimits_b \mu_{nb}^d(t))^2$$
$$+ \sum\nolimits_{nd} \theta_n^d (A_n^d(t) + \sum\nolimits_a \mu_{an}^d(t))^2 - 2\sum\nolimits_{nd} \theta_n^d Q_n^d(t)$$
$$\times \left(\sum\nolimits_b \mu_{nb}^d(t) - A_n^d(t) - \mu_{an}^d(t)\right).$$

It is not difficult to show that

$\Delta L(Q) \leq 2BN - 2\sum_{nd} \theta_n^d \varepsilon Q_n^d(t)$,where $B \triangleq \frac{1}{2N}\sum_{n\varepsilon N} \theta_{\max}[(\mu_{\max,n}^{\text{out}})^2 + (A_n^{\max} + \mu_{\max,n}^{in})^2]$.

Using the above, we can rewrite drift inequality as follows:

$\Delta L(Q) \leq 2B'N\theta_{\max} - 2\sum_{nd} \theta_n^d \varepsilon Q_n^d(t)$, where $B' \triangleq \frac{1}{2N^2}\sum_{n\varepsilon N}[(\mu_{\max,n}^{\text{out}})^2 + (A_n^{\max} + \mu_{\max,n}^{in})^2]$.

This drift inequality is in the exact form for application of the Lyapunov drift lemma, proving the stability of the algorithm.

The weighted sum of all queues is as follows:

$\limsup_{t\to\infty} \frac{1}{t}\sum_{\tau=0}^{t} E\{\sum_{n,d} \theta_n^d Q_n^d(\tau)\} \leq \frac{NB'\theta_{\max}}{\varepsilon_{\max}}$, which proves the optimality.

30.5.1 The Enhanced Layered Backpressure Policy

Routing protocols must be dynamic in order to cope with mobility of nodes in modern wireless networks. Widely varying mobility characteristics are expected to have a significant impact on the performance of routing protocols that are based on node grouping (like ***CB-BP***, ***LayBP***) in order to route packets even if links among nodes are updated. In case of grouping-based routing protocols, high mobility of nodes which lead them to change groups degrades the performance of the methods since this "wrong" information is used in the routing procedure. Although ***LayBP*** does not use gateways, it still suffers from this behavior if the layer that the moving nodes belong to is not updated. The differential backlog of each link is computed according to the difference between current and destination's node layers. It is clear that ***LayBP*** behavior can be affected by "misplaced" nodes. In this case, packets may be forwarded to layers different than the desired, making the method inappropriate.

In order to cope with node mobility, we incorporate in ***LayBP*** algorithm another step in which moving nodes and all the one-hop neighbors recalculate their cluster according to their neighborhood. In the initiation phase, every node has a counter $C0_{nl}$ for every layer ID, indicating how many neighbors belong to it, and a variable Layer_n indicating the layer that node n belongs to. For every time slot t, the following actions are performed:

- Calculate C_{nl}, the total number of neighbors that belong to every detected layer.
- if $C_{nl} >= C0_{nl}$ for $l = \text{Layer}_n$, then moving node remains in the same layer.
- else calculate the layer with the most neighbors $M_{nl} = \max C_{nl}$. if $M_{nl} > C_{nl}$, then moving node belongs to layer M_{nl}.
- assign for every layer l, $C0_{nl} = C_{nl}$ as new initial values for next time slot.

This algorithm can be executed every k time slots according to how fast we want the system to adapt to node mobility. Only moving nodes and their one-hop neighbors update the information on their counters in order to find the appropriate layer. Also, one-hop neighbors of the moving nodes need to update their information more rarely, than the moving nodes do, since a certain number of neighbors must be replaced in order to affect them. The algorithm does not perform reclustering, but only "helps" layers incorporate moving nodes.

30.6 Further Reading

Recent interesting work on the backpressure family of algorithms comprises the study of the trade-off between throughput optimality and delay [1, 4, 5, 7] and attempts to decouple the routing and scheduling components of the algorithm [2].

References

1. Alresaini M, Sathiamoorthy M, Krishnamachari B, Neely MJ (2012) Backpressure with adaptive redundancy (BWAR). In: Proceedings of IEEE INFOCOM, pp 2300–2308
2. Athanasopoulou E, Bui L, Ji T, Srikant R, Stoylar A (2013) Back-pressure-based packet-by-packet adaptive routing in communication networks. IEEE/ACM Trans Networking 21(1):244–257
3. Bui L, Srikant R, Stolyar A (2009) Novel architectures and algorithms for delay reduction in back-pressure scheduling and routing. In: Proceedings of IEEE INFOCOM mini conference.
4. Cui Y, Lau VKN, Wang R, Huang H, Zhang S (2012) A survey on delay-aware resource control for wireless systems—Large deviation theory, stochastic Lyapunov drift, and distributed stochastic learning. IEEE Trans Inf Theory 58(3):1677–1701
5. Huang L, Moeller S, Neely MJ, Krishnamachari B (2013) LIFO-Backpressure achieves near optimal utility-delay tradeoff. IEEE/ACM Trans Networking 21(3):831–844
6. Moeller S, Sridharan A, Krishnamachari B, Gnawali O (2010) Routing without routes: the backpressure collection protocol. In: Proceedings of IPSN
7. Si W, Starobinski D (2013) On the channel-sensitive delay behavior of lifo-backpressure. In: Proceedings of Allerton, pp 715–722
8. Tassiulas L, Ephremides A (1992) Stability properties of constrained queueing systems and scheduling policies for maximum throughput in multihop radio networks. IEEE Trans Autom Control 37(12):1936–1948
9. Ying L, Shakkotai S, Reddy A (2009)On combining shortest path and back-pressure routing over multihop wireless networks.In: Proceedings of IEEE INFOCOM
10. Ying L, Srikant R, Towsley DF (2008) Cluster-based backpressure routing algorithm. In: Proceedings of IEEE INFOCOM, pp 484–492

Chapter 31
Cache Control Issues in Pub–Sub Networks and Wireless Sensor Networks

Dimitrios Katsaros

31.1 Introduction

Applications that exploit the publish–subscribe (pub–sub) paradigm are organized as a collection of clients which interact by publishing events and by subscribing to the events they are interested in. In a pub–sub network, any message is guaranteed to reach all interested active clients whose subscriptions are known at publish time. However, in a dynamic distributed environment, clients join and leave the network during time, and it is possible that a client joins the network after the publishing of an interesting message. In current pub–sub systems, it is not possible for a new subscriber to retrieve previously published messages that match her subscription. Therefore, enabling the retrieval of previously published content by means of storing is one of the most challenging problems in pub–sub networks. Wireless sensor networks (WSNs) consist of wirelessly interconnected devices that can interact with their environment by controlling and sensing "physical" parameters. Although there is no single realization of a WSN to support all applications, there are some common characteristics of these networks that need to be efficiently addressed in all these applications: (a) lifetime, (b) scalability, and (c) data-centric networking (whereas the target of a conventional communication network is to move bits from one machine to another, the purpose of a sensor network is to provide information and answers). Therefore, techniques of temporary caching of information at various places in the sensor network is a challenging issue that can achieve all three requirements. The caching decisions are strongly dependent on the network topology; therefore, analysis of the topology by discovering which nodes are located in "central" positions of the network can improve the caching algorithms.

D. Katsaros (✉)
University of Thessaly, Volos, Greece
e-mail: dkatsar@inf.uth.gr

J.H. van Schuppen and T. Villa (eds.), *Coordination Control of Distributed Systems*,
Lecture Notes in Control and Information Sciences 456,
DOI 10.1007/978-3-319-10407-2_31

31.2 Motivation

To support archival retrieval in a pub–sub network, data storage servers replicate the whole content of a given server. When a client is interested in the content of that server, his/her request is redirected to one of the existing storages (i.e., the closest one). Since storages serve only a portion of the total requests and are placed closer to the client, clients are served faster. Therefore, the objective function is to minimize client's response latency subject to installing the minimum number of storages.

On the other hand, the vast majority of applications running over WSNs require the optimization of the communication among the sensors so as to serve the requested data in short latency and with minimal energy consumption. The battery lifetime can be extended if the "amount" of communication is reduced, which in turn can be done by caching useful data for each sensor either in its local store or in the near neighborhood. Additionally, caching can be very effective in reducing the need for network-wide transmissions, thus reducing the interference and overcoming the variable channel conditions. The cooperative data caching is an effective and efficient technique to achieve these goals. The fundamental aspect in every cooperative caching schemes for sensor networks is the identification of the nodes which will implement the aspects of the cooperation concerning the caching decisions. Therefore, we need to define nodes that will run control decisions usually without complete knowledge of the state of neighboring nodes, and also to define a cache admission/replacement policy for the contents of each sensor node cache.

31.3 Examples

Several industrial and academic pub–sub systems such as IBM's Gryphon, Siena, Elvin, and REDS develop an overlay event notification service. An event notification service is an infrastructure that facilitates the construction of event-based systems, whereby producers of events publish event notifications to the infrastructure, and consumers of events subscribe with the infrastructure to receive relevant notifications. The two primary services that should be provided by the infrastructure are the determination of which notifications match which subscriptions, and routing notifications from producers to consumers. However, these systems are lacking support of archival retrieval.

In the area of WSNs, the necessity of cooperative caching can be exemplified via the following example scenario. Consider, a sensor network deployed in a modern battlefield, with sensor nodes dispersed in a large area; each sensor node is equipped with a micro-camera that can take a photograph of a very narrow angle around its position. The sensors update the photographs they take (storing a prespecified number of the immediate past images, so as to be able to respond to historic queries), and share (on demand) with each other the new photographs, in order to built a more complete view of the region that is being monitored. The sharing is necessary because every micro-camera can capture a limited view of the whole region, either due to the sensor node's position or because of the obstacles that exist nearby the sensor node. Therefore, every sensor may request and receive a large number of photographs taken by some other sensor(s) through multihop communication. Afterward, each sensor is able to respond to queries about "high-level" events, e.g., enemy presence. Apparently, sensor battery recharging might be infeasible or difficult due to the limited access to the field. Also, the location of the sensors has not been decided by some placement algorithm (the sensors were dropped by an unmanned aircraft), and the communication is strictly multihop.

31.4 Problem Formulation

Even though the caching problems in pub–sub and WSNs are not identical, they bear many similarities, and thus, we will only provide the formulation of the online cooperative caching problem in WSNs. This is a control problem combining the cache admission and the cache replacement control policies, which continuously try to optimize the cache contents in a way that optimizes a performance measure, e.g., the percentage of requests serviced by each cache.

In the online cooperative caching problem, there are several goals that need to be optimized, such as energy consumption and access latency. These goals are often conflicting. Therefore, it is unfeasible to formulate the online cooperative caching problem using a single equation that would encompass all these factors. We express it here as an optimization problem with the goal of optimizing one of these metrics only, i.e., access latency. Thus, we provide the following formulation for the cooperative caching problem.

Given an ad hoc network of sensor nodes $G(V, E)$ with p equisized data items $D_1, D_2, \ldots, D_p$, where data item D_j can be served by a sensor SN_j, a sensor may act as a server of multiple data items. Each sensor SN_i has a capacity of m_i units of storage, e.g., bits. We use a_{ij} to denote the access frequency with which a sensor SN_i requests the data item D_j and d_{il} to denote the distance (in hops) between sensors i and l. The *cooperative caching problem* is an online problem, with the goal being the selection of a set of sets $M = \{M_1, M_2, \ldots, M_p\}$, where M_j is a set of sensors that store a copy of D_j, to minimize the total access cost:

$$\tau(G, M) = \sum_{i \in V} \sum_{j=1}^{p} a_{ij} \times min_{l \in (\{SN_j\} \cup M_j)} d_{il} \tag{31.1}$$

and fulfilling the memory capacity constraint that:

$$|\{M_j | i \in M_j\}| \leq m_i \qquad \text{for all } i \in V,$$

which means that each network node SN_i appears in at most m_i sets of M.

31.5 Theory and Concepts

Since earlier work [4] suggested that the smart selection of the so-called "mediator" nodes is a crucial factor in addressing energy and latency considerations, one of the aims was to design centrality measures [8] to help us in the selection of the mediator nodes which will be robust and easy to compute (without the need of complex calculations or many rounds of message exchanges). The "central" nodes are able to control the communication among others: for instance, (a) in routing protocols for sensor networks, such nodes can be selected to forward the packets because, due to their position, they will succeed in reducing the routing latency, (b) in disconnection-tolerant mobile sensor networks, such nodes can be selected as data mules to carry messages, until they find the chance to pass these messages to sensors which are closer to the packets' final destination, and so no. Therefore, the significance of such central sensors varies depending on the application and the protocol, and thus, we use the word "influence" to depict the ability of the central nodes to affect (usually for optimization purposes) the communication among other sensors.

31.6 Overview of Research Contributions

The sensor degree, i.e., the number of its 1-hop neighbors, has been used as a measure of centrality. Looking at Fig. 31.1, we see that the nodes 3, 4, 7, 6 are equally central with respect to their degree. If we compute the betweenness centrality for each sensor—the percentage of shortest paths among all pairs of sensors that pass via this sensor—in the whole graph, then node 7 is the most "central," followed by nodes 3, 4 and then comes node 6. This is somehow counter-intuitive, since node 6 has *all network nodes* at its vicinity.

Starting from this observation, we propose a new centrality measure, called the μ-*Power Community Index*, defined as follows [12]:

Definition 31.1 The μ-*Power Community Index* of a sensor v is equal to k, such that there are up to $\mu \times k$ sensors in the μ-hop neighborhood of v with degree greater than or equal to k, and the rest of the sensors with in that neighborhood have a degree less than or equal to k.

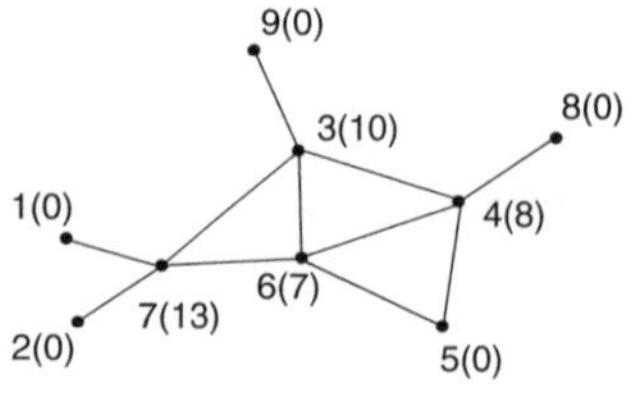

Fig. 31.1 Betweenness centrality (the numbers in parentheses) for a graph

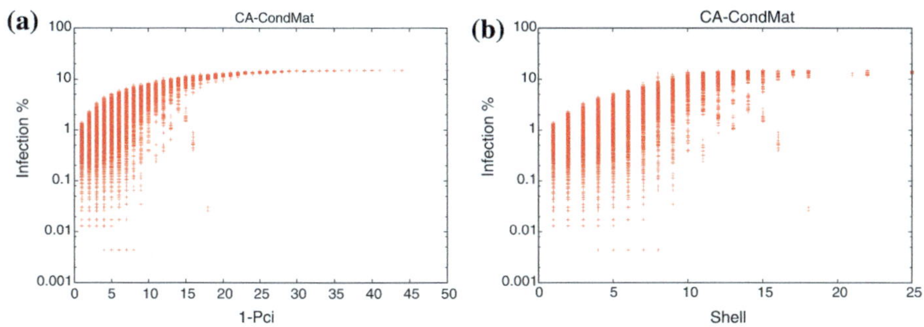

Fig. 31.2 Spreading capability of nodes in the ca-CondMat network with a single original spreader according to **a** 1-PCI and **b** k-shell index. There are nodes with high k-shell indices, some of which infect a large portion of the network, as well as nodes with the same k-shell index (16) that infect a significantly smaller part of the network. On the other hand, only nodes with very small 1-PCI exhibit such behavior

The calculation of this measure is completely local involving only communication among neighboring nodes without knowledge of the complete network topology. Having defined such "controller" sensors, the task of providing a solution to the online cooperative caching problem can be done along the lines proposed in [4] and [5].

The μ*-Power Community Index* has been used also to address the problem of influential spreader identification in complex networks [2], which is yet another network control problem that aims at finding the nodes in complex networks that can spread a message rapidly among other nodes or finding nodes that can control the rest of the nodes. So far, the k-shell decomposition was the champion method; if from a given graph, we recursively delete all vertices and edges incident with them of degree less than k, the remaining graph is the k-shell.

Figure 31.2 shows all nodes' spreading capability according to their 1-PCIs and k-shell indices for the ca-CondMat collaboration networks from the e-print arXiv covering condensed matter physics. The 1-PCI method results in a more monotonic distribution than k-shell decomposition, providing a clearer ranking of spreading capabilities. It converges to an approximately straight line, where maximum influence lies, more steeply than the k-shell method in all the studied cases. Choosing a spreader with, say, 1-PCI $>$ 23 will yield the maximum influence, whereas choosing one from the core or from the high shells might not be optimal because in some cases nodes within the same shell have different spreading capabilities.

31.7 Further Reading

Relevant research on the use of social network analysis for improving the performance of networks includes the analysis of time-varying networks [1, 16], the detection of communities [7, 13, 15], the proposal of new centrality measures

[6, 11, 14, 17], the discovery of influential spreaders [2], and the social-based routing [3, 9, 10, 18].

References

1. Acer UG, Drineas P, Abouzeid A (2010) Random walks in time-graphs. In: Proceedings of MOBIOPP, pp 93–100
2. Basaras P, Katsaros D, Tassiulas L (2013) Detecting influential spreaders in complex, dynamic networks. IEEE Comput mag 46(4):26–31
3. Daly EM, Haahr M (2007) Social network analysis for routing in disconnected delay-tolerant MANETs. In: Proceedings of ACM MOBIHOC, pp 32–40
4. Dimokas N, Katsaros D, Manolopoulos Y (2008) Cooperative caching in wireless multimedia sensor networks. ACM Mob Netw Appl 13(3–4):337–356
5. Dimokas N, Katsaros D, Tassiulas L, Manolopoulos Y (2011) High performance, low complexity cooperative caching for wireless sensor networks. ACM Wireless Netw 17(3):717–737
6. Dolev S, Elovici Y, Puzis R (2010) Routing betweenness centrality. J ACM 57
7. Fortunato S (2010) Community detection in graphs. Phys Rep 486(3–5):75–174
8. Freeman LC (1977) A set of measures of centrality based on betweenness. Sociometry 40(1):35–41
9. Gao W, Li Q, Zhao B, Cao G (2009) Multicasting in delay tolerant networks: a social network perspective. In: Proceedings of ACM MOBIHOC, pp 299–308
10. Hui P, Crowcroft J, Yoneki E (2008) Bubble rap: social-based forwarding in delay tolerant networks. In: Proceedings of ACM MOBIHOC, pp 241–250
11. Hwang W, Kim T, Ramanathan M, Zhang A (2008) Bridging centrality: graph mining from element level to group level. In: Proceedings of ACM SIGKDD, pp 336–344
12. Katsaros D, Dimokas N, Tassiulas L (2010) Social network analysis concepts in the design of wireless ad hoc network protocols. IEEE Netw mag 24(6):23–29
13. Li F, Wu J (2009) LocalCom: a community-based epidemic forwarding scheme in disruption-tolerant networks. In: Proceedings of IEEE SECON
14. Nanda S, Kotz D (2008) Localized bridging centrality for distributed network analysis. In: Proceedings of 17th international conference on IEEE ICCCN, pp 1–6
15. Newman NEJ (2006) Modularity and community structure in networks. Proc National Acad Sci 103(23):8577–8582
16. Pallis G, Katsaros D, Dikaiakos MD, Loulloudes N, Tassiulas L (2009) On the structure and evolution of vehicular networks. In: Proceedings of IEEE MASCOTS, pp 502–511
17. Perra N, Fortunato S (2008) Spectral centrality measures in complex networks. Phys Rev E 78(036107):036107-1-03107-10
18. Pujol JM, Toledo AL, Rodriguez P (2009) Fair routing in delay tolerant networks. In: Proceedings of IEEE INFOCOM, pp 837–845

Chapter 32
Detecting Influential Nodes in Complex Networks with Range Probabilistic Control Centrality

Dimitrios Katsaros and Pavlos Basaras

32.1 Introduction and Motivation

Real-world entities often interconnect with each other through explicit or implicit relationships, by transient and continuous ways to form a complex network. Such networks are studied in many fields of science like bioinformatics, statistical mechanics, sociology, and computer science [1]. Complex networks have provided a wealth of evidence for their ability to disseminate information rapidly among other node users [2]. Rumors and fashion, but also social unrest or the spreading of infectious diseases among those people networks, highlight the need for identifying those entities to either boost or hinder spreading. A great deal of research into the structure of complex systems has focused on trying to identify such entities in an attempt to efficiently control complex systems. For the identification of such entities, traditional centrality measures have been proposed such as the shortest-path betweenness centrality or spectral centrality measures, e.g., PageRank [3]. More sophisticated methods for the detection of influentials are reported in [4, 5]

The common characteristic of the aforementioned efforts is that they all deal with static complex networks, i.e., they apply for a specific instance of the network's lifetime; at that specific instance, a link between a specific pair of nodes either exists or not. However, many of the real-world complex networks are continuously evolving, and their links rapidly appear and disappear. Examples of such complex systems are vehicular ad hoc networks [6] whose links live for only a few seconds and are usually characterized by a link quality parameter, ranging from a zero value indicating an absent link, to a value of one, indicating a perfect communication link. Moreover,

D. Katsaros (✉) · P. Basaras
Electrical and Computer Engineering Department,
University of Thessaly, Volos, Greece
e-mail: dkatsar@inf.uth.gr

P. Basaras
e-mail: pabasara@uth.gr

J.H. van Schuppen and T. Villa (eds.), *Coordination Control of Distributed Systems*,
Lecture Notes in Control and Information Sciences 456,
DOI 10.1007/978-3-319-10407-2_32

many evolving complex networks are examined from an 'aggregated' perspective, associating to each link the percentage of time that this link existed.

The study of influentials in complex networks with probabilistic links is a challenging, new task, because apart from the 'neighborhoods' that an influential can exert influence, we should also take into account the 'strength' of the links. We could resort to the old ideas finding stochastic shortest paths and computing analogous betweenness centralities, but these centralities have already been shown that they do not perform well for static networks either [5].

In this article, we develop a semi-local centrality measure for dynamic, complex networks with probabilistic links, the *range probabilistic control centrality* (RPCC), which considers both the 'strength' of links emanating from each node, and it additionally estimates the influence region of the node based on ideas from the literature of control theory. In the absence of relevant methods, the proposed centrality measure is compared against a baseline method, namely the localized weighted-degree centrality [7], for a couple of networks with various distributions for the probabilities of the links.

32.2 Utility Examples

Consider the vehicular ad hoc network (VANET) where the existence and quality of connections between vehicles (e.g., time of active connection, signal strength) is a factor of several parameters such as the vehicle's direction, acceleration–deceleration of vehicles, the underlying road network topology, possible obstacles or interference, and so on. The aggregate effect of these factors results in having a temporal network. A vital operation in a VANET is that of locating the nodes that can disseminate a safety message to as many vehicles as possible within the whole network or focused parts of it, e.g., safety geocasting messages.

Apart from these ad hoc communication networks, a wide variety of complex systems in nature, society, and technology can be represented as graphs with entities linked by probabilistic edges. A couple of other examples include a transportation or airline network [8], where schedules of transportations vary or change, and examples of phone, email, or social networks, depicting contacts as entities and the amount of time of their interaction as their links strength [9], and we need to determine the entities that can exert the maximum influence over the network. Earlier works such as betweenness centralities based on stochastic shortest paths suffer from the inability to detect influential spreaders [5]. Recent efforts using positive and negative links [10] are not rich enough to address the present problem.

The present article investigates the issue of detecting influential nodes in temporal networks with probabilistic links and makes the following contributions:

- Investigates the issue of influential spreaders in complex networks with probabilistic links.

- Extends the concept of control centrality [11] and proposes an adjustable centrality measure, the range probabilistic control centrality (RPCC), based on control theory, to help identify such nodes.
- Evaluates this centrality measure across a range of complex networks and distributions of probabilities over the links, and compares it with a baseline method, namely weighted-degree centrality [7].

32.3 Range Probabilistic Control Centrality

The concept of *control centrality* was introduced in [11] based on the work of [12]. Their motivation was to detect the nodes of the network that can control a directed network, i.e., to drive, based on specific inputs, the 'controlled' nodes to the state required by the control goal. They described the notion of a *stem*, which is a directed path starting from an initial node, such that no nodes appear more than once along the path, e.g., $j \rightarrow k \rightarrow l \rightarrow m$. A *stem-cycle disjoint subgraph* of G is the subgraph of G consisting of stems and cycles with no common nodes. The control centrality of a node is defined as the largest number of edges among all possible stem-cycle disjoint subgraphs.

Whereas their definition of centrality is very interesting from a control-theoretic perspective, our needs for addressing the requirements of all the aforementioned application fields demand two major reconsiderations. The first one concerns the fact that our *links are probabilistic*, and this must be incorporated in the definition of a control-theoretic type of centrality. Additionally, it does not make sense, for a VANET for instance, to demand from a single vehicle to be able to 'control' the whole ad hoc network; we need to redefine the centrality measure in a way that it can be *defined for both the entire network, and for neighborhoods around each node*.

Following on these requirements, we define the generic concept of *stem significance* (ssf), as the product of two scalar terms:

$$: \mathrm{sf} = \mathrm{sizeOfStem}\ \ x\ \ \mathrm{weight\,OfStem} \tag{32.1}$$

where sizeOfStem is the number of edges of the stem and weightOfStem is the product of its weights.

Based on this, we build two approaches for defining centrality measures over probabilistic graphs for range-limited neighborhoods. In the first approach, we adjust appropriately the ideas of [11], but in the second approach, we depart significantly from them and rely on the graph-theoretic concept of the influence range of a node, which is defined as the set of nodes reachable from a specific node.

32.3.1 RPCC with Cycle Extraction ($RPCC_{CE}$)

In our fist attempt to identify the most influential users following the idea of stem-cycle disjoint subgraphs, we denote the *cycle significance, csf,* in a similar way as *ssf*:

$$\text{csf} = \text{cyclePointer} * \text{weightOfCycle} * (\text{sizeOfCycle} + 1) \tag{32.2}$$

where cyclePointer is the weight of the edge through which we visit a node of the cycle, weightOfCycle is the product of the weights of the edges that form it and sizeOfCycle is the number of its edges.

We compute the k-RPCC of a node i as the sum of the significances of the disjoint stems and cycles within the k-specified range. The pseudocode for the algorithm is as follows:

Step 1: Remove all incoming links of node i.
Step 2: Perform the Cycle Extraction procedure.
Step 3: Calculate and sum: cycle significances.
Step 4: Calculate and sum: stem significances of the remaining graph.
Step 5: Sum results of Steps 4 and 5.

For $n = 1$, this method is identical to the weighted-degree centrality. For $n = 2$, we exclude Steps 2 and 3, since there can be no cycles within such range. For $n \geq 3$, the computation is as given. The procedure *Cycle Extraction* is described in the next paragraphs.

32.3.1.1 Cycle Extraction

In each step, one cycle is removed from the graph. The first identified cycle becomes a candidate for extraction. Cycle Weight is the average sum of the weights of the cycle and Cycle Size i the number of edges that form it. When multiple cycles exist, the criteria to change candidates are as follows:

1. CycleWeight > CandCycleWeight and CycleSize/CandCycleSize > 0.7
2. CycleSize > CandCycleSize and CycleWeight/CandCycleWeight > 0.7.

CandCycleWeight and CandCycleSize denote the characteristics of the previous cycle candidate, and CycleWeight and CycleSize are the characteristics of the newly found one. The first criterion is to prevent small-sized cycles with high weights to be chosen over larger ones with high-quality links, due to their small number of edges. We use the second criterion to account for cases where the significance of a newly found cycle might be lower than that of the candidates, but if its number of edges is greater, and their significances are not far off (e.g., are more than 70 % equal), then the new cycle may be a better choice. If at least one of the above criteria is true, then the newly found cycle becomes the candidate. Finally, the candidate is removed and the process is repeated until there are no cycles in the graph. The choice 0.7

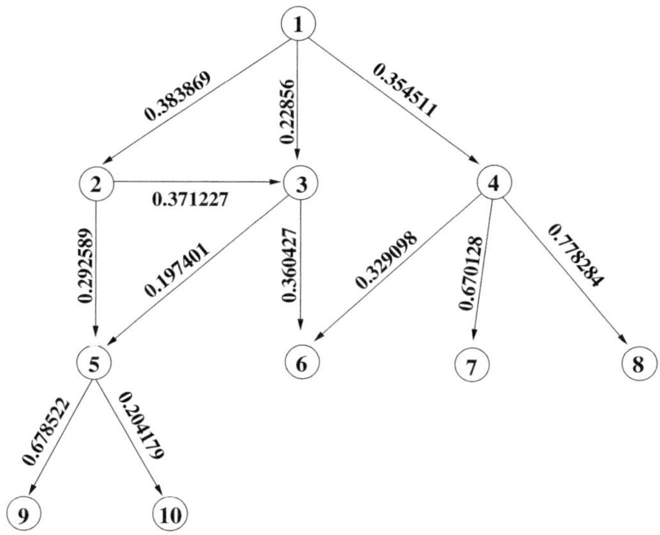

Fig. 32.1 A small complex network with probabilistic links

of the relative importance might seem arbitrary, but it does not have a significant algorithmic impact, as long as is larger than 0.4.

Figure 32.1 illustrates a small graph with probabilities on edges. The weighted degree of node 4 indicates that this node is the most influential one, however as illustrated, through 4 only three nodes are potentially accessible. From our point of view, node 1 becomes the better choice. Its RPCC value is equal to 2.02493, whereas node 1's RPCC value equals to 1.77751.

32.3.2 RPCC Without Cycle Extraction ($RPCC_C$)

In our second approach to calculate the importance of a node, we use only the sum of the significances of the stems within a specified range, leaving the cycles of the graph intact. The calculation of RPCC for each node is as previously, but now without Steps 2 and 3.

With this approach, we test the quality of paths through which a node i sees the rest of the network within the specified range. Since a path may be accessed by more than one nodes (e.g., $i \rightarrow j \rightarrow k \rightarrow l \rightarrow m$ and $i \rightarrow t \rightarrow k \rightarrow l \rightarrow m$), this approach also takes into account with how many ways a certain portion of the network can be controlled by i.

This approach targets the elimination of the burden of cycle calculation that can become significant in large networks and when k becomes relatively large. In principle, it does not differentiate significantly the performance of the method with respect to the previous method.

32.4 Simulation Parameters and Experimentation

For evaluation purposes, we had to select appropriate competitor methods, use networks with probabilistic edges, and also propagation models. As already mentioned, in the absence of competitors designed specifically for our problem, we used the weighted degree [7]. It is a straightforward generalization of the traditional unweighted degree as used in [5] for the evaluation of the spreading capabilities of a node in complex networks. Also, despite the wealth of real datasets that concern complex networks, it is hard to find appropriate input networks with probabilistic links. Therefore, we had to resort to the solution of using real complex networks and annotate their links with probabilities drawn from various distributions (uniform, zipfian, exponential, gaussian). Specifically, in the present article, we present results from a social network, namely *Wiki-Vote* which is part of the Stanford Network Analysis Platform [13]. As far as the propagation model is concerned, there is a wealth of such models in the literature, and it is worth examining the performance of the methods for each one of them. In this article, we confined ourselves to the SIR model with the characteristic that an infection originates from a single spreader, which is quite popular and has been used in similar studies [5]. We use relatively small values of infection probability to highlight the importance of influential spreaders.

The proposed centrality methods k-RPCC can be calculated for regions around the node of interest, and the whole network, as well. We experimented for values of $k = 2$, $k = 3$, and $k = 5$, where k is the distance in hops. Similar to [5, 14], we used the average size of the network's infected area as the performance measure.

For the experiment presented here due to lack of space, the probabilities of the edges are assigned based on uniform distribution and $k = 2$ for the (RPCC_C) approach. The probabilities range from 0.1 to 1. As said, these values depict the probability of an edge to be active on the graph. Links with values close to 1 are mostly active in our inspection time, whereas values near 0.1 are mostly inactive. According to these probabilities, we take 10 snapshots of the input graph resulting in 10 temporal graphs. To obtain statistically unbiased results, we repeated the computation 1000 times for each vertex in every temporal graph, i.e., 10,000 spreading processes.

32.5 Evaluation and Overview of Research Contributions

Figure 32.2 illustrates the results of the comparison of RPCC_C versus weighted degree for $k = 2$. The y-axis corresponds to the portion of the temporal network that got infected in percentage, and the x-axis depicts the values of the respective centrality measure. An ideal performance curve would be a very 'slim' one; in this curve, a very small number of infection percentages (values at y-axis) correspond to the same centrality value (value at x-axis). This would mean that the centrality measure would be able to divide the nodes on non-overlapping classes based on the

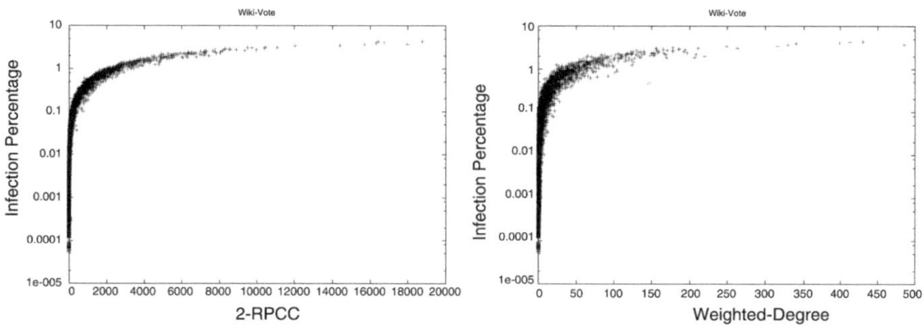

Fig. 32.2 2-RPCC$_C$ versus weighted degree

network percentage that can infect. For our competitor, there is no such observation since nodes with value of 100 are equal spreaders to those of 200, as depicted in (b). For a fixed k-RPCC value, there is a small deviation in the spreading capabilities, converging to a thinner curve, whereas for the weighted-degree, the deviation is much wider. Overall, the performance curve of the proposed method is much closer to the ideal one, than the competitor's curve.

In general, we expect that the network topologies and link probability distributions will affect the algorithms' performance, but for any influential spreader detection algorithm in order to be characterized as an efficient one, it is important that the algorithm has a steep ascending curve, which is 'thin,' especially as we move to larger values along the x-axis.

The study of complex, dynamic networks with probabilistic links arises naturally in some application fields, such as vehicular ad hoc networks, and aggregated descriptions of evolving complex networks. The identification of influential nodes in such networks is a new and interesting topic of investigation. This article takes a first step toward exploring this area and develops a measure of significance for the nodes of such complex network that quantifies whether each node is the starting point of 'strong' (i.e., almost permanent) paths that subsequently can 'control' the rest of the nodes. For the future, it is interesting to investigate the RPCC from a control theory perspective, instead of a pure engineering aspect, as it was done in the present article.

References

1. Newman MEJ, (2010) Networks: aN iNtroductioN. Oxford UNiversity Press, Cambridge
2. Doerr B, Fouz M, Friedrich T, (2011) Social networks spread rumors in sublogarithmic time, In Proceedings of ACM STOC, 2011, pp 21–30
3. Langville A, Meyer C (2006) Google's PageRank and beyond: the science of search engine rankings. Princeton University Press, New Jersey
4. Basaras P, Katsaros D, Tassiulas L (2013) Detecting influential spreaders in complex, dynamic networks. IEEE Comput Mag 46(4):26–31

5. Kitsak M, Gallos LK, Havlin S, Liljeros F, Muchnik L, Stanley HE, Makse HA (2010) Identification of influential spreaders in complex networks. Nat Phys 6:888–893
6. Pallis G, Katsaros D, Dikaiakos MD, Loulloudes N, Tassiulas L (2009) On the structure and evolution of vehicular networks. In: Proceedings of IEEE/ACM MASCOTS, pp 502–511
7. Fountalis I, Bracco A, Dovrolis C (2013) Spatio-temporal network analysis for studying climate patterns. Climate Dynamics, 2013, (to appear)
8. Barrat A, Barthelemy M, Pastor-Satorras R, Vespignani A (2004) The architecture of complex weighted networks. Proc Nat Acad Sci 101(11):3747–3752
9. Lambiotte R, Blondel VD, de Kerchove C, Huens E, Prieur C, Smoreda Z, Van Dooren P (2008) Geographical dispersal of mobile communication networks. Physica, 1(1)
10. Li Y, Chen W, Wang Y, Zhang Z.-L (2013) Influence diffusion dynamics and influence maximization in social networks with friend and foe relationships. In Proceedings of ACM WSDM, 2013
11. Y.-Y. Liu, J.-J. Slotine, and A.-L. Barabasi (2012) Control centrality and hierarchical structure in complex networks. PLOS One, 7(9)
12. Hosoe S (1980) Determination of generic dimensions of controllable subspaces and its applications. IEEE Trans Autom Control 25(6):1192–1196
13. The Stanford Network Analysis Project, Available at http://snap.stanford.edu/data
14. Pei S, Muchnik L, Andrade Jr J, Zheng JSZ, Makse HA (2014) Searching for superspreaders of information in real-world social media, 2014, Available at http://arxiv.org/abs/1405.1790

Chapter 33
Model Predictive Controllers over Differentiated Services Packet Networks

Riccardo Muradore, Davide Quaglia and Paolo Fiorini

33.1 Problem Statement and Motivation

In this work, we theoretically analyze the design of a MPC over a differentiated services network which provides two different "virtual wires" featuring different packet loss rates (Fig. 33.1) or different constant delays (Fig. 33.2). The MPC approach is used to choose at the same time the command value and the service class according to the predicted state of the plant and the knowledge of network condition. When the plant output is far from the desired reference and the network condition is bad, then a more guaranteed transmission is needed: This means that the high-quality virtual wire is used which guarantees a lower packet loss rate or communication delay. The DiffServ mechanism assumes a parsimonious use of the high-quality service class since its guarantee is obtained at the cost of communication resources taken from the low-quality class. For this reason, the high-quality transmission "costs" more than the low-quality one and so it should be used only when it is strictly needed. On the other hand, when the output is close to the reference or the network condition is good, the controller should avoid the waste of network resources if the degradation featured by the low-quality virtual wire still guarantees an acceptable control performance.

The plant $P(z)$ is assumed to be a linear time-invariant (LTI) discrete-time full-information system

$$P(z) : \begin{cases} x_{k+1} &= Ax_k + Bu_k \\ y_k &= x_k + n_k \end{cases} \tag{33.1}$$

R. Muradore (✉) · D. Quaglia · P. Fiorini
Department of Computer Science, University of Verona, Strada Le Grazie 15, Ca' Vignal 2, 37134 Verona, Italy
e-mail: riccardo.muradore@univr.it

D. Quaglia
e-mail: davide.quaglia@univr.it

P. Fiorini
e-mail: paolo.fiorini@univr.it

J.H. van Schuppen and T. Villa (eds.), *Coordination Control of Distributed Systems*, Lecture Notes in Control and Information Sciences 456,
DOI 10.1007/978-3-319-10407-2_33

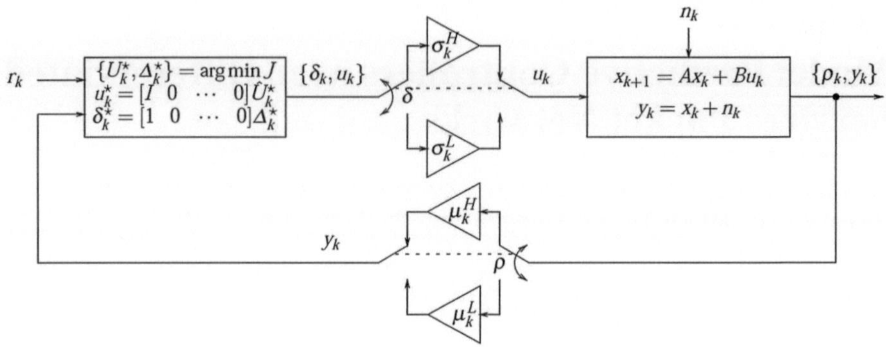

Fig. 33.1 Block diagram of a NCS over differentiated services lossy network

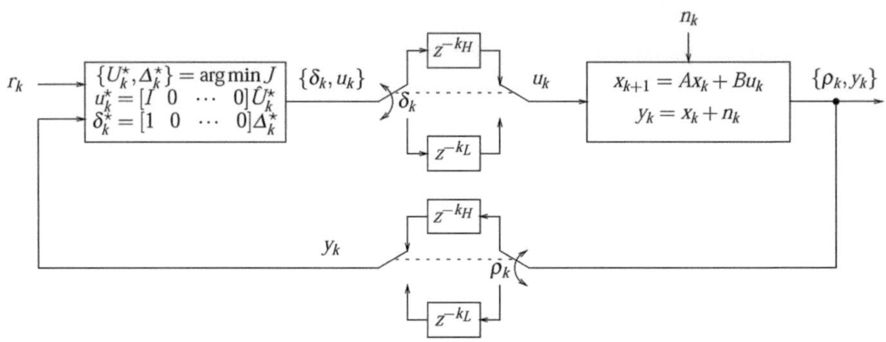

Fig. 33.2 Block diagram of a NCS over delay-based differentiated services network

where x_k is the vector state, u_k is the command, and n_k is the measurement noise. The controller $C(z)$ is located on the other side of a packet-based communication network. The commands u_k and measurements y_k may be affected by packet loss or delay which may compromise NCS stability and performance. The differentiated services approach is based on the selection of the service class on a packet-to-packet basis, [2, 4–7]. As already stated, we analyze two different network scenarios:

- **Lossy Networks**: the DiffServ architecture can choose between two channels, H and L, characterized by different loss probabilities. Let δ_k be the binary variable representing the marker strategy for the command u_k, whereas ρ_k is the binary variable for the marker strategy of the measurement. When $\delta_k = 1$, the channel behavior at time k is described by the binary Bernoulli variable σ_k^H (i.e., high-priority service); otherwise, σ_k^L is considered (i.e., low-priority service). When the Bernoulli variable is equal to 1, the packet reaches the other side of the network; otherwise, it gets lost. The same holds for the plant-to-controller channel. Mathematically, the probabilities of successful reception are

$$P\{\sigma_k^L = 1\} = \bar{\sigma}^L, \quad P\{\sigma_k^H = 1\} = \bar{\sigma}^H$$
$$P\{\mu_k^L = 1\} = \bar{\mu}^L, \quad P\{\mu_k^H = 1\} = \bar{\mu}^H$$

where the constants $\bar{\sigma}^L, \bar{\sigma}^H$ refer to the virtual wires from the controller-to-plant path, and $\bar{\mu}^L, \bar{\mu}^H$ refer to the virtual wires from the plant-to-controller path. Since the H service is more reliable than service L, we have $\bar{\sigma}^H > \bar{\sigma}^L$ and $\bar{\mu}^H > \bar{\mu}^L$. The effect of packet loss on the transmitted data can be modeled by multiplying the packets by binary independent and identically distributed Bernoulli variables, [8, 10]. Therefore, the state equation of the model can be rewritten as

$$x_{k+1} = Ax_k + \left[(1 - \delta_k)\sigma_k^L + \delta_k \sigma_k^H\right] Bu_k, \tag{33.2}$$

whereas the measurement equation will be

$$y_k = \left[(1 - \rho_k)\mu_k^L + \rho_k \mu_k^H\right] [x_k + n_k] \tag{33.3}$$

where n_k is the measurement noise.

- **Delay-based Networks**: the DiffServ architecture can still choose between two channels, H and L characterized, in this scenario, by different constant delays. Let δ_k (ρ_k) be the binary variable representing the marker strategy for the command u_k (measurement y_k). When $\delta_k = 1$ (or $\rho_k = 1$), the channel behavior at time k is described by a constant delay equal to $k_H T_s$ (i.e., high-priority service); otherwise, $k_L T_s$ is considered (i.e., low-priority service), where $k_H, k_L \in \mathbb{N}$. In this network scenario, we have $k_H < k_L$, i.e., a smaller delay. The state equation of the model and the measurement equation take the form

$$x_{k+1} = Ax_k + (1 - \delta_k)Bu_{k-k_L} + \delta_k Bu_{k-k_H}, \tag{33.4}$$
$$y_k = (1 - \rho_k)x_{k-k_L} + \rho_k x_{k-k_H} + n_k. \tag{33.5}$$

To simplify the analysis, in this work, we make the following assumptions:

(A1) we deal with two transmission priorities;
(A2) the marking strategy is limited to the controller-to-plant path, while the plant-to-controller path is assumed to be reliable, i.e., $y_k = x_k + n_k$.

Remark 33.1 Even though the state Eqs. (33.2) and (33.4) look very similar, they are different in nature. The equation for the lossy scenario is stochastic because the packet loss rate is modeled by two Bernoulli random variables. This means that the MPC problem formulations are slightly different. In particular, we need to introduce the expectation operator within the performance index in the lossy network scenario.

This work aims at jointly designing the optimal controller based on MPC techniques and the DiffServ marking strategy. The performance index defined in standard MPC methods [3, 9] has now to be modified by:

1. penalizing the use of the high-quality service,
2. inserting the conditional expectation based on the present and past states.

Let MPC-QoS be the name of the following problems summarizing the above considerations:

Problem 33.1 *(Model Predictive Control problem over lossy networks)* Given the system (33.2), find the optimal control $u_k^\star$ and the optimal transmission strategy $\delta_k^\star$ at time k by solving the stochastic MPC-QoS problem

$$\begin{cases} \{u_k^\star, \delta_k^\star\} = \arg\min \quad J_{\text{MPC-QoS}}(k) = \mathbf{E}_{x_k} \sum_{i=0}^{N} \left[\|\hat{x}_{k+i|k}\|^2_{Q_i} + \|\hat{u}_{k+i|k}\|^2_{R_i} + \|\delta_{k+i}\|^2_{W_i} \right] \\ \qquad \text{subject to } m_u \leq \hat{u}_{k+i|k} \leq M_u \\ \qquad m_x \leq \hat{x}_{k+i|k} \leq M_x \\ \qquad \delta_k \in \{0, 1\} \\ \qquad \sigma_k^L, \sigma_k^H \text{ i.i.d. Bernoulli.} \end{cases}$$

Problem 33.2 *(Model Predictive Control problem over delayed-based networks)* Given the system (33.4), find the optimal control $u_k^\star$ and the optimal transmission strategy $\delta_k^\star$ at time k by solving the deterministic MPC-QoS problem

$$\begin{cases} \{u_k^\star, \delta_k^\star\} = \arg\min \quad J_{\text{MPC-QoS}}(k) = \sum_{i=0}^{N} \left[\|\hat{x}_{k+i|k}\|^2_{Q_i} + \|\hat{u}_{k+i|k}\|^2_{R_i} + \|\delta_{k+i}\|^2_{W_i} \right] \\ \qquad \text{subject to } m_u \leq \hat{u}_{k+i|k} \leq M_u \\ \qquad m_x \leq \hat{x}_{k+i|k} \leq M_x \\ \qquad \delta_k \in \{0, 1\} \end{cases}$$

33.2 Theory

The solutions of the deterministic/stochastic MPC-QoS problems are obtained by rewriting the minimization problem as a programming problem where the state equation and the constraints are reformulated as mixed logical dynamical (MLD) systems, [1].

33.2.1 Solution of Problem 33.1

By defining the auxiliary variable $a_k = \delta_k u_k$, the state equation takes the form

$$x_{k+1} = \text{A}x_k + \sigma_k^L \text{B}u_k + \left[-\sigma_k^L + \sigma_k^H \right] \text{B}a_k.$$

The decision variable "disappears," and an auxiliary variable shows up. The two state equations are equivalent if and only if a set of inequalities is satisfied, [1]. Collecting the terms $\hat{x}_{k+i|k}$, $\hat{u}_{k+i|k}$, $\hat{u}_{k+i|k}\sigma^L_{k+i}$, $\delta_{k+i|k}$, $\hat{a}_{k+i|k}$, $\hat{a}_{k+i|k}(-\sigma^L_{k+i}+\sigma^H_{k+i})$ for $i=0,\ldots,N$ in the vectors $\hat{X}(k)$, $\hat{U}(k)$, $\hat{U}_\sigma(k)$, $\hat{\Delta}(k)$, $\hat{A}(k)$, $\hat{A}_\sigma(k)$, and defining the other matrices accordingly, the matrix notations of the state equation, of the index, and of the constraints are

$$\hat{X}(k) = \mathscr{A}x_k + \mathscr{B}\hat{U}_\sigma(k) + \mathscr{B}\hat{A}_\sigma(k), \tag{33.6}$$

$$J_{\text{MPC-QoS}} = \hat{X}(k)^T\mathscr{Q}\hat{X}(k) + \hat{U}^T(k)\mathscr{R}\hat{U}(k) + \Delta^T(k)\mathscr{W}\Delta(k), \tag{33.7}$$

$$\mathscr{E}_u\hat{U}(k) + \mathscr{E}_a\hat{A}(k) + \mathscr{E}_\delta\Delta(k) \le \mathscr{E}, \tag{33.8}$$

with $\mathscr{Q}=\text{diag}\{Q_i\}_{i=0}^N$, $\mathscr{R}=\text{diag}\{R_i\}_{i=0}^N$, $\mathscr{W}=\text{diag}\{W_i\}_{i=0}^N$. The constrained minimization can then be rewritten as a mixed integer quadratic programming (MIQP) problem by explicitly computing the conditional expectations, see [5] for the details. Due to the binary variables and the constraints on the state and command, it is not possible to solve the above problem in a recursive way. The choice of N should then be a trade-off between computational time and performance.

Theorem 33.1 *The stochastic MPC-QoS* Problem 33.1 *is equivalent to the deterministic MIQP problem*

$$\begin{cases} V^\star(k) = \arg\min_{V(k)} & V^T(k)\mathscr{H}V(k) + \mathscr{P}^T V(k) \\ \qquad\qquad\text{subject to} & \mathscr{C}V(k) \le \mathscr{D} \\ & \Delta_f(k) \in \{0,1\}^{N_u} \end{cases}$$

where $V(k) := \begin{bmatrix}\hat{U}(k) & \hat{A}(k) & \Delta(k)\end{bmatrix}^T$ *and the matrices* $\mathscr{H}$ *and* $\mathscr{P}$ *are obtained by computing the conditional expectation in (33.7). The optimal control and the optimal transmission strategy at time k are* $u^\star(k) = \begin{bmatrix}I & 0 & \cdots & 0\end{bmatrix}\hat{U}^\star(k)$ *and* $\delta^\star(k) = \begin{bmatrix}1 & 0 & \cdots & 0\end{bmatrix}\Delta^\star(k)$.

33.2.2 Solution of Problem 33.2

Let $k_H, k_L \in \mathbb{N}$ be the delays in the controller-to-plant path with $k_H < k_L$. For sake of simplicity, we assume $k_H = 0$, $k_L = 1$. The state Eq. (33.4) can be rewritten as

$$\begin{bmatrix}x_{k+1}\\ u_k\end{bmatrix} = \begin{bmatrix}A & B\\ 0 & 0\end{bmatrix}\begin{bmatrix}x_k\\ u_{k-1}\end{bmatrix} + \begin{bmatrix}-B\\ 0\end{bmatrix}\delta_k u_{k-1} + \begin{bmatrix}B\\ -I\end{bmatrix}\delta_k u_k + \begin{bmatrix}0\\ I\end{bmatrix}u_k. \tag{33.9}$$

By defining the new state vector $\mathbf{x}_k := \begin{bmatrix}x_k\\ u_{k-1}\end{bmatrix}$ and the auxiliary variable $a_k := \delta_k\begin{bmatrix}\mathbf{x}_k\\ u_k\end{bmatrix}$, the model can be rewritten in a more compact way as

$$\mathbf{x}_{k+1} = \underbrace{\begin{bmatrix} A & B \\ 0 & 0 \end{bmatrix}}_{F} \mathbf{x}_k + \underbrace{\begin{bmatrix} 0 & -B & B \\ 0 & 0 & -I \end{bmatrix}}_{G_a} a_k + \underbrace{\begin{bmatrix} 0 \\ I \end{bmatrix}}_{G_u} u_k. \tag{33.10}$$

In the general case with $k_L > k_H \geq 0$, the matrix F and the enhanced state vector in (33.10) take the form

$$F = \left[\begin{array}{cccccc|cccc} A & B & 0 & \cdots & 0 & & 0 & 0 & \cdots & 0 \\ 0 & 0 & I & \cdots & 0 & & 0 & 0 & \cdots & 0 \\ \vdots & \vdots & \vdots & \ddots & \vdots & & 0 & 0 & \cdots & 0 \\ 0 & 0 & 0 & \cdots & I & & 0 & 0 & \cdots & 0 \\ 0 & 0 & 0 & \cdots & 0 & & I & 0 & \cdots & 0 \\ \hline 0 & 0 & 0 & \cdots & 0 & & 0 & I & \cdots & 0 \\ 0 & 0 & 0 & \cdots & 0 & & \vdots & \vdots & \ddots & \vdots \\ 0 & 0 & 0 & \cdots & 0 & & 0 & 0 & \cdots & I \\ 0 & 0 & 0 & \cdots & 0 & & 0 & 0 & \cdots & 0 \end{array}\right], \qquad \mathbf{x}_k = \left[\begin{array}{c} x_k \\ u_{k-k_L} \\ u_{k-k_L+1} \\ \vdots \\ u_{k-k_L+k_H} \\ \hline u_{k-k_L+k_H+1} \\ \vdots \\ u_{k-1} \end{array}\right]$$

where the upper-left block of F models the "common delay," $D = k_H$, whereas the bottom-right block models the "relative delay," $d = k_L - k_H$. The matrices G_a, G_u in (33.10) have to be modified in the same way. It is worth remarking that when a command sent on the low-delay class is received, it overtakes other commands traveling on the un-protected class: More precisely, it overtakes $k_L - k_H$ commands. Since we assume to apply always the newest command received, all the $k_L - k_H$ overtaken commands are discharged.

At the end, we use the MLD formalism and the matrix formulations of the state equation and the performance index to derive a deterministic MIQP problem equivalent to the MPC-QoS problem. We refer the reader to [6] for the mathematical details.

33.3 Applications

33.3.1 Lossy Network Scenario

The MPC-QoS controller is used to steer to zero the state of a discrete-time LTI system with sample time 50 ms, control and prediction horizons both equal to $N = 10$ and with the constraint on the input command to belong to the interval $[-5, 5]$. The measurements are affected by Gaussian white noise with zero mean and variance 0.05, and two network scenarios are compared: $\bar{\sigma}^H = 0.9, \bar{\sigma}^L = 0.5$, and $\bar{\sigma}^H = 0.9, \bar{\sigma}^L = 0.65$. The goal of these simulation experiments is to show that the control strategy adopts the high-priority service when the commands to be sent to the plant are "important," i.e., when the noise is bringing the state far away from zero or the network condition is bad. The plots in Fig. 33.3 show the time series for the two

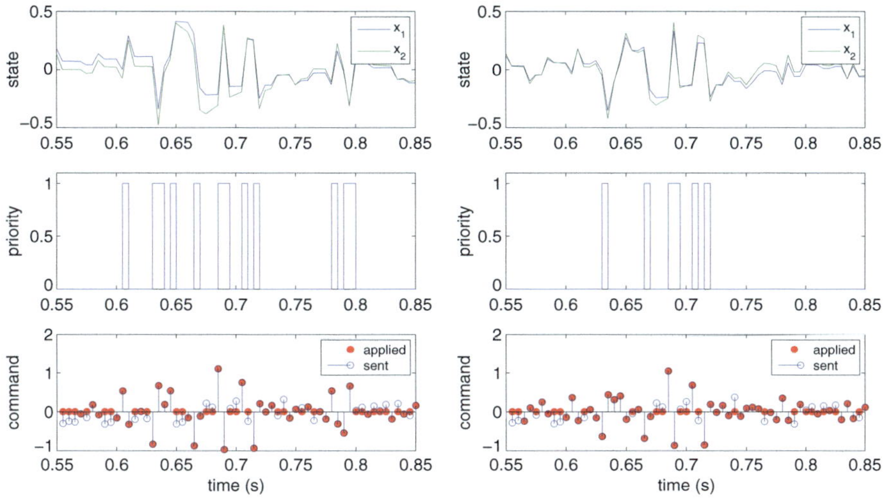

Fig. 33.3 MPC-QoS-based regulation with lossy channels. *Left* $\bar{\sigma}^H = 0.9, \bar{\sigma}^L = 0.5$. *Right* $\bar{\sigma}^H = 0.9, \bar{\sigma}^L = 0.65$

components of the state variable x_k, the optimal marking strategy $\delta_k^\star$ and the sent and applied commands. The command computed by the MPC-QoS controller $u_{k|k}^\star$ is sent to the network through the service H if $\delta_k^\star$ is set to 1, and through the service L if $\delta_k^\star$ is set to 0. The bottom plots in the figure show with empty blue circles the sent commands and with full red circles the applied commands ($\sigma_k^H u_{k|k}^\star$ or $\sigma_k^L u_{k|k}^\star$). If there are no losses ($\sigma_k^H = 1$ or $\sigma_k^L = 1$), these values are exactly the same; otherwise, ($\sigma_k^H = 0$ or $\sigma_k^L = 0$) the applied command is zero. The control architecture uses more the H service in the first case because the packet loss probability of the L service is higher than in the second case.

Remark 33.2 In the above examples, we also took into account measurement noise. This means that the constraints on the state (i.e., $m_x \leq \hat{x}_{k+i|k} \leq M_x$) in Problem 33.1 are no longer *hard* constraints.

33.3.2 Delayed-Based Network Scenario

In this second scenario, we compare the same delay configuration $k_H = 0, k_L = 2$ but with two different weighting matrices $W = 10$ and $W = 20$ to highlight the impact of the cost of the high-quality channel in the choice of the communication strategy (the weighting matrices for the state and the command are kept constant). Moreover, we force the command to stay in the range $[-25, 25]$ and we assume that a disturbance is active in the interval between 0.3 and 035 s.

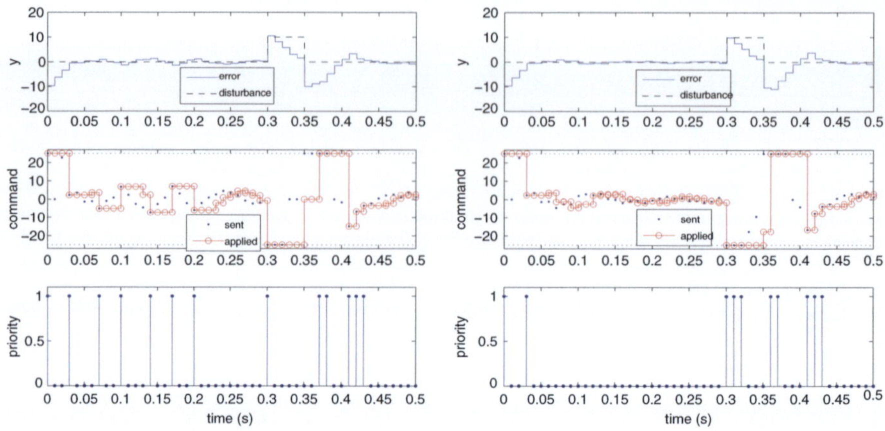

Fig. 33.4 MPC-QoS-based regulation with delayed channels. *Left* $k_H = 0$, $k_L = 2$ and $W = 10$. *Right* $k_H = 0$, $k_L = 2$ and $W = 20$

Figure 33.4 shows that when W increases (i.e., the task is less important), the number of commands sent on the H service class is smaller and located on the most critical interval, i.e., when the disturbance step arrives.

Remark 33.3 Even though not explicitly discussed in the examples above, it is clear that (1) large values for k_L and k_H worsen the performance, (2) the larger the difference $k_L - k_H$, the more likely the controller will select the high-quality channel when the performance is poor.

References

1. Bemporad A, Morari M (1999) Control of systems integrating logic, dynamics, and constraints. Automatica 35:407–428
2. J.C. De Martin and D. Quaglia. Distortion-based packet marking for MPEG video transmission over DiffServ networks. In IEEE International Conference on Multimedia and Expo, 2001.
3. J.M. Maciejowski. Predictive control: with constraints. Pearson education, 2002.
4. R. Muradore, D. Quaglia, and P. Fiorini. Adaptive LQ Control over Differentiated Service Lossy Networks. In Proc. World Congress of the International Federation of Automatic Control (IFAC), August 28 - September 2, 2011.
5. R. Muradore, D. Quaglia, and P. Fiorini. Predictive control of networked control systems over differentiated services lossy networks. In Proceedings of the Conference on Design, Automation and Test in Europe, pages 1245–1250, 2012.
6. R. Muradore, D. Quaglia, and P. Fiorini. Model predictive control over delay-based differentiated services control networks. In Design, Automation & Test in Europe Conference & Exhibition (DATE), 2013, pages 1117–1122, 2013.
7. K. Nichols, V. Jacobson, and L. Zhang. A Two-bit Differentiated Services Architecture for the Internet. RFC 2638, July 1999.
8. A. Papoulis. Probability, random variables, and stochastic processes. McGraw-Hill New York, 1991.

9. Primbs JA, Sung CH (2009) Stochastic receding horizon control of constrained linear systems with state and control multiplicative noise. IEEE Transactions on Automatic Control 54(2):221–230
10. Schenato L, Sinopoli B, Franceschetti M, Poolla K, Sastry SS (2007) Foundations of control and estimation over lossy networks. Proceedings of the IEEE 95(1):163–187

Chapter 34
A SystemC/MATLAB Co-simulation Tool for Networked Control Systems

Davide Quaglia, Riccardo Muradore and Paolo Fiorini

34.1 Motivation and Problem Statement

Networked control systems (NCS) are feedback control systems in which the control loop is closed through a shared digital communication network rather than by an ideal point-to-point connection. In NCS, the communication channel can significantly affect the quality of the control due to communication delay and packet loss. Many solutions have been proposed to address this issue [3]. The simulation of NCS plays a crucial role in the verification, validation, and fine-tuning of these solutions. A simulator should capture and represent both the control and communication aspects. For instance, the control aspects include signal generation and analysis as well as plant/controller specification, whereas the communication aspects include channel and protocol specification as well as packet flow generation and routing.

This essay addresses the problem of *building an accurate simulator for NCSs*. For example, MATLAB/Simulink is one of the most used tools to design and simulate dynamic systems. Concerning the network, this tool provides low-level propagation models, which slow down simulation, and abstract queuing models which do not describe the network architecture in terms of nodes, links, and competing packet flows. There are well-known tools for network simulation [4, 7], but they do not address the simulation of dynamic systems in a native way. A possible solution to handle the heterogeneity of NCSs is the *co-simulation approach* to make different tools working together to simulate different parts of the overall system. The tools

D. Quaglia(✉) · R. Muradore · P. Fiorini
Department of Computer Science, University of Verona, Strada Le Grazie 15, Ca' Vignal 2, 37134 Verona, Italy
e-mail: davide.quaglia@univr.it

R. Muradore
e-mail: riccardo.muradore@univr.it

P. Fiorini
e-mail: paolo.fiorini@univr.it

J.H. van Schuppen and T. Villa (eds.), *Coordination Control of Distributed Systems*, Lecture Notes in Control and Information Sciences 456,
DOI 10.1007/978-3-319-10407-2_34

have to run *simultaneously* (to reduce the simulation time on a multiprocessor system) and in a *synchronized way* (to provide correct results). Therefore, different problems must be solved to achieve this objective.

The first problem to be solved is the *interconnection of the simulation tools*. Assuming that each tool is executed by a specific process in the host operating system, simulation data should be exchanged by using inter-process communications, e.g., shared memory or network sockets. The transfer of simulation data between tools should be efficient and independent of the complexity of the simulated model.

Another problem consists in the introduction of *new modeling entities* in each tool to represent the connection of the standard entities provided by the tool with the other components modeled by the other tool. For instance, in MATLAB/Simulink workspace, new blocks are needed to represent the fact that the controller and the plant are connected together through a component modeled outside MATLAB to simulate the network.

The third issue is the creation of the *same time domain* for the global simulation. This implies that tools should perform simulation in a synchronized way and that cause–effect relationship between events belonging to different tools should be preserved.

To show how these issues can be solved, we refer to an actual co-simulation platform in which MATLAB/Simulink is connected to the SystemC Network Simulation Library (SCNSL) [1, 2]. However, the explained concepts are quite general and they can be adapted to other tools.

34.2 Framework Key Concepts

Regarding the *interconnection of the simulation tools*, network sockets are used to handle data transfer and synchronization between MATLAB and SystemC. As depicted in Fig. 34.1, socket management is separated from system/network modeling, thus making it independent of the complexity of the NCS model. In SystemC simulator, socket communications have been implemented in the simulation kernel, while in MATLAB, they are addressed by a special Simulink block named *MATLAB Wrapper*, developed as Level-2 m-file s-function. The use of sockets, instead of shared memory mechanisms, allows to distribute simulation not only among different CPUs but also among different hosts to enable load balancing or remote on-demand simulation services.

Concerning the introduction of *new modeling entities*, the MATLAB Wrapper plays this role in MATLAB/Simulink, while special objects named *registers* have been introduced in SystemC (Fig. 34.1). The MATLAB Wrapper can be connected to a user-defined number of scalar and vector input and output ports. Each port has a unique identification number and a given update frequency. MATLAB executes the Wrapper at the highest of these frequency values and, for each input port, creates a co-simulation message bearing data, the port identification number, and the simulation timestamp. Messages coming from SystemC with a given identification number

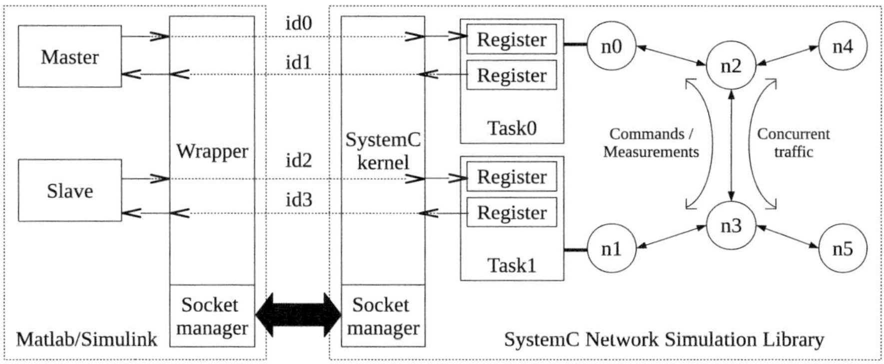

Fig. 34.1 Relationship between the new entities in MATLAB and SystemC and example of teleoperation system sharing the network with concurrent traffic

are decoded, and data are written on the corresponding output port. The SystemC simulator, born to model HW components, represents input/output ports as registers, which have been extended to let the model send/receive data to/from MATLAB. Each instance of these ports has a unique identification number which is used to associate it to the corresponding port in the MATLAB Wrapper.

Concerning the creation of the *same time domain*, a synchronization protocol has been created by using the blocking read primitive of the socket library [5]. MATLAB and SystemC exchange the time information about the next co-simulation event and then perform local simulation until this time is reached. Then, one of the peers waits for a co-simulation message from the other peer and the protocol is repeated. MATLAB simulation advances time according to a given sampling frequency, which allows to determine the time of the next co-simulation event. SystemC kernel is an event-driven simulator which manages a list of time-sorted simulation events; the next co-simulation event can be either the next event in the queue or a periodic synchronization point whose frequency is set by the designer to trade-off between time accuracy and simulation speed.

34.3 SystemC Network Simulation Library

SystemC Network Simulation Library (SCNSL) is an extension of SystemC to allow modeling packet-based networks such as wireless networks, ethernet, fieldbus, etc. As done by basic SystemC for signals on the bus, SCNSL provides primitives to model packet transmission, reception, propagation, and contention on the channel and wireless path loss. The use of SCNSL allows the easy and complete modeling of distributed applications of networked embedded systems such as wireless sensor networks, routers, and distributed plant controllers. The network scenario is described in an object-oriented way by instantiating tasks, nodes, and channels. *Tasks* are used to model node functionality in terms of packet transmission and reception; in the context of NCS, tasks are used to interface with the MATLAB Wrapper through the already mentioned registers; when data arrive from MATLAB, the corresponding task transmits them on the network and vice versa. Tasks are hosted by *Nodes*, which represent physical devices. Thus, tasks deployed on different nodes shall communicate through the channel. In case of wireless channel, some transmission properties are bound to the nodes, i.e., position in a 3D space, transmission power, and bitrate. Such properties are used by the simulator to reproduce mobility, propagation delay, loss of signal strength as a function of distance, and collisions. The *channel* represents the transmission medium; SCNSL provides models for both point-to-point (full-duplex, half-duplex, and unidirectional) and shared channels as described below. In this work, instances of point-to-point full-duplex channel are used to model a wired network, while the shared channel is used to model a wireless network.

Point-to-Point Channel Models

Point-to-point channels are used to connect node pairs. A point-to-point channel is characterized by its *capacity* and *delay*; the former represents the amount of bits that can be delivered in the time unit, while the latter represents the propagation delay; both are assumed constant during the simulation. Full-duplex channels allow transmission in both directions at the same time while half-duplex in different time intervals.

Shared Channel Models

Shared channels are used to connect more than two nodes. For each transmitting node, signal power and distance are considered with respect to all the other nodes to evaluate whether the transmitted packet can be reached and whether it generates collisions with other packets. A shared channel is characterized by its *attenuation exponent* which is applied to the distance to compute the power attenuation. Simple shared channels are also characterized by a constant *propagation delay*, while delayed shared channels are characterized by a *propagation speed* which allows to compute the propagation delay as a function of the distance between the transmitter and the receiver.

34.4 Applications

In this section, two NCS applications are presented and modeled by using the co-simulation tool to show its potentiality. For each application, the network model is detailed and some experimental results are presented.

Bilateral Teleoperation System

A particular example of NCS is the bilateral teleoperation system [5] shown in Fig. 34.1, which consists of the *master device* through which the operator controls the *remote slave robot* and a *packet-based network* which delivers all the signals (e.g., commands and measurements). `Task0` and `Task1` are the counterparts of master and slave devices in the network simulator. They are hosted by nodes `n0` and `n1`.

Nodes are connected by point-to-point full-duplex channels to create the so-called bottleneck topology; peripheral nodes are connected through dedicated links to a common backbone link. Nodes have queues to store packets exiting toward a busy link; since the backbone capacity is shared among the different traffic flows, queue level may vary during simulation and congestion may happen. Over this topology, two end-to-end application flows have been defined between applications endpoints; in Fig. 34.1, they are represented by curved arrows. The packet flow between master and slave sides in the teleoperation application (in general, between controller and plant in a NCS) has a constant bit rate since samples of commands and measurements are taken at constant rate and put in packets. A concurrent flow has been modeled between nodes `n4` and `n5`. It features an ON/OFF behavior with constant bit rate during ON periods. Teleoperation has been simulated in both uncongested and congested network conditions.

Figure 34.2a has been generated by MATLAB, and it shows the tracking error for one of the joints of the slave robot. The dashed black line refers to the uncongested scenario, whereas the red continuous line refers to the congested scenario in which control performance are affected by packet delays and losses. Figure 34.2b, c have been generated by the network simulator, and they show the packet loss rate and the communication delay, respectively, of the path from the master to the slave. In all the figures, the vertical lines separate the ON and OFF intervals of the concurrent traffic. During OFF periods, the delay is minimum and equal to the propagation delay. When the concurrent source is switched on, queues at the edges of the bottleneck link start to grow and the delay increases. When the queues are full, arriving packets are dropped. When the concurrent traffic is switched off, the number of enqueued packets decreases as well as the delay. These results show that the co-simulation tool works as expected. More complex network scenarios and concurrent traffic models (e.g., probabilistic, actual recorded traffic, etc.) can be easily implemented to model all possible kinds of working conditions.

Formation Control

Formation control is a traditional control problem in which autonomous vehicles should adapt their trajectory and speed to keep relative position with respect to each other [6]. In our scenario (Fig. 34.3a), each vehicle (except the formation leader) has only one leader and zero or more followers. We assume that each vehicle knows its position and speed and periodically broadcasts this information by using wireless messages so that followers can know it. Therefore, each vehicle receives position and speed of neighbors but considers only the information coming from its leader and changes its trajectory and speed accordingly. As depicted in Fig. 34.3b, each vehicle can be considered a NCS where the dynamic model of the vehicle represents the plant which receives directly the command u_i from the controller; the output y_i (position and speed of the vehicle) is sent back to the controller and to the neighbors by using wireless messages; the output of a given vehicle is the reference signal r_i for all its followers; each vehicle (i.e., the corresponding plant) is affected by a perturbation signal (e.g., due to wind, water flow, obstacles) which alters its position and speed.

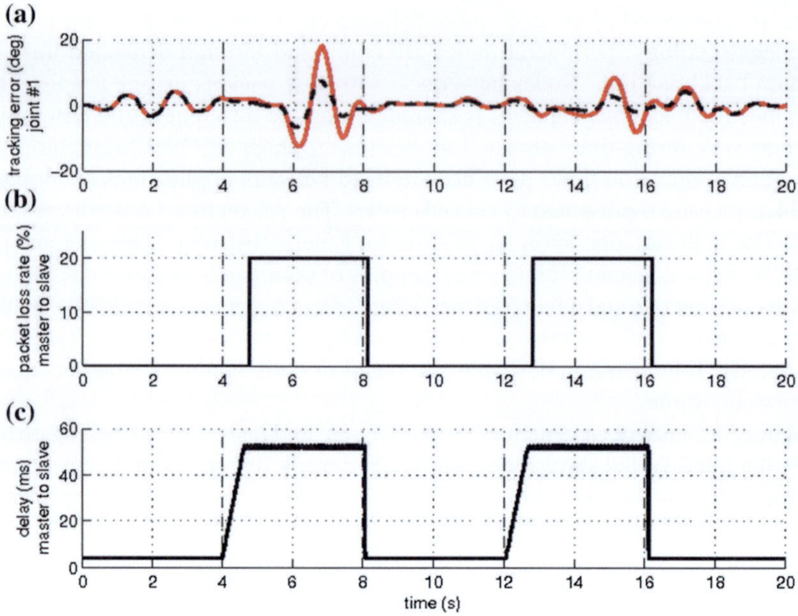

Fig. 34.2 Tracking error, packet loss rate, and delay from the teleoperation simulation

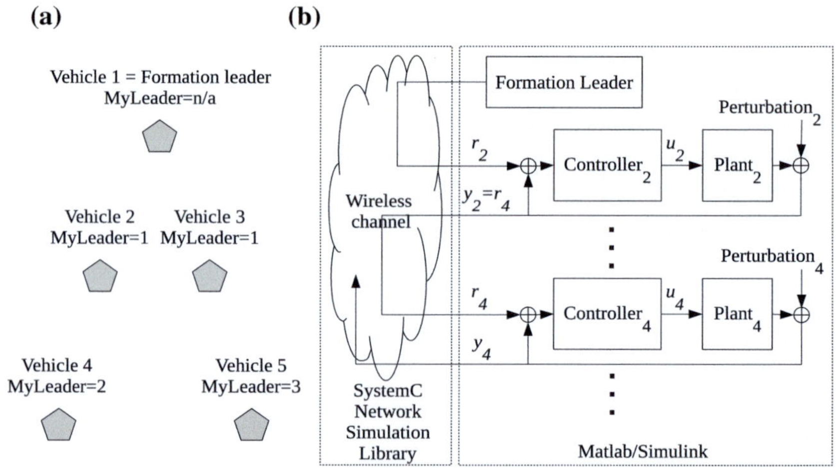

Fig. 34.3 Formation control scenario (**a**) and architecture of each NCS with tool mapping (**b**)

One of the most critical issues of this scenario is related to wireless transmission on the shared channel. Messages may not arrive to the followers because of packet collisions (i.e., overlapping of more transmitted signals at the receiver side) and out-of-range transmission. If the position of the leader is not heard, then the follower cannot react promptly to trajectory changes and perturbations so that colli-

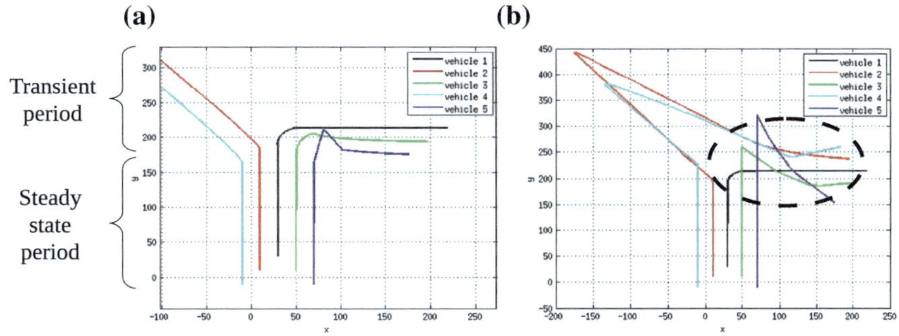

Fig. 34.4 Vehicle trajectories in two different communication scenarios: minimum (**a**) and maximum (**b**) transmission power

sions between vehicles and loss of vehicles may occur. Reaction delay depends on the timely reception of reference messages which depends on channel arbitration (to avoid or compensate collisions) and propagation delay as a function of distance. Analytical approaches to study channel behavior are not scalable with the number of vehicles, and therefore, simulation is crucial to identify problems and to validate solutions before the actual deployment. As depicted in Fig. 34.3b, we used the proposed co-simulation tool in which MATLAB simulates the different vehicles (i.e., controllers, plants, and perturbations), while SystemC reproduces the behavior of the wireless channel in between and the communication protocol.

Figure 34.4 shows the simulated behavior of the vehicles when the leader changes trajectory in the presence of perturbations as a function of two different transmission power settings. In the scenario shown in Figure 34.4.a, the transmission power of each vehicle is set to the minimum required to reach the followers when the distance requirements are satisfied. In the transient period, one vehicle (and its follower) gets lost since it cannot hear the reference signal of its leader. In the scenario shown in Figure 34.4.b, the transmission power of each vehicle is set to the maximum so that messages can be heard over a great distance. In the transient period, the perturbed vehicle increases the distance from the leader but, after a delay, rejoins the group since messages continue to be heard even if the distance is great. The drawback of this approach is the higher number of ripples in the trajectories (see dashed region) due to problems in the reception of reference messages caused by an increased number of collisions; in fact, nodes interfere with each other over a larger area due to the higher transmission power. The results in Table 34.1 confirm this conjecture; the higher variability of position error is related to the higher number of packet losses depending on message collisions. In the table, the position error is also compared to the one obtained with pure MATLAB simulation in which inter-node communications are modeled as constant delay blocks. It is worth noting that the purpose of this table is not to assess the performance of a particular control strategy, protocol, or simulation tool but to show that no verification is possible without the accurate modeling of the network provided by a suitable co-simulation tool.

Table 34.1 Relationship between control performance and network behavior as a function of the simulation strategy

Simulation strategy	Position error (m)		Packet loss rate
	Mean	Std. deviation	
Co-sim. with minimum TX power	2.0	5.5	29 %
Co-sim. with maximum TX power	2.3	9.4	72 %
MATLAB only	0.4	1.0	N/A

34.5 Further Research

In general, co-simulation has some computational overhead due to synchronization. To avoid this, a new trend in this context consists in representing both the discrete- and the continuous-time components of the system by a single hybrid model. The application of this approach to NCS should be still investigated in detail.

34.6 Further Reading

This co-simulation tool can be fruitfully used to fine-tune joint control/network design techniques as those proposed in Chap. 33.

References

1. Accellera. IEEE Std 1666–2005 IEEE Standard SystemC Language Reference Manual. IEEE Std 1666–2005, pages 1–423, 2006.
2. SystemC Network Simulation Library - version 2, 2012. URL: http://sourceforge.net/projects/scnsl
3. Hespanha JP, Naghshtabrizi P, Xu Y (2007) A survey of recent results in networked control systems. Proceedings of the IEEE 95(1):138–162
4. S. McCanne and S. Floyd. NS Network Simulator - version 2. URL: http://www.isi.edu/nsnam/ns
5. Quaglia D, Muradore R, Bragantini R, Fiorini P (2012) A SystemC/Matlab co-simulation tool for networked control systems. Simulation Modelling Practice and Theory 23:71–86
6. Wang PKC (1991) Navigation Strategies for Multiple Autonomous Mobile Robots Moving in Formation. J. Robot. Syst. 8(2):177–195
7. Zhu C, Yang OWW, Aweya J, Ouellette M, Montuno DY (Jun. 2002) A comparison of active queue management algorithms using the OPNET Modeler. IEEE Communications Magazine 40(6):158–167

Chapter 35
On Shannon's Duality of a Source and a Channel and Nonanticipative Communication and Communication for Control

Charalambos D. Charalambous, Christos K. Kourtellaris and Photios A. Stavrou

35.1 Communication System Model

Shannon in a seminal paper "A Mathematical Theory of Communication" published in 1948 [1] initiated the mathematical area known today as "Information Theory." Shannon states, "The fundamental problem of communication is that of reproducing at one point either exactly or approximately a message selected at another point." The block diagram is shown in Fig. 35.1. A message generated randomly by an information source is encoded by the encoder, transmitted over a noisy channel, and the received signal at the output of the channel is decoded by the decoder. The fundamental problem is the design of an encoder and a decoder. This fundamental problem often is separated into [2]

- Sub-problem 1. What information should be transmitted?
- Sub-problem 2. How the information should be transmitted?

Sub-problem 2 is concerned with encoding messages from a set of messages so that reliable communication over a noisy channel is possible. This amounts to establishing coding theorems, to characterize the maximum rate of information transmission reliably, called channel capacity. Sub-problem 1 is exclusively addressed by Shannon himself in another seminal paper published in 1959 [3] "Coding Theorems for a Discrete Source with Fidelity." Sub-problem 1 is concerned with selecting the set of possible messages to be transmitted reliably. This amounts to compressing source messages with respect to a fidelity or reconstruction criterion, by establishing coding

C.D. Charalambous (✉) · C.K. Kourtellaris · P.A. Stavrou
Department of Electrical and Computer Engineering (ECE), University of Cyprus,
75 Kallipoleos, P.O. Box 20537, CY-1678 Nicosia, Cyprus
e-mail: chadcha@ucy.ac.cy

C.K. Kourtellaris
e-mail: kourtellaris.christos@ucy.ac.cy

P.A. Stavrou
e-mail: stavrou.fotios@ucy.ac.cy

J.H. van Schuppen and T. Villa (eds.), *Coordination Control of Distributed Systems*,
Lecture Notes in Control and Information Sciences 456,
DOI 10.1007/978-3-319-10407-2_35

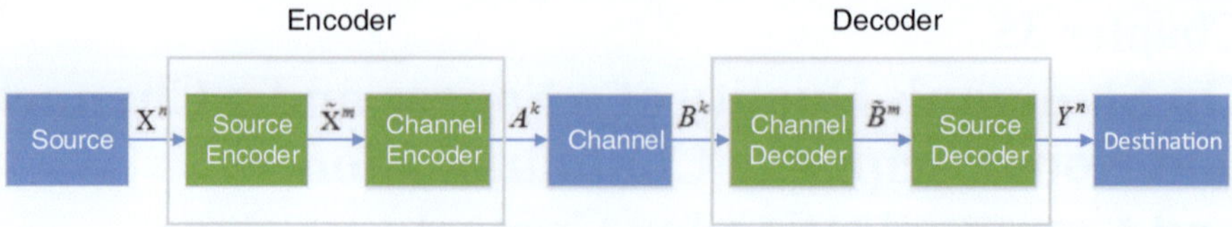

Fig. 35.1 Diagram of a communication system displaying the various subsystems

theorems, to characterize the minimum rate of compressing a source, called optimal performance theoretically attainable (OPTA) by noncausal codes, and given by the rate distortion function (RDF) of the source [3].

This separation into sub-problems 1 and 2 enabled the communication community to develop source and channel codes [4], often with very large codeword lengths to achieve the OPTA and the channel capacity, even for network communication problems. Some of the fundamental concepts often used are entropy, mutual information, and Marko's [5] bidirectional information of sequences of random variables, and weak law of large numbers (asymptotic equipartition property).

Since Shannon seminal paper [1], the field of information theory has grown considerably and found applications in other areas of science and engineering. Many excellent textbooks on information theory have been published extending Shannon's original tools and coding theorems in various directions, such as [2, 6–12].

Sub-problem 1. For capacity of channels with memory and feedback, recent approaches utilize Marko's bidirectional information [5], distinguishing the direction of information flow between two statistically dependent processes. This concept is further elaborated by Massey [13] in the context of channels with feedback, and developed by Kramer [14], including networks communication problems. A small sample for point-to-point communication, when capacity is defined using directed information between the channel input and channel output sequences, is [15–18], and for network communication [19]. In the past, such problems of capacity for channels with memory and feedback are dealt with using mutual information between the source and the channel output sequences, under the assumption that the source is not affected causally by channel outputs [12, 20].

Sub-problem 2. For sources with memory, the OPTA by noncausal and causal codes (not necessarily delayless) given by the RDF and entropy rate of the reproduction symbols subject to fidelity [21], respectively, is less developed compared to channel capacity, possibly due to the difficulty of computing the RDF and the optimal compression channel distribution when sources have memory.

The separation of the fundamental problem into sub-problems 1 and 2 offers many advantages both in analysis and design of communication systems. There is, however, one drawback with respect to the understanding how encoders are designed to control the channel output process, that of determining, simultaneously, what information should be transmitted, and how the information should be transmitted, and whether separating these sub-problems, compromises the optimality and complexity of the overall design. The problem of designing jointly the encoders and decoders, without separating them into source–channel encoders and decoders, is often called joint source–channel coding (JSCC) or source–channel matching [22].

Shannon himself made the following remarks [3]:

> **Duality of a Source and a Channel**. *There is a curious and provocative duality between the properties of a source with a distortion measure and those of a channel. This duality is enhanced if we consider channels in which there is a "cost" associated with the different input letters, and it is desired to find the capacity subject to the constraint that the expected cost not exceed a certain quantity. The solution of this problem leads to a capacity cost function $C(a)$ for the channel. Solving this problem corresponds, in a sense, to finding a source that is just right for the channel and the desired cost.*
>
> *In a somewhat dual way, evaluating the rate distortion function $R(d)$ for the source amounts, to 1...*
>
> *Solving this problem corresponds to finding a channel that is just right for the source and allowed distortion level. This duality can be pursued further and is related to a duality between past and future and the notions of control and knowledge.*

Unfolding the duality relation between the two fundamental limits of communication, that of data compression, and of data transmission is of fundamental importance. Two fascinating examples are the independent identically distributed (IID) binary source with Hamming distortion transmitted uncoded over a symmetric memoryless channel, and the IID Gaussian source with average squared-error distortion, transmitted over an additive white Gaussian noise (AWGN) channel, with the encoder and decoder scaling their inputs [6, 22, 23]. These examples demonstrate the simplicity of JSCC systems when a "duality of a source and a channel" is at work, in operating optimally with zero delay, in complexity, when compared to the asymptotic noncausal performance of optimally separating sub-problem 1 and sub-problem 2. Most research in JSCC is focused on point-to-point communication, memoryless sources, and channels, and in showing that source–channel separation is optimal, using long codes, without any emphasis on real-time communication.

Understanding the notion of "duality of a source and a channel" and how this can be extended beyond these two examples, and beyond memoryless sources and memoryless channels [8] is believed to be related to the fundamental problem of communication, and communication for control, in which the optimal transmission is indeed real-time transmission. It presupposes further understanding of the structural properties of capacity-achieving encoders for channels with memory with and without feedback, and further development of nonanticipative rate distortion theory, to deal with questions of realizing optimal compression conditional distributions by nonanticipative or real-time communication systems, that of processing at the encoder, channel, and decoder information causally without any dependence on future inputs (i.e., zero-delay processing). Such schemes are important in delay-sensitive communication and communication for control applications.

The purpose of this article is twofold: (1) to provide a brief introduction to the fundamental problem of communication, and the main tools on this topic; (2) to identify topics for further research related to sub-problems 1 and 2, and to nonanticipative or real-time communication, and communication for control.

Emphasis will be given on raising questions to prepare the ground on unfolding Shannon's remarks on "duality of a source and a channel," with respect to nonanticipative communication, and how this benefits communication for control.

This unfolding is further discussed by the authors in the subsequent chapters on this topic. An example is presented in the last chapter, illustrating *duality of a source with memory and a channel with/without feedback, operating optimally in real time*.

Mathematical Model of Real-Time Communication Systems.
The alphabet spaces of the source output, channel input, channel output, and decoder output are sequences of finite cardinality spaces, $\{\mathscr{X}_t : t = 1, \ldots, n\}$, $\{\mathscr{A}_t : t = 1, \ldots, n\}$, $\{\mathscr{B}_t : t = 1, \ldots, n\}$, and $\{\mathscr{Y}_t : t = 1, \ldots, n\}$; their corresponding product spaces are $\mathscr{X}^n \triangleq \times_{k=1}^n \mathscr{X}_k$, $\mathscr{A}^n \triangleq \times_{k=1}^n \mathscr{A}_k$, $\mathscr{B}^n \triangleq \times_{k=1}^n \mathscr{B}_k$, $\mathscr{Y}^n \triangleq \times_{k=1}^n \mathscr{Y}_k$, and the finite-valued discrete-time random processes are $X^n \triangleq \{X_t : t = 1, \ldots, n\}$, $A^n \triangleq \{A_t : t = 1, \ldots, n\}$, $B^n \triangleq \{B_t : t = 1, \ldots, n\}$, $Y^n \triangleq \{Y_t : t = 1, \ldots, n\}$, respectively. The conditional independence of Random Variables (RVs) X, B given the RV A is denoted by the Markov chain (MC) $X \leftrightarrow A \leftrightarrow B$ [8, 9].

In nonanticipative communication, the source-encoder–channel-decoder in Fig. 35.1 processes information causally, while in communication for control, the source is a controlled process, causally affected by the channel outputs (or decoder outputs), as defined below, using probability mass functions (PMF).

Information Source. The information source is a sequence of conditional PMF

$$P_{X_i|X^{i-1},A^{i-1},B^{i-1}}(x_i|x^{i-1}, a^{i-1}, b^{i-1}), \qquad i = 1, \ldots, n. \tag{35.1}$$

If the source symbols are independent of the previous encoder outputs and channel outputs, then $P_{X_i|X^{i-1},A^{i-1},B^{i-1}}(x_i|x^{i-1}, a^{i-1}, b^{i-1}) = P_{X_i|X^{i-1}}(x_i|x^{i-1})$, and if the source is memoryless, then $P_{X_i|X^{i-1},A^{i-1},B^{i-1}}(x_i|x^{i-1}, a^{i-1}, b^{i-1}) = P_X(x_i)$, $i = 1, \ldots, n$.

Encoder with and without Feedback. An encoder with feedback is a sequence of conditional PMF

$$P_{A_i|A^{i-1},B^{i-1},X^i}(a_i|a^{i-1}, b^{i-1}, x^i), \qquad i = 1, \ldots, n. \tag{35.2}$$

An encoder without feedback is $P_{A_i|A^{i-1},X^i}(a_i|a^{i-1}, x^i), i = 1, \ldots, n$. A deterministic encoder with feedback is a sequence of measurable functions $\{e_i : \mathscr{A}^{i-1} \times \mathscr{B}^{i-1} \times \mathscr{X}^i \mapsto \mathscr{A}_i : a_i = e_i(a^{i-1}, b^{i-1}, x^i), i = 1, \ldots, n\}$, and without feedback $\{e_i : \mathscr{A}^{i-1} \times \mathscr{X}^i \mapsto \mathscr{A}_i : a_i = e_i(a^{i-1}, x^i), i = 1, \ldots, n\}$.

Communication Channel. A communication channel is a sequence of PMF

$$P_{B_i|B^{i-1},A^i,X^i}(b_i|b^{i-1}, a^i, x^i), \qquad i = 1, \ldots, n. \tag{35.3}$$

If the channel is conditional independent of the source, then $P_{B_i|B^{i-1},A^i,X^i}(b_i|b^{i-1}, a^i, x^i) = P_{B_i|B^{i-1},A^i}(b_i|b^{i-1}, a^i)$, and if the channel is memoryless, known as discrete memoryless channel (DMC), then $P_{B_i|B^{i-1},A^i,X^i}(b_i|b^{i-1}, a^i, x^i) = P_{B|A}(b_i|a_i), i = 1, \ldots, n$.

Decoder. A decoder is a sequence of PMF

$$P_{Y_i|Y^{i-1},B^i}(y_i|y^{i-1},b^i), \qquad i=1,\ldots,n. \tag{35.4}$$

A deterministic decoder is a sequence of measurable functions $\{d_i \colon \mathscr{Y}^{i-1} \times \mathscr{B}^i \mapsto \mathscr{Y}_i \colon y_i = d_i(y^{i-1}, b^i), i = 1, \ldots, n\}$.

Problems of Communication. When sub-problems 1 and 2 are separated, fundamental progress in point-to-point communication and network communication (often by considering memoryless sources and channels) is made addressing

1. What is the best compression, the OPTA by noncausal and causal (as defined by Neuhoff and Gilbert [21], not necessarily delayless) lossy codes, and lossless codes, of a random source?
2. What is the maximum rate of communicating information over a noisy channel?

There is, however, a need for further progress in addressing the following questions.

1. How does one design source codes to achieve the OPTA of a given source?
2. How does one design channel codes to achieve the capacity of a given channel?
3. How does one design JSCC systems, and what is the performance?

Problems of Nonanticipative Communication and Communication for Control. The current research is at its infancy, when (a) each subsystem is nonanticipative or real-time communication is imposed, (b) the source is a controlled process, controlled using nonanticipative communication, and (c) control and communication subsystems are integrated. Very little is known on how to modify the existing theory to account for real-time communication and communication for control?. Past research on control over limited rate communication channels, although offered significant insights, did not address the above questions [15, 24, 25].

The rest of the chapter is focused on a brief discussion of current tools, and how these can be modified to account for nonanticipative communication and communication for control, based on the authors' views.

35.2 Information Theoretic Measures, MCs, AEP, Fano's Inequality

Source, channel coding theorems, and JSCC theorems are often derived using the asymptotic equipartition property (AEP), MCs, and Fano's inequality [8, 10]. These are briefly described below.

Information Measures (IM). Consider any joint PMF $P_{X^n,Y^n}(x^n, y^n) \equiv P(x^n, y^n)$ induced by arbitrary processes $\{(X_i, Y_i) \colon i = 1, \ldots, n\}$.

Entropy. Self-entropy of a sequence $X^n = x^n$ and self-conditional entropy of $X^n = x^n$ given $Y^n = y^n$ are defined by

$$i(x^n) \triangleq -\log P(x^n) = -\sum_{i=1}^{n} \log P(x_i|x^{i-1}), \quad i(x^n|y^n) \triangleq -\log P(x^n|y^n). \tag{35.5}$$

Their average values called conditional entropy and entropy are

$$H(X^n|Y^n) \triangleq \mathbb{E}\{i(X^n|Y^n)\}, \quad H(X^n) \triangleq \mathbb{E}\{i(X^n)\} = \sum_{i=1}^{n} H(X_i|X^{i-1}). \tag{35.6}$$

The self-entropy of a sequence $X^n = x^n$ is the amount of uncertainty or information of the event. This is often measured in bits or nats (depending on the choice of the base of the logarithm used). The average entropy is the average uncertainty of the processes.

Mutual Information. Self-mutual information between two sequences $X^n = x^n$ and $Y^n = y^n$ is defined by

$$i(x^n; y^n) \triangleq \log \frac{P(y^n, x^n)}{P(y^n)P(x^n)} = \log \frac{P_{Y^n|X^n}(y^n|x^n)}{P(y^n)} = -\log P(y^n) + \log P(y^n|x^n). \tag{35.7}$$

Its average value, called mutual information, is

$$I(X^n; Y^n) \triangleq \mathbb{E}\{i(X^n; Y^n)\} = H(Y^n) - H(Y^n|X^n) = H(X^n) - H(X^n|Y^n). \tag{35.8}$$

Directed Information. Following Marko [5], an application of Bayes' theorem to (35.7), yields $i(x^n; y^n) = i(x^n \to y^n) + i(x^n \leftarrow y^n)$, where $i(x^n \to y^n)$ is the self-directed information from sequence $X^n = x^n$ to the sequence $Y^n = y^n$, and $i(x^n \leftarrow y^n)$ the self-directed information from sequence $Y^n = y^n$ to the sequence $X^n = x^n$, defined by

$$i(x^n \to y^n) \triangleq \sum_{i=1}^{n} \log \frac{P(y_i|y^{i-1}, x^i)}{P(y_i|y^{i-1})}, \quad i(x^n \leftarrow y^n) \triangleq \sum_{i=1}^{n} \log \frac{P(x_i|x^{i-1}, y^{i-1})}{P(x_i|x^{i-1})}. \tag{35.9}$$

Their average values called directed information are

$$\begin{aligned} I(X^n \to Y^n) \triangleq \mathbb{E}\Big\{i(X^n \to Y^n)\Big\} &= \sum_{i=1}^{n} I(X^i; Y_i|Y^{i-1}) \\ &= \sum_{i=1}^{n} \Big(H(Y_i|Y^{i-1}) - H(Y_i|Y^{i-1}, X^i)\Big), \end{aligned}$$

$$I(X^n \leftarrow Y^n) \triangleq \mathbb{E}\Big\{i(X^n \leftarrow Y^n)\Big\} = \sum_{i=1}^{n} I(Y^{i-1}; X_i|X^{i-1})$$
$$= \sum_{i=1}^{n} \Big(H(X_i|X^{i-1}) - H(X_i|X^{i-1}, Y^{i-1})\Big).$$

Properties of directed information is given in Chap. 36.

Note that $I(X^n; Y^n) = I(X^n \rightarrow Y^n) + I(X^n \leftarrow Y^n)$, and for network and feedback communication problems, the relevant information communicated from sources to destinations is directional. On the other hand, if $P(x_i|x^{i-1}, y^{i-1})$ is not affected by the process Y^{i-1} i.e., $P(x_i|x^{i-1}, y^{i-1}) = P(x_i|x^{i-1}), i = 1, \ldots, n$, then $I(X^n; Y^n) = I(X^n \rightarrow Y^n) \equiv \mathbb{I}_{X^n \rightarrow Y^n}(P_{X^n}, \{P_{Y_i|Y^{i-1},X^i}, i = 1, \ldots, n\})$, and the joint PMF is $P_{X^n,Y^n} = \prod_{i=1}^n P_{Y_i|Y^{i-1},X^i} P_{X^n}$. This is the approach taken in [12, 20] to discuss capacity of channels with memory and feedback, prior to Marko's directed information influence.

Markov Chains. Using Fig. 35.1, the following MCs (or variations of them) are important in separating the fundamental problem of communication into sub-problems 1 and 2, identifying the proper information measures of channel capacity, OPTA by noncausal, causal, and zero-delay codes, JSCC, while special cases of them are important in nonanticipative communication and communication for control.

MC1. If $X^t \leftrightarrow (A^i, B^{i-1}) \leftrightarrow B_i$ for $i = 0, 1, \ldots, n,\ t \leq n$, then

$$I(X^t; B^n) \leq I(A^n \rightarrow B^n), \qquad t \leq n. \tag{35.10}$$

MC2. If $X_i \leftrightarrow (X^{i-1}, B^n) \leftrightarrow Y^t$ for $i = 0, 1, \ldots, t,\ t \leq n$, then

$$I(X^t; Y^t) \leq I(X^t; B^n), \qquad t \leq n. \tag{35.11}$$

MC3. If the conditions of statements MC1 and MC2 hold then

$$I(X^t \rightarrow Y^t) \leq I(X^t; Y^t) \leq I(A^n \rightarrow B^n) \leq I(A^n; B^n), \qquad t \leq n. \tag{35.12}$$

For source–encoders–channels defined by (35.1)–(35.3) (i.e., channels or encoders with feedback, and sources affected by channel outputs), Marko's directional information flow is utilized, implying that capacity should be defined via $I(X^n \rightarrow B^n)$. On the other hand, if the channel is not affected by the source, i.e., the MC holds, $X^t \leftrightarrow (B^{i-1}, A^i) \leftrightarrow B_i, i = 1, \ldots, n, t \leq n$, capacity is defined via $I(A^n \rightarrow B^n)$. Most paper in the literature assumes MC1 (see bibliography and references within). If in addition the channel is used without feedback, since $B^{i-1} \leftrightarrow A^{i-1} \leftrightarrow A_i, i = 1, \ldots, n$ if and only if $I(A^n \leftarrow B^n) = 0$, then capacity is defined via $I(A^n; B^n)$.

For nonanticipative communication and communication for control, the conditions for MC1 and MC2 are $X^i \leftrightarrow (A^i, B^{i-1}) \leftrightarrow B_i$ and $X_i \leftrightarrow (X^{i-1}, B^i) \leftrightarrow Y^i, i = 0, 1, \ldots, n$, and $t = n$ in (35.11)–(35.12).

Fano's Inequality. In information theory, the converse part of channel coding theorems often employs Fanon's inequality to identify an upper bound on capacity, which is tight; that is, violating this bound can result in probability of decoding error arbitrary close to 1. The direct part establishes achievability of this bound.

Define the average error probability over any sequence $X^n = x^n$ reproduced by $Y^n = y^n$, according to the joint PMF by $P_e^{(n)} = \frac{1}{n}\sum_{i=1}^{n} \text{Prob}\{\text{X}_\text{i} \neq \text{Y}_\text{i}\}$. Fano's inequality states that if the sample spaces $\mathscr{X}_i, \mathscr{Y}_i$ are of finite cardinality M, $i = 1, \ldots, n$, then

$$\frac{1}{n}H(X^n|Y^n) \leq P_e^{(n)} \log M + h(P_e^{(n)}), \quad h(z) = -z\log z - (1-z)\log(1-z). \tag{35.13}$$

When inequality (35.13) and MCs are applied to Fig. 35.1, they relate $P_e^{(n)}$, $H(X^n)$ and $I(A^n; B^n)$, and $I(X^n; Y^n)$ to $I(A^n; B^n)$, to obtain tight bounds in coding theorems, and relate channel capacity and RDF. Analogous relations can be obtained for nonanticipative communication and communication for control.

Asymptotic Equipartition Property (AEP). In information theory, coding theorems are typically derived, in the limit, as the length of the codes tends to infinity. The AEP is often used to divide all sequences into sets of high and low probability of occurrence, and it is applied in the direct or achievability part of coding theorems, using random coding arguments, to show existence of at least one encoder and decoder, with probability of decoding error which can be made arbitrary small, as the codeword length n tends to infinity. The definitions which make use of the AEP of increasing generality are summarized below.

1. *Jointly Independent and Identically Distributed (IID)*. Let $\{(X_i, Y_i) : i = 1, 2, \ldots, n\}$ be jointly IID with joint distribution $P(x^n, y^n) = \prod_{i=1}^{n} P(x_i, y_i), n = 1, 2, \ldots$. Then, by the weak law of large numbers, the following hold.

$$\lim_{n\longrightarrow\infty} -\frac{1}{n}\log P(X^n) = H(X), \quad \lim_{n\longrightarrow\infty} -\frac{1}{n}\log P(X^n, Y^n) = H(X,Y) \quad \text{in prob.} \tag{35.14}$$

$$\lim_{n\longrightarrow\infty} -\frac{1}{n}\log P_{Y^n}(Y^n) = H(Y), \quad \lim_{n\longrightarrow\infty} \frac{1}{n}\log\frac{P(y^n|x^n)}{P(y^n)} = I(X;Y) \quad \text{in prob.} \tag{35.15}$$

The AEP states that the set of jointly typical sequences (x^n, y^n) with respect to the distribution $P(x, y)$ defined by

$$A_\varepsilon^{(n)} = \{(x^n, y^n) \in \mathcal{X}^n \times \mathcal{Y}^n : | -\frac{1}{n}\log P(x^n) - H(X)| < \varepsilon,$$
$$| -\frac{1}{n}\log P(y^n) - H(Y)| < \varepsilon, | -\frac{1}{n}\log P(x^n, y^n) - H(X,Y)| < \varepsilon\}, \quad \varepsilon > 0. \tag{35.16}$$

is such that, given any $\varepsilon > 0$, n can be chosen big enough so that

- $Prob\{((X^n, Y^n) \notin A_\varepsilon^{(n)}\} < \varepsilon$ for sufficiency large n;
- The number of elements in $A_\varepsilon^{(n)}$ is at most $|A_\varepsilon^{(n)}| \le 2^{n(H(X,Y)+\varepsilon)}$ for every n;
- If $(\tilde{X}^n, \tilde{Y}^n) \sim P(x^n)P(y^n)$ (i.e., $\tilde{X}^n$ and $\tilde{Y}^n$ are independent with the same marginals as $P_{X^n,Y^n}(x^n, y^n)$), then there exists sufficiently large n such that

$$(1-\varepsilon)2^{-n(I(X;Y)+3\varepsilon)} \le P\{(\tilde{X}^n, \tilde{Y}^n) \in A_\varepsilon^{(n)}\} \le 2^{-n(I(X;Y)-3\varepsilon)}. \tag{35.17}$$

2. *Stationary Ergodic Processes.* Let $\{Z_i \triangleq (X_i, Y_i) : i = 1, 2, \ldots, n\}$ be stationary ergodic with joint PMF $P(x^n, y^n), n = 1, 2, \ldots$. Then, by the Shannon–McMillan–Breiman theorem, the following hold.

$$\lim_{n\longrightarrow\infty} -\frac{1}{n}\log P(X^n) = \lim_{n\longrightarrow\infty} \mathbb{E}\Big\{ -\log P(X_n|X^{n-1})\Big\} = \lim_{n\longrightarrow\infty} \frac{1}{n}\sum_{i=1}^{n} H(X_i|X^{i-1}) \text{ w.p. } 1.$$

$$\lim_{n\longrightarrow\infty} -\frac{1}{n}\sum_{i=1}^{n}\log P(Z_i|Z^{i-1}) = \lim_{n\longrightarrow\infty} \mathbb{E}\Big\{ -\log P(Z_n|Z^{n-1})\Big\}$$
$$= \lim_{n\longrightarrow\infty} \frac{1}{n}\sum_{i=1}^{n} H(Z_i|Z^{i-1}) \text{ w.p. } 1.$$

3. *Information Stability.* For each $\varepsilon > 0$, define the ε-typical set by

$$J_\varepsilon^{(n)} \triangleq \{(x^n, y^n) \in \mathcal{X}^n \times \mathcal{Y}^n : \frac{1}{n}|i(x^n; y^n) - I(X^n; Y^n)| \le \varepsilon\}.$$

The process $\{(X_i, Y_i) : i = 1, \ldots, n\}$ is called information stable if $\lim_{n\longrightarrow\infty} \mathrm{Prob}\{(X^n, Y^n) \in J_\varepsilon^{(n)}\} = 1, \forall \varepsilon > 0$. For stationary ergodic processes and information stable processes, there is an analog of the AEP of IID processes, which is standard in deriving the direct part of coding theorems, for sub-problem 2 and sub-problem 1 [2, 10].

4. *Information Spectrum.* For nonstationary, nonergodic processes, a generalization of AEP of mutual information is based on information spectrum methods [10]. Most results derived in information theory related to sub-problems 1 and 2, and JSCC employs the AEP corresponding to the above notions of stationary ergodicity, information stability, and information spectrum methods, and random code generation to show that a specific tight bound coming from Fano's inequality corresponds to the extremum of all achievable rates. For nonanticipative communication or communication for control, in which real-time communication is imposed this is not a choice, due to requirement of real-time processing, this part will be explained further shortly.

35.3 Channel Capacity and Rate Distortion Functions

Channel Capacity and Coding Theorems. The operational definition of reliable communication over noisy channels is based on encoding long source sequences, into channel sequences, and then decoding the channel output sequences. A channel code and the operational definition of capacity is defined below.

Channel Code. An $\{(n, M_n, \epsilon_n): n = 1, \dots\}$ code sequence for a channel without feedback consists of the following.

1. A set of messages $\mathscr{M}_n \triangleq \{1, 2, \dots, M_n\}$, encoder mappings $\{\varphi_i: \mathscr{M}_n \times \mathscr{A}^{i-1} \mapsto \mathscr{A}_i: i = 1, \dots, n\}$ that transform each message $X \in \mathscr{M}_n$ into a channel input $A^n \in \mathscr{A}^n$ of length n. The codeword for $x \in \mathscr{M}_n$ is $u_x \in \mathscr{A}^n$, $u_x = (\varphi_1(x), , \dots, \varphi_n(x, a^{n-1}))$, $\mathscr{C}_n = (u_1, u_2, \dots, u_{M_n})$ is the code for the message set $\mathscr{M}_n$, generated independently according to $P(a^n)$ (random code). When the transmitter wishes to send the message $x \in \mathscr{M}_n$, it transmits the codeword u_x of the current message x.
2. Decoder measurable mappings $d^n: \mathscr{B}^n \mapsto \mathscr{M}_n$, $Y^n = d^n(B^n)$, such that the average probability of decoding error satisfies $P_e^n \triangleq \frac{1}{M_n} \sum_{x \in \mathscr{M}_n} Prob(Y^n \neq x | X = x) = \epsilon_n$.

R is an achievable rate if there exists such a code sequence satisfying $\lim_{n\to\infty} \epsilon_n = 0$ and $\liminf_{n\to\infty} \frac{1}{n} \log M_n \geq R$. The channel capacity is $C \triangleq \sup\{R: R \text{is achievable}\}$. $r_n \triangleq \frac{1}{n} \log M_n$ is the coding rate (or transmission rate).
Feedback encoder mappings are $\{\varphi_i : \mathscr{M}_n \times \mathscr{A}^{i-1} \times \mathscr{B}^{i-1} \mapsto \mathscr{A}_i: i = 1, \dots, n\}$.

In channel coding problems, the objective is to find the encoder and decoder mappings which maximize the coding rate while keeping the probability of error small.

The identification of the information measure which corresponds to channel capacity, C, is based on the converse part of the coding theorem, which aims at identifying a tight upper bound on the achievable rates via Fano's inequality.

The converse part of the coding theorem is illustrated below, to bring out some of the silent features and assumptions imposed.

Converse Part. Suppose R is achievable. Then, there exists an (n, M_n, ϵ_n) code $\mathscr{C}_n = (u_1, u_2, \dots, u_{M_n})$ such that $\lim_{n\to\infty} \epsilon_n = 0$ and $\liminf_{n\to\infty} \frac{1}{n} \log M_n \geq R$. For each n, then $P_e^n = \varepsilon_n$, and by Fano's inequality [8] then

$$\begin{aligned}
\log M_n &= H(X) = H(X|B^n) + I(X; B^n) \\
&\leq h(\varepsilon_n) + \varepsilon_n \log M_n + I(X; B^n) = h(\varepsilon_n) + \varepsilon_n \log M_n + H(B^n) \\
&\quad - \sum_{i=1}^{n} H(B_i | B^{i-1}, X) \\
&= h(\varepsilon_n) + \varepsilon_n \log M_n + H(B^n)
\end{aligned}$$

$$
\begin{aligned}
&- \sum_{i=1}^{n} H(B_i|B^{i-1}, A_1(X), \ldots, A^{n-1}(X, A^{n-1}), X) \\
&\leq \frac{h(\varepsilon_n) + \sum_{i=1}^{n} I(X, A^i; B_i)}{1-\varepsilon_n}.
\end{aligned}
$$

Since $\varepsilon_n \rightarrow 0, h(\varepsilon_n) \rightarrow 0$, as, $n \rightarrow \infty$, the upper bound is obtained.

$$
R \leq \liminf_{n\rightarrow\infty} \frac{1}{n} \log M_n \leq \liminf_{n\rightarrow\infty} \frac{1}{n} \sum_{i=1}^{n} I(X, A^i; B_i|B^{i-1}). \tag{35.18}
$$

If the MC $X \leftrightarrow (B^{i-1}, A^i) \leftrightarrow B_i, i = 1, \ldots, n$ **holds,** then

- $\sum_{i=1}^{n} I(X, A^i; B_i|B^{i-1}) = \sum_{i=1}^{n} I(A^i; B_i|B^{i-1}) \equiv I(A^n \rightarrow B^n)$, and the upper bound in (35.18) is independent of the source;
- any achievable rate R satisfies

$$
\text{Feedback.} \quad R \leq C^{FB,-} \triangleq \sup_{\{P(a_i|a^{i-1},b^{i-1}):\, i=1,\ldots\}} \liminf_{n\longrightarrow\infty} \frac{1}{n} I(A^n \rightarrow B^n), \tag{35.19}
$$

$$
\text{No feedback.} \quad R \leq C^{NFB,-} \triangleq \sup_{\{P(a_i|a^{i-1}):\, i=1,\ldots\}} \liminf_{n\longrightarrow\infty} \frac{1}{n} I(A^n; B^n), \tag{35.20}
$$

$$
\text{DMC with or without Feedback.} \quad R \leq C^{DMC} \triangleq \sup_{P(a)} I(A; B); \tag{35.21}
$$

- the tightness of the upper bound is often shown via the direct part using the AEP.

Direct Part. Often, using the AEP and random coding, for stationary ergodic or information stable processes, it can be shown that any rate R satisfying $R < C^{FB,-}$, $R < C^{NFB,-}$ is achievable, and for DMC that $R < C^{DMC}$; hence, $C^{FB,-}$, $C^{NFB,-}$, C^{DMC} are the corresponding channel capacities. For nonstationary, nonergodic processes, similar results are shown via information spectrum methods (outage capacity) [10].

Capacity with Transmission Cost. Proving the supremum over all channel input distributions with transmission cost set $\mathscr{P}_n(P)$ is achieved, for abstract spaces, $\mathscr{A}^n$, $\mathscr{B}^n$ (such as, continuous) and channels with feedback, defined by

$$
C^{FB} = \lim_{n\longrightarrow\infty} \frac{1}{n} C_{0,n}^{FB}, \quad C_{0,n}^{FB}(P) \triangleq \sup_{\{P(a_i|a^{i-1},b^{i-1}):\, i=1,\ldots,n\}\in\mathscr{P}_n(P)} I(A^n \rightarrow B^n), \tag{35.22}
$$

is a challenging problem. Proving the supremum over all achievable rates given by the limiting expression $C^{FB}(P)$ is also challenging. Progress in this direction for finite-state unit memory channels defined by $\{P_{B_i|B_{i-1},A_i}(b_i|b_{i-1}, a_i): i = 1, \ldots, n\}$

without transmission cost, under the assumption that the capacity-achieving distribution has the important property $\{P_{A_i|A^{i-1},B^{i-1}}(a_i|a^{i-1},b^{i-1}) = P_{A_i|B_{i-1}}(a_i|b_{i-1})\colon i=1,\ldots,n\}$, which then implies $I(A^n \rightarrow B^n) = \sum_{i=1}^n I(A_i; B_i|B_{i-1})$, is found in [16] (a proof of the encoder property has not been located in the literature).
If the MC $X^i \leftrightarrow (B^{i-1}, A^i) \leftrightarrow B_i, i=1,\ldots,n$ does not hold, and the source is causally affected by channel input–outputs, $\{P_{X_i|X^{i-1},A^{i-1},B^{i-1}}\colon i=1,\ldots,n\}$, to show $C^{GFB} = \lim_{n\longrightarrow\infty} \frac{1}{n} C_{0,n}^{GFB}$, where

$$C_{0,n}^{GFB}(P) \triangleq \sup_{\{P(x_i|x^{i-1},a^{i-1},b^{i-1}),a_i=g_i(a^{i-1},b^{i-1},x^i)\}_{i=1}^n \in \mathscr{P}_n(P)} I(X^n \rightarrow B^n), \tag{35.23}$$

is the supremum of all achievable rates is a challenging problem.
For nonanticipative communication, and communication for control, the additional challenge is to show achievability of C^{FB} and C^{GFB}, using real-time transmission. Whether real-time encoders can give as good performance as classical noncausal encoders is a fundamental question.

Rate Distortion Function. The above methodology applies to sub-problem 1. For stationary ergodic processes or information stable processes, the OPTA by noncausal codes subject to fidelity is the RDF of the source, defined by

$$R(D) \triangleq \lim_{n\longrightarrow\infty} \frac{1}{n} R_{0,n}(D), \quad R_{0,n}(D) \triangleq \inf_{P_{Y^n|X^n}\in\mathscr{Q}_{0,n}(D)} I(X^n; Y^n), \tag{35.24}$$

where the fidelity set is defined by $\mathscr{Q}_{0,n}(D) \triangleq \{P_{Y^n|X^n}\colon \frac{1}{n}\mathbb{E}\{d_{0,n}(X^n, Y^n)\} \leq D\}$. The optimal conditional distribution is given by the implicit expression [2]

$$P^*_{Y^n|X^n}(y^n|x^n) = \frac{e^{sd_{0,n}(x^n,y^n)} P^*_{Y^n}(y^n)}{\sum_{y^n\in\mathscr{Y}_{0,n}} e^{sd_{0,n}(x^n,y^n)} P^*_{Y^n}(y^n)}, \qquad s \leq 0, \tag{35.25}$$

where $s \in (-\infty, 0]$ is the Lagrange multiplier associated with $\mathscr{Q}_{0,n}(D)$. However, $R(D)$ has the following important limitations.

- The exact expressions of $R_{0,n}(D)$, $P^*_{Y^n|X^n}(dy^n|x^n)$, for finite n, and as, $n \longrightarrow \infty$, are known for a small class of sources, which are memoryless or Gaussian [2].
- (35.25) is generally noncausal or anticipative, and often cannot be used in JSCC using nonanticipative transmission.

Hence, for sources with memory, attempting to identify any duality of a source and a channel is rather difficult, with the current definition of RDF. An alternative approach is via the nonanticipatory $\epsilon-$entropy and message generation rates introduced by Gorbunov and Pinsker [26] defined by $R^{\varepsilon}(D) = \lim_{n\longrightarrow\infty} \frac{1}{n} R^{\varepsilon}_{0,n}(D)$, where

$$R_{0,n}^{\varepsilon}(D) \triangleq \inf_{P_{Y^n|X^n} \in \mathcal{Q}_{0,n}(D) \cap \{X_{i+1}^n \leftrightarrow X^i \leftrightarrow Y_i, i=1,\ldots,n-1\}} I(X^n; Y^n). \qquad (35.26)$$

This nonanticipative RDF, although suitable for communication problems, because the source is not affected causally by past reproductions, is not suitable for controlled sources, because the source is affected by past reproductions. Hence, a nonanticipative generalization of rate distortion theory, based on Marko's [5] directed information $I(X^n \rightarrow Y^n)$, appears appropriate.

Understanding *duality of a source and a channel* for nonanticipative communication (and control) reduces to the properties of solutions of capacity expressions (35.22), (35.23) and (35.26) (and its generalizations).

Examples of JSCC Systems. Coding over large block lengths, achieves, under certain conditions, the optimal performance, yet it is not the only choice. The following two examples of source–channel pairs introduced by Shannon reinforce the belief that real-time processing in jointly designed encoders–decoders is optimal.

IID Bernoulli Source-Hamming distortion. The RDF of an IID Bernoulli source with single letter Hamming distortion is given by $R(D) = 1 - H(D)$. The capacity of a binary symmetric DMC with crossover probability ϵ, called BSC(ϵ), is $C = 1 - H(\epsilon)$. Optimal JSSC design is then achieved by setting the average distortion $D = \epsilon$, and this is obtained by uncoded transmission; that is, the encoder and the decoder are unitary maps on their inputs, as shown in Fig. 35.2.

IID Gaussian source-mean-square-error (MSE) distortion. The RDF of an IID Gaussian RV, $N(0, \sigma_X^2)$, with MSE distortion is $R(D) = \frac{1}{2}\log_2(\frac{\sigma_X^2}{D})$ for $D \leq \sigma_X^2$, and zero otherwise, while the capacity of an AWGN memoryless channel, with noise $N(0, \sigma^2)$, and average power constraint P is $C(P) = \frac{1}{2}\log_2(1 + \frac{P}{\sigma^2})$. Optimal JSSC design is then achieved by setting $D = \frac{\sigma^2\sigma_X^2}{\sigma^2+P}$ and realizing the optimal reproduction distribution by scaling the encoder/decoder inputs, as illustrated in Fig. 35.3. These two examples illustrate the simplicity in complexity of the optimal JSCC system, processing information in real time.

Whether the simplicity of the optimal encoder/decoder pair of these examples is the rule, rather than an exception remains, however, to be answered.

35.4 Further Research

The fundamental problem of communication is often separated into sub-problems 1 and 2. Often, such separation compromises optimality compared to that of JSCC

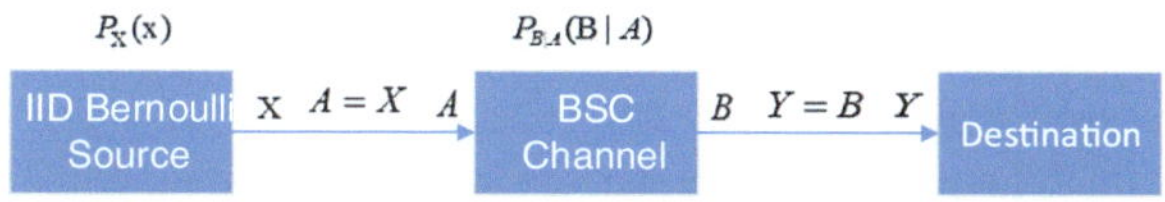

Fig. 35.2 JSCC system of a IID Bernoulli source with Hamming distortion over a BSC

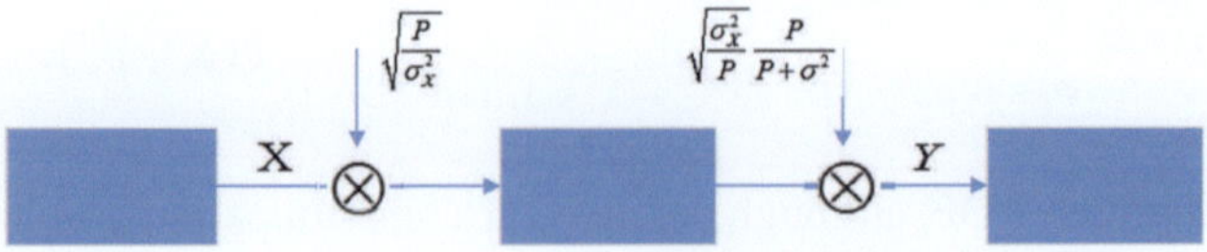

Fig. 35.3 JSCC system of an IID Gaussian source with MSE distortion over an AWGN memoryless channel

design. The understanding of how optimal systems operate can benefit by addressing the following issues.

1. Sub-problem 1. What is the RDF or OPTA by noncausal, causal, and real-time codes, and what are the structural properties of OPTA quantizers?
2. Sub-problem 2. What are the structural properties of capacity-achieving encoders? What is the encoder's role in controlling the channel output process?
3. What are the trade-offs of separating sub-problems 1 and 2 compared to JSCC?
4. What is a proper framework of reliable communication, and communication for control based on real-time information processing?

The direct analogy between a control process and a controlled process is yet to be exploited in problems of information theory. It appears the encoder output is the control process, the channel output is the controlled process, and the role of an encoder is to control channel outputs. Whether control theory can help remove some of the standard assumptions, such as stationary ergodic and information stability, or account for the luck of them, and aid the analysis of finite-horizon communication using JSCC based on a duality of a source and a channel, in real-time, remains to be seen.

Acknowledgments The authors are grateful to Jan H. van Schuppen for stimulating discussions and technical suggestions, during the preparation of this manuscript and throughout the years.

References

1. Shannon CE (1948) A mathematical theory of communication. Bell Sys Tech J 27:379–423, 1948.
2. Berger T (1971) Rate Distortion Theory: a mathematical basis for data compression. Prentice-Hall, Englewood Cliffs
3. Shannon CE (1959) Coding theorems for a discrete source with a fidelity criterion. IRE Nat Conv Rec 4:142–163
4. Richardson T, Urbanke R (2008) Modern coding theory. Cambridge University Press, Cambridge
5. Marko H (1973) The bidirectional communication theory-A generalization of information theory. IEEE Trans Commun 21(12):1345–1351
6. Csiszár I, Körner J (1981) Information theory: coding theorems for discrete memoryless systems. Academic Press, Waltham
7. Gallager RG (1968) Information theory and reliable communication. Wiley, New York
8. Cover TM, Thomas JA (1991) Elements of information theory. Wiley-Interscience, New York

9. Gray RM (1990) Entropy and information theory. Springer, Berlin
10. Han TS (2003) Information-spectrum methods in information theory. Springer, Berlin
11. Gamal AE, Kim YH (2011) Network information theory. Cambridge University Press, Cambridge
12. Ihara S (1993) Information theory for continuous systems. World-Scientific, Singapore
13. Massey JL (1990) Causality, feedback and directed information. In: International symposium on information theory and its applications (ISITA). Nov 27–30:303–305
14. Kramer G (1998) Directed information for channels with feedback. Ph.D thesis, Swiss Federal Institute of Technology, Zurich, Switzerland
15. Tatikonda S (2000) Control under communication constraints. Ph.D thesis, Massachusetts Institute of Technology (MIT), MA, USA
16. Chen J, Berger T (2005) The capacity of finite-state channels with feedback. IEEE Trans Inf Theory 51(3):780–798
17. Tatikonda S, Mitter S (2009) The capacity of channels with feedback. IEEE Trans Inf Theory 55(1):323–349
18. Permuter HH, Weissman T, Goldsmith A (2009) Finite state channels with time-invariant deterministic feedback. IEEE Trans Inf Theory 55(2):644–662
19. Kramer G (2008) Topics in multi-user information theory. Found Trends Inf Theory 4(4–5):265–444
20. Cover TM, Pombra S (1989) Gaussian feedback capacity. IEEE Trans Inf Theory 35(1):37–43
21. Neuhoff DL, Gilbert R (1982) Causal source codes. IEEE Trans Inf Theory 28(5):701–713
22. Berger T (2003) Shannon lecture: living information theory. IEEE Inf Theory Soc Newslett 53(1)
23. Gastpar M, Rimoldi B, Vetterli M (2003) To code or not to code: lossy source-channel communication revisited. IEEE Trans Inf Theory 49(5):1147–1158
24. Wong W, Brockett R (1997) Systems with finite communication bandwidth constraints I: state estimation problems. IEEE Trans Autom Control 42(9):1294–1299
25. Nair GN, Evans RJ, Mareels IMY, Moran W (2004) Topological feedback entropy and nonlinear stabilization. IEEE Trans Autom Control 49(9):1585–1597
26. Gorbunov AK, Pinsker MS (1973) Nonanticipatory and prognostic epsilon entropies and message generation rates. Probl Inf Transm 9(3):184–191

Chapter 36
Directed Information on Abstract Spaces: Properties and Extremum Problems

Charalambos D. Charalambous, Photios A. Stavrou and Christos K. Kourtellaris

36.1 Motivation

In information theory, directed information [1, 2] or its variants is used extensively to characterize capacity of channels with memory and feedback, sequential and nonanticipative lossy compression, and their extensions to networks [3], while in biology, it is used as a measure of causality (alternative to Granger's causality). Directed information is particularly suitable in developing realizable filters, and in addressing problems of reliable control over limited rate channels (noisy or noiseless), when the unobserved processes or control process are affected by the output of the channel [4, 5]. Almost all such applications of directed information involve extremum problems on the space of conditional distributions, with directed information representing the payoff.

The analysis of such extremum problems necessitates the understanding of functional and topological properties of directed information, which is the main scope of this chapter. For readers unfamiliar with Shannon's information transmission theorems, we suggest [6] and Chap. 35 which serves as an introduction.

C.D. Charalambous (✉) · P.A. Stavrou · C.K. Kourtellaris
Department of Electrical and Computer Engineering (ECE),
University of Cyprus, 75 Kallipoleos, P.O. Box 20537, 1678 Nicosia, Cyprus
e-mail: chadcha@ucy.ac.cy

P.A. Stavrou
e-mail: stavrou.fotios@ucy.ac.cy

C.K. Kourtellaris
e-mail: kourtellaris.christos@ucy.ac.cy

J.H. van Schuppen and T. Villa (eds.), *Coordination Control of Distributed Systems*,
Lecture Notes in Control and Information Sciences 456,
DOI 10.1007/978-3-319-10407-2_36

36.2 Problems and Issues

The main issues of extremum problems of directed information are the following.

1. Is directed information concave, respectively, convex with respect to the encoder distribution, respectively, the channel distribution?
2. Is directed information upper and/or lower semicontinuous with respect to the encoder distribution or the channel distribution?
3. Are the encoder admissible set of distributions or the distortion (fidelity) admissible set of distributions compact?
4. What are the appropriate function spaces on which existence of maximizing and minimizing distributions over the encoder admissible set of distributions and the distortion admissible set of distributions, respectively, can be sought?
5. Is it possible to express directed information via variational equalities involving minimization/maximization operations over appropriate sets on the space of measures?

In this chapter, we address the above questions (see [7] for derivations). Applications of these results to information nonanticipative rate-distortion function (RDF), and nonanticipative or real-time communication and communication for control are found in the other two chapters of this book by the same authors (see also [4, 8]).

36.3 Nonanticipative Equivalent Channels on Abstract Spaces

In this section, we present the necessary mathematical constructs to define directed information using relative entropy, as a functional of two consistent families of conditional distributions that uniquely define two families of conditional distributions, $\{P_{X_i|X^{i-1},Y^{i-1}}(\cdot|\cdot,\cdot) : i = 0, 1, \ldots\}$ and $\{P_{Y_i|Y^{i-1},X^i}(\cdot|\cdot,\cdot) : i = 0, 1, \ldots\}$, respectively, and vice versa following [7]. These constructions are vital for the derivation of convexity and concavity of directed information. Throughout the paper, Random Variables $\{(X_n, Y_n) : n = 0, 1, \ldots\}$ take values in Polish spaces (complete separable metric spaces), to include both finite alphabet and continuous alphabet applications (such as Gaussian processes and general function spaces).

Notation. Let $\mathbb{N} \triangleq \{0, 1, 2, \ldots\}$, and $\mathbb{N}^n \triangleq \{0, 1, 2, \ldots, n\}$. Introduce two sequence of measurable spaces $\{(\mathscr{X}_n, \mathscr{B}(\mathscr{X}_n)) : n \in \mathbb{N}\}$ and $\{(\mathscr{Y}_n, \mathscr{B}(\mathscr{Y}_n)) : n \in \mathbb{N}\}$, where $\mathscr{B}(\mathscr{X}_n)$ and $\mathscr{B}(\mathscr{Y}_n)$ are Borel $\sigma-$algebras of subsets of $\mathscr{X}_n$ and $\mathscr{Y}_n$, respectively, which are Polish spaces. Points in $\mathscr{X}^{\mathbb{N}} \triangleq \times_{n\in\mathbb{N}} \mathscr{X}_n$, $\mathscr{Y}^{\mathbb{N}} \triangleq \times_{n\in\mathbb{N}} \mathscr{Y}_n$ are denoted by $\mathbf{x} \triangleq \{x_0, x_1, \ldots\} \in \mathscr{X}^{\mathbb{N}}$, $\mathbf{y} \triangleq \{y_0, y_1, \ldots\} \in \mathscr{Y}^{\mathbb{N}}$, and their restrictions to finite coordinates by $x^n \triangleq \{x_0, x_1, \ldots, x_n\} \in \mathscr{X}_{0,n}$, $y^n \triangleq \{y_0, y_1, \ldots, y_n\} \in \mathscr{Y}_{0,n}$, for $n \in \mathbb{N}^n$. Let $\mathscr{B}(\mathscr{X}^{\mathbb{N}}) \triangleq \odot_{i\in\mathbb{N}} \mathscr{B}(\mathscr{X}_i)$, $\mathscr{B}(\mathscr{Y}^{\mathbb{N}}) \triangleq \odot_{i\in\mathbb{N}} \mathscr{B}(\mathscr{Y}_i)$ denote the $\sigma-$algebras on $\mathscr{X}^{\mathbb{N}}$, $\mathscr{Y}^{\mathbb{N}}$, respectively, generated by cylinder sets, and $\mathscr{B}(\mathscr{X}_{0,n})$, $\mathscr{B}(\mathscr{Y}_{0,n})$ the $\sigma-$algebras of cylinder sets in $\mathscr{X}^{\mathbb{N}}$, $\mathscr{Y}^{\mathbb{N}}$ with basis over $\{0, 1, \ldots, n\}$, respectively. The set of stochastic kernels on $\mathscr{Y}$ given $\mathscr{X}$ is denoted by $\mathscr{Q}(\mathscr{Y}; \mathscr{X})$.

Feedback Channel. Suppose for each $n \in \mathbb{N}$, the distributions $\{p_n(\mathrm{d}x_n; x^{n-1}, y^{n-1}) : n \in \mathbb{N}\}$ satisfy the following conditions.

(i) For $n \in \mathbb{N}$, $p_n(\cdot; x^{n-1}, y^{n-1})$ is a probability measure on $\mathscr{B}(\mathscr{X}_n)$;
(ii) For every $A_n \in \mathscr{B}(\mathscr{X}_n)$, $n \in \mathbb{N}$, $p_n(A_n; x^{n-1}, y^{n-1})$ is a $\odot_{i=0}^{n-1}\big(\mathscr{B}(\mathscr{X}_i) \odot \mathscr{B}(\mathscr{Y}_i)\big)$—measurable function of $x^{n-1} \in \mathscr{X}_{0,n-1}$, $y^{n-1} \in \mathscr{Y}_{0,n-1}$.

Every cylinder set $C \in \mathscr{B}(\mathscr{X}_{0,n})$ has the form $C \triangleq \big\{\mathbf{x} \in \mathscr{X}^{\mathbb{N}} : x_0 \in C_0, \ldots, x_n \in C_n\big\}$, $C_i \in \mathscr{B}(\mathscr{X}_i)$, $i \in \mathbb{N}^n$. Define $C_{0,n} = \times_{i=0}^{n} C_i$. Define a family of measures $\mathbf{P}(\cdot|\mathbf{y})$ on $\mathscr{B}(\mathscr{X}^{\mathbb{N}})$ by

$$\mathbf{P}(C|\mathbf{y}) \triangleq \int_{C_0} p_0(\mathrm{d}x_0) \cdots \int_{C_n} p_n(\mathrm{d}x_n; x^{n-1}, y^{n-1}) \equiv \overleftarrow{P}_{0,n}(C_{0,n}|y^{n-1}). \tag{36.1}$$

The notation $\overleftarrow{P}_{0,n}(\cdot|y^{n-1})$ denotes the restriction of the measure $\mathbf{P}(\cdot|\mathbf{y})$ on cylinder sets $C \in \mathscr{B}(\mathscr{X}_{0,n})$, for $n \in \mathbb{N}$. Thus, if conditions (**i**) and (**ii**) hold then for each $\mathbf{y} \in \mathscr{Y}^{\mathbb{N}}$, the right-hand side (RHS) of (36.1) defines a consistent family of finite-dimensional distribution on $(\mathscr{X}^{\mathbb{N}}, \mathscr{B}(\mathscr{X}^{\mathbb{N}}))$, and hence, there exists a unique measure on $(\mathscr{X}^{\mathbb{N}}, \mathscr{B}(\mathscr{X}^{\mathbb{N}}))$, from which $p_i(\cdot; \cdot, \cdot)$, $i = 1, \ldots, n$ are obtained. An alternative, equivalent definition of a feedback channel is established by considering a family of measures $\mathbf{P}(\cdot|\mathbf{y})$ on $(\mathscr{X}^{\mathbb{N}}, \mathscr{B}(\mathscr{X}^{\mathbb{N}}))$ satisfying the following consistency condition.

C1: If $E \in \mathscr{B}(\mathscr{X}_{0,n})$, then $\mathbf{P}(E|\mathbf{y})$ is $\mathscr{B}(\mathscr{Y}_{0,n-1})$—measurable function of $\mathbf{y} \in \mathscr{Y}^{\mathbb{N}}$. The set of such measures is denoted by $\mathscr{Q}^{\mathbf{C1}}(\mathscr{X}^{\mathbb{N}}; \mathscr{Y}^{\mathbb{N}})$.

For Polish spaces, it can be shown that for any family of measures $\mathbf{P}(\cdot|\mathbf{y})$ satisfying **C1** there exists a collection of conditional distributions $\{p_n(\cdot; \cdot, \cdot) : n \in \mathbb{N}\}$ satisfying conditions (**i**) and (**ii**) which are connected with $\mathbf{P}(\cdot|\mathbf{y})$ via relation (36.1) [7].

Feedforward Channel. The previous methodology applies to the collection of distributions $\{q_n(\mathrm{d}y_n; y^{n-1}, x^n) : n \in \mathbb{N}\}$ which satisfy similar conditions to (**i**) and (**ii**).

Define a family of measures $\mathbf{Q}(\cdot|\mathbf{x})$ on $(\mathscr{Y}^{\mathbb{N}}, \mathscr{B}(\mathscr{Y}^{\mathbb{N}}))$ for $D \in \mathscr{B}(\mathscr{Y}^{\mathbb{N}})$, $D_{0,n} \triangleq \times_{i=0}^{n} D_i \in \mathscr{B}(\mathscr{Y}_{0,n})$, by

$$\mathbf{Q}(D|\mathbf{x}) \triangleq \int_{D_0} q_0(\mathrm{d}y_0; x_0) \ldots \int_{D_n} q_n(\mathrm{d}y_n; y^{n-1}, x^n) \equiv \overrightarrow{Q}_{0,n}(D_{0,n}|x^n). \tag{36.2}$$

Then, (36.2) is a unique measure on $(\mathscr{Y}^{\mathbb{N}}, \mathscr{B}(\mathscr{Y}^{\mathbb{N}}))$ from which $\{q_n(\mathrm{d}y_n; y^{n-1}, x^n) : n \in \mathbb{N}\}$ is obtained.

An equivalent definition of a feedforward channel is a family of measures $\mathbf{Q}(\cdot|\mathbf{x})$ on $(\mathscr{Y}^{\mathbb{N}}, \mathscr{B}(\mathscr{Y}^{\mathbb{N}}))$ satisfying the following consistency condition.

C2: If $F \in \mathscr{B}(\mathscr{Y}_{0,n})$, then $\mathbf{Q}(F|\mathbf{x})$ is $\mathscr{B}(\mathscr{X}_{0,n})$—measurable function of $\mathbf{x} \in \mathscr{X}^{\mathbb{N}}$. The set of such measures is denoted by $\mathscr{Q}^{\mathbf{C2}}(\mathscr{Y}^{\mathbb{N}}; \mathscr{X}^{\mathbb{N}})$.

In feedback capacity problems, the distributions $\{p_n(\mathrm{d}x_n; x^{n-1}, y^{n-1}) : n \in \mathbb{N}\}$ are the feedback encoder distributions and $\{q_n(\mathrm{d}y_n; y^{n-1}, x^n) : n \in \mathbb{N}\}$ are the channel with memory distributions, in which $\{x_n : n \in \mathbb{N}\}$ are the channel inputs and $\{y_n : n \in \mathbb{N}\}$ are the channel outputs. In control problems, $\{p_n(\mathrm{d}x_n; x^{n-1}, y^{n-1}) : n \in \mathbb{N}\}$ are the controlled process and $\{q_n(\mathrm{d}y_n; y^{n-1}, x^n) : n \in \mathbb{N}\}$ are the control process distributions.

36.4 Directed Information and its Properties

Next, we define directed information $I(X^n \to Y^n)$ using $\mathbf{P}(\cdot|\mathbf{y}) \in \mathscr{Q}^{\mathbf{C1}}(\mathscr{X}^{\mathbb{N}}; \mathscr{Y}^{\mathbb{N}})$ and $\mathbf{Q}(\cdot|\mathbf{x}) \in \mathscr{Q}^{\mathbf{C2}}(\mathscr{Y}^{\mathbb{N}}; \mathscr{X}^{\mathbb{N}})$. Define the following measures.
P1: The joint distribution on $\mathscr{X}^{\mathbb{N}} \times \mathscr{Y}^{\mathbb{N}}$

$$P_{0,n}(\times_{i=0}^{n} A_i \times B_i) = (\overleftarrow{P}_{0,n} \otimes \overrightarrow{Q}_{0,n})(\times_{i=0}^{n} A_i \times B_i), \quad A_i \in \mathscr{B}(\mathscr{X}_i),\ B_i \in \mathscr{B}(\mathscr{Y}_i).$$

P2: The marginal distributions on $\mathscr{X}^{\mathbb{N}}$

$$\mu_{0,n}(\times_{i=0}^{n} A_i) = (\overleftarrow{P}_{0,n} \otimes \overrightarrow{Q}_{0,n})(\times_{i=0}^{n}(A_i \times \mathscr{Y}_i)), \quad A_i \in \mathscr{B}(\mathscr{X}_i),\ i = 0, 1, \ldots, n.$$

P3: The marginal distributions on $\mathscr{Y}^{\mathbb{N}}$

$$\nu_{0,n}(\times_{i=0}^{n} B_i) = (\overleftarrow{P}_{0,n} \otimes \overrightarrow{Q}_{0,n})(\times_{i=0}^{n}(\mathscr{X}_i \times B_i)), \quad B_i \in \mathscr{B}(\mathscr{Y}_i),\ i = 0, 1, \ldots, n.$$

P4: The measure $\overrightarrow{\Pi}_{0,n} : \mathscr{B}(\mathscr{X}_{0,n}) \odot \mathscr{B}(\mathscr{Y}_{0,n}) \longmapsto [0, 1]$

$$\overrightarrow{\Pi}_{0,n}(\times_{i=0}^{n}(A_i \times B_i)) \stackrel{\triangle}{=} (\overleftarrow{P}_{0,n} \otimes \nu_{0,n})(\times_{i=0}^{n}(A_i \times B_i)), \quad A_i \in \mathscr{B}(\mathscr{X}_i),\ B_i \in \mathscr{B}(\mathscr{Y}_i).$$

Directed information [7] is expressed via relative entropy as follows.

$$I(X^n \to Y^n) \stackrel{\triangle}{=} \sum_{i=0}^{n} I(X^i; Y_i|Y^{i-1}) = \mathbb{D}(\overleftarrow{P}_{0,n} \otimes \overrightarrow{Q}_{0,n} || \overrightarrow{\Pi}_{0,n}) \tag{36.3}$$

$$= \int \log\Big(\frac{\overrightarrow{Q}_{0,n}(\mathrm{d}y^n|x^n)}{\nu_{0,n}(\mathrm{d}y^n)}\Big)(\overleftarrow{P}_{0,n} \otimes \overrightarrow{Q}_{0,n})(\mathrm{d}x^n, \mathrm{d}y^n) \equiv \mathbb{I}_{X^n \to Y^n}(\overleftarrow{P}_{0,n}, \overrightarrow{Q}_{0,n}). \tag{36.4}$$

$\mathbb{I}_{X^n \to Y^n}(\cdot, \cdot)$ indicates the functional dependence of $I(X^n \to Y^n)$ on $\{\overleftarrow{P}_{0,n}, \overrightarrow{Q}_{0,n}\}$.

The next results are important in establishing existence of solutions to extremum problems of directed information, and several of their properties, similar to extremum problems of mutual information.

Convexity and Concavity of Directed Information. Let $\mathscr{Q}^{\mathbf{C1}}(\mathscr{X}_{0,n}; \mathscr{Y}_{0,n-1})$ denote the projection of $\mathscr{Q}^{\mathbf{C1}}(\mathscr{X}^{\mathbb{N}}; \mathscr{Y}^{\mathbb{N}})$ to a finite number of coordinates, and similarly for $\mathscr{Q}^{\mathbf{C2}}(\mathscr{Y}_{0,n}; \mathscr{X}_{0,n})$. The following convexity/concavity of directed information is analogous to that of mutual information.

Theorem 36.1 [7] *Assume $\{(\mathscr{X}_n, \mathscr{Y}_n) : n \in \mathbb{N}\}$ are Polish spaces. Then,*

(1) $\mathscr{Q}^{\mathbf{C1}}(\mathscr{X}_{0,n}; \mathscr{Y}_{0,n-1})$, $\mathscr{Q}^{\mathbf{C2}}(\mathscr{Y}_{0,n}; \mathscr{X}_{0,n})$ are convex sets.
(2) $\mathbb{I}_{X^n \to Y^n}(\overleftarrow{P}_{0,n}, \overrightarrow{Q}_{0,n})$ is a convex functional of $\overrightarrow{Q}_{0,n} \in \mathscr{Q}^{\mathbf{C2}}(\mathscr{Y}_{0,n}; \mathscr{X}_{0,n})$ for a fixed $\overleftarrow{P}_{0,n} \in \mathscr{Q}^{\mathbf{C1}}(\mathscr{X}_{0,n}; \mathscr{Y}_{0,n-1})$, and a concave functional of $\overleftarrow{P}_{0,n} \in \mathscr{Q}^{\mathbf{C1}}(\mathscr{X}_{0,n}; \mathscr{Y}_{0,n-1})$ for a fixed $\overrightarrow{Q}_{0,n} \in \mathscr{Q}^{\mathbf{C2}}(\mathscr{Y}_{0,n}; \mathscr{X}_{0,n})$

Lower Semicontinuity–continuity of directed information. Next, we discuss the lower semicontinuity and continuity of directed information. Let $\mathbb{Z}_+ \stackrel{\triangle}{=} \{1, 2, \ldots\}$. The notation $P^\alpha \stackrel{w}{\Longrightarrow} P^o$ denotes weak convergence of the sequence of probability measures $\{P^\alpha : \alpha \in \mathbb{Z}_+\}$ to P^o. The following theorem has fundamental implication in extremum problems of directed information, in point to point and network communications.

Theorem 36.2 [7] **Part A**: *Let $\mathscr{Y}_{0,n}$ be a compact Polish space and $\mathscr{X}_{0,n}$ a Polish space. Assume $\overleftarrow{P}_{0,n}(\cdot|y^{n-1}) \in \mathscr{Q}^{\mathbf{C1}}(\mathscr{X}_{0,n}; \mathscr{Y}_{0,n-1})$ satisfy the following condition.*
CA: *For all $g(\cdot) \in BC(\mathscr{X}_n)$*

$$(x^{n-1}, y^{n-1}) \longmapsto \int_{\mathscr{X}_n} g(x) p_n(dx; x^{n-1}, y^{n-1}) \in \mathbb{R} \tag{36.5}$$

is jointly continuous in $(x^{n-1}, y^{n-1}) \in \mathscr{X}_{0,n-1} \times \mathscr{Y}_{0,n-1}$.
Then, the following weak convergence results hold.

(A1) Let $\overleftarrow{P}_{0,n}(\cdot|y^{n-1}) \in \mathscr{Q}^{\mathbf{C1}}(\mathscr{X}_{0,n}; \mathscr{Y}_{0,n-1})$ and $\{\overrightarrow{Q}^{\alpha}_{0,n}(\cdot|x^n) : \alpha \in \mathbb{Z}_+\} \in \mathscr{Q}^{\mathbf{C2}}(\mathscr{Y}_{0,n}; \mathscr{X}_{0,n})$. Then, the joint measure $(\overleftarrow{P}_{0,n} \otimes \overrightarrow{Q}^{\alpha}_{0,n})(dx^n, dy^n) \stackrel{w}{\Longrightarrow} (\overleftarrow{P}_{0,n} \otimes \overrightarrow{Q}^{o}_{0,n})(dx^n, dy^n)$, where $\overrightarrow{Q}^{o}_{0,n}(\cdot|x^n) \in \mathscr{Q}^{\mathbf{C2}}(\mathscr{Y}_{0,n}; \mathscr{X}_{0,n})$.
(A2) Let $\overleftarrow{P}_{0,n}(\cdot|y^{n-1}) \in \mathscr{Q}^{\mathbf{C1}}(\mathscr{X}_{0,n}; \mathscr{Y}_{0,n-1})$ and $\{\overrightarrow{Q}^{\alpha}_{0,n}(\cdot|x^n) : \alpha \in \mathbb{Z}_+\} \in \mathscr{Q}^{\mathbf{C2}}(\mathscr{Y}_{0,n}; \mathscr{X}_{0,n})$ and define the family of joint measures $\{(\overleftarrow{P}_{0,n} \otimes \overrightarrow{Q}^{\alpha}_{0,n})(dx^n, dy^n) : \alpha \in \mathbb{Z}_+\}$ having marginals $\{\nu^\alpha_{0,n} : \alpha \in \mathbb{Z}_+\}$ on $\mathscr{Y}_{0,n}$ and $\{\mu^\alpha_{0,n} : \alpha \in \mathbb{Z}_+\}$ on $\mathscr{X}_{0,n}$. Then, $\nu^\alpha_{0,n}(dy^n) \stackrel{w}{\Longrightarrow} \nu^o_{0,n}(dy^n)$ and $\mu^\alpha_{0,n}(dx^n) \stackrel{w}{\Longrightarrow} \mu^o_{0,n}(dx^n)$, where $\nu^o_{0,n} \in \mathscr{M}_1(\mathscr{Y}_{0,n})$ and $\mu^o_{0,n} \in \mathscr{M}_1(\mathscr{X}_{0,n})$ are the marginals of $\overleftarrow{P}_{0,n} \otimes \overrightarrow{Q}^{o}_{0,n})(dx^n, dy^n)$.

(A3) *The sets of measures* $\mathcal{Q}^{\mathbf{C1}}(\mathcal{X}_{0,n};\mathcal{Y}_{0,n-1})$, *and* $\mathcal{Q}^{\mathbf{C2}}(\mathcal{Y}_{0,n};\mathcal{X}_{0,n})$ *are weakly compact.*

(A4) *Let* $\overleftarrow{P}_{0,n}(\cdot|y^{n-1}) \in \mathcal{Q}^{\mathbf{C1}}(\mathcal{X}_{0,n};\mathcal{Y}_{0,n-1})$, $\{\overrightarrow{Q}^{\alpha}_{0,n}(\cdot|x^n) : \alpha \in \mathbb{Z}_+\} \in \mathcal{Q}^{\mathbf{C2}}(\mathcal{Y}_{0,n};\mathcal{X}_{0,n})$, *and* $\{\nu^{\alpha}_{0,n} : \alpha \in \mathbb{Z}_+\}$ *the marginals of* $\{(\overleftarrow{P}_{0,n} \otimes \overrightarrow{Q}^{\alpha}_{0,n})(dx^n, dy^n) : \alpha \in \mathbb{Z}_+\}$. *Then,* $\overrightarrow{\Pi}^{\alpha}_{0,n}(dx^n, dy^n) \equiv \overleftarrow{P}_{0,n}(dx^n|dy^{n-1}) \otimes \nu^{\alpha}_{0,n}(dy^n) \overset{w}{\Longrightarrow} \overleftarrow{P}_{0,n}(dx^n| dy^{n-1}) \otimes \nu^{o}_{0,n}(dy^n) \equiv \overrightarrow{\Pi}^{o}_{0,n}(dx^n, dy^n)$, *where* $\nu^{o}_{0,n} \in \mathcal{M}_1(\mathcal{Y}_{0,n})$ *is the weak limit of* $\nu^{\alpha}_{0,n} \in \mathcal{M}_1(\mathcal{Y}_{0,n})$.

Part B: *Let* $\mathcal{X}_{0,n}$ *be a compact Polish space and* $\mathcal{Y}_{0,n}$ *a Polish space. Assume* $\overrightarrow{Q}_{0,n}(\cdot|x^n) \in \mathcal{Q}^{\mathbf{C2}}(\mathcal{Y}_{0,n};\mathcal{X}_{0,n})$ *satisfy the following condition.*
CB: *For all* $h(\cdot) \in BC(\mathcal{Y}_n)$

$$(x^n, y^{n-1}) \longmapsto \int_{\mathcal{Y}_n} h(y) q_n(dy; y^{n-1}, x^n) \in \mathbb{R} \tag{36.6}$$

is jointly continuous in $(x^n, y^{n-1}) \in \mathcal{X}_{0,n} \times \mathcal{Y}_{0,n-1}$.
The statements of **Part A** *hold by interchanging* $\overleftarrow{Q}_{0,n}$ *with* $\overleftarrow{P}_{0,n}$, $\nu_{0,n}$ *with* $\mu_{0,n}$, $\overrightarrow{\Pi}_{0,n}$ *with* $\overleftarrow{\Pi}_{0,n} \overset{\triangle}{=} (\mu_{0,n} \otimes \overrightarrow{Q}_{0,n})$.

By using Theorem 36.2 we can show lower semicontinuity of directed information $I(X^n \rightarrow Y^n) \equiv \mathbb{I}_{X^n \rightarrow Y^n}(\overleftarrow{P}_{0,n}, \overrightarrow{Q}_{0,n})$.

Theorem 36.3 [7]

(1) *Suppose the conditions in* Theorem 36.2, **Part A**, *hold. Then,* $\mathbb{I}_{X^n \rightarrow Y^n}(\overleftarrow{P}_{0,n}, \overrightarrow{Q}_{0,n})$ *is lower semicontinuous on* $\overrightarrow{Q}_{0,n} \in \mathcal{Q}^{\mathbf{C2}}(\mathcal{Y}_{0,n};\mathcal{X}_{0,n})$ *for fixed* $\overleftarrow{P}_{0,n} \in \mathcal{Q}^{\mathbf{C1}}(\mathcal{X}_{0,n};\mathcal{Y}_{0,n-1})$.

(2) *Suppose the conditions in* Theorem 36.2, **Part B**, *hold. Then,* $\mathbb{I}_{X^n \rightarrow Y^n}(\overleftarrow{P}_{0,n}, \overrightarrow{Q}_{0,n})$ *is lower semicontinuous on* $\overleftarrow{P}_{0,n} \in \mathcal{Q}^{\mathbf{C1}}(\mathcal{X}_{0,n};\mathcal{Y}_{0,n-1})$ *for fixed* $\overrightarrow{Q}_{0,n} \in \mathcal{Q}^{\mathbf{C2}}(\mathcal{Y}_{0,n};\mathcal{X}_{0,n})$.

Sufficient conditions for continuity of $\mathbb{I}_{X^n \rightarrow Y^n}(\overleftarrow{P}_{0,n}, \overrightarrow{Q}_{0,n})$ as a function of $\overleftarrow{P}_{0,n}$ for fixed $\overrightarrow{Q}_{0,n}$ are given in [7, Theorem III.9].

36.5 Extremum Problems and Variational Equalities

Next, we discuss two extremum problems of information theory, and we present two variational equalities.

Existence of Capacity Achieving Distribution. Consider the finite horizon information capacity of channels with memory and feedback defined by

$$C^{fb}_{0,n} \overset{\triangle}{=} \sup_{\overleftarrow{P}_{0,n}(\cdot|y^{n-1}) \in \overleftarrow{\mathcal{P}}_{0,n}(P) \text{ or } \mathcal{Q}^{\mathbf{C1}}(\mathcal{X}_{0,n};\mathcal{Y}_{0,n-1})} \mathbb{I}_{X^n \rightarrow Y^n}(\overleftarrow{P}_{0,n}, \overrightarrow{Q}_{0,n}) \tag{36.7}$$

where $\overleftarrow{\mathscr{P}}_{0,n}(P)$ is the power constraint defined by

$$\overleftarrow{\mathscr{P}}_{0,n}(P) \triangleq \Big\{ \overleftarrow{P}_{0,n}(\cdot|y^{n-1}) \in \mathscr{Q}^{\mathbf{C1}}(\mathscr{X}_{0,n}; \mathscr{Y}_{0,n-1}) : \int g_{0,n}(x^n, y^{n-1})(\overleftarrow{Q}_{0,n} \otimes \overrightarrow{P}_{0,n})(dx^n, dy^n) \leq P \Big\}$$

and $g_{0,n} : \mathscr{X}_{0,n} \times \mathscr{Y}_{0,n-1} \longmapsto [0, \infty]$ is Borel measurable. Existence of the maximizing distribution is given in [9, Theorem 5], under very general assumptions.

Existence of Nonanticipative rate-distortion achieving distribution. Consider a reconstruction channel $\overrightarrow{Q}_{0,n}(\cdot|x^n) \in \mathscr{Q}^{\mathbf{C2}}(\mathscr{Y}_{0,n}; \mathscr{X}_{0,n})$, a fixed source $\mu_{0,n}(dx^n) \in \mathscr{M}_1(\mathscr{X}_{0,n})$, and define the fidelity set by

$$\overrightarrow{Q}_{0,n}(D) \triangleq \Big\{ \overrightarrow{Q}_{0,n}(\cdot|x^n) \in \mathscr{Q}^{\mathbf{C2}}(\mathscr{Y}_{0,n}; \mathscr{X}_{0,n}) : \int d_{0,n}(x^n, y^n)(\mu_{0,n} \otimes \overrightarrow{Q}_{0,n})(dx^n, dy^n) \leq D \Big\}$$

where $D \geq 0$, and $d_{0,n} : \mathscr{X}_{0,n} \times \mathscr{Y}_{0,n} \longmapsto [0, \infty]$ is Borel measurable.

The finite horizon information nonanticipative RDF is defined by

$$R^{na}_{0,n}(D) \triangleq \inf_{\overrightarrow{Q}_{0,n}(\cdot|x^n) \in \overrightarrow{Q}_{0,n}(D)} \mathbb{I}_{X^n \rightarrow Y^n}(\mu_{0,n}, \overrightarrow{Q}_{0,n}). \tag{36.8}$$

Existence of the minimizing distribution is given in [9, Theorem 6].

Variational Equalities of Directed Information. Many extremum problems in information theory such as capacity and rate distortion are often calculated using the Blahut–Arimoto algorithm (BAA), which uses variational equalities of mutual information. Here, we describe two such variational equalities of directed information.

Let $\mathbf{S}(\cdot|\mathbf{x})$ be any measure on $(\mathscr{Y}^{\mathbb{N}}, \mathscr{B}(\mathscr{Y}^{\mathbb{N}}))$ satisfying consistency condition

C3: If $F \in \mathscr{B}(\mathscr{Y}_{0,n})$, then $\mathbf{S}(F|\mathbf{x})$ is a $\mathscr{B}(\mathscr{X}_{0,n-1})$—measurable.

Denote this family by $\mathbf{S}(\cdot|\mathbf{x}) \in \mathscr{Q}^{\mathbf{C3}}(\mathscr{Y}^{\mathbb{N}}; \mathscr{X}^{\mathbb{N}})$. Then, for $D_{0,n} \triangleq \times_{i=0}^{n} D_i \in \mathscr{B}(\mathscr{Y}_{0,n})$,

$$\mathbf{S}(D|\mathbf{x}) = \int_{D_0} s_0(dy_0) \cdots \int_{D_n} s_n(dy_n; y^{n-1}, x^{n-1}) \equiv \overleftarrow{S}_{0,n}(D_{0,n}|x^{n-1}).$$

Unlike $\overrightarrow{Q}_{0,n}(\cdot|x^n)$, the measure $\overleftarrow{S}_{0,n}(\cdot|x^{n-1})$ is conditioned on $x^{n-1} \in \mathscr{X}_{0,n-1}$. Let $\mathbf{R}(\cdot|\mathbf{y})$ be any family of measures on $(\mathscr{X}^{\mathbb{N}}, \mathscr{B}(\mathscr{X}^{\mathbb{N}}))$ satisfying condition

C4: If $E \in \mathscr{B}(\mathscr{X}_{0,n})$, then $\mathbf{R}(E|\mathbf{y})$ is a $\mathscr{B}(\mathscr{Y}_{0,n})$—measurable.

Denote this family by $\mathbf{R}(\cdot|\mathbf{y}) \in \mathscr{Q}^{\mathbf{C4}}(\mathscr{X}^{\mathbb{N}}; \mathscr{Y}^{\mathbb{N}})$. Then, for $G_{0,n} \triangleq \times_{i=0}^{n} G_i \in \mathscr{B}(\mathscr{X}_{0,n})$

$$\mathbf{R}(G|\mathbf{y}) = \int_{G_0} r_0(dx_0; y_0) \cdots \int_{G_n} r_n(dx_n; x^{n-1}, y^n) \equiv \overrightarrow{R}_{0,n}(G_{0,n}|y^n).$$

Unlike $\overleftarrow{P}_{0,n}(\cdot|y^{n-1})$, the measure $\overrightarrow{R}_{0,n}(\cdot|y^n)$ is conditioned on $y^n \in \mathcal{Y}_{0,n}$. Define another joint distribution on $\left(\mathcal{X}^{\mathbb{N}} \times \mathcal{Y}^{\mathbb{N}}, \odot_{n\in\mathbb{N}}\mathcal{B}(\mathcal{X}_n)\odot\mathcal{B}(\mathcal{Y}_n)\right)$ by $(\overleftarrow{S}_{0,n}\otimes \overrightarrow{R}_{0,n})(\mathrm{d}x^n, \mathrm{d}y^n)$. The next theorem gives the two variational equalities.

Theorem 36.4 [7] **Part A**. *For any arbitrary measure $\bar{\nu}_{0,n} \in \mathcal{M}_1(\mathcal{Y}_{0,n})$, then*

$$\mathbb{I}_{X^n\to Y^n}(\overleftarrow{P}_{0,n}, \overrightarrow{Q}_{0,n}) \triangleq \mathbb{D}(P_{0,n}||\overrightarrow{\Pi}_{0,n}) = \inf_{\bar{\nu}_{0,n}\in\mathcal{M}_1(\mathcal{Y}_{0,n})} \mathbb{D}(\overleftarrow{P}_{0,n}\otimes\overrightarrow{Q}_{0,n}||\overleftarrow{P}_{0,n}\otimes\bar{\nu}_{0,n})$$

and the infimum is achieved at $\bar{\nu}^*_{0,n}(dy^n) = \int_{\mathcal{X}_{0,n}}(\overleftarrow{P}_{0,n}\otimes\overrightarrow{Q}_{0,n})(dx^n, dy^n) \equiv \nu_{0,n}(dy^n)$.
Part B. *For any $\mathbf{S}(\cdot|\mathbf{x}) \in \mathcal{Q}^{\mathbf{C3}}(\mathcal{Y}^{\mathbb{N}}; \mathcal{X}^{\mathbb{N}})$ and $\mathbf{R}(\cdot|\mathbf{y}) \in \mathcal{Q}^{\mathbf{C4}}(\mathcal{X}^{\mathbb{N}}; \mathcal{Y}^{\mathbb{N}})$, then*

$$\begin{aligned}\mathbb{I}_{X^n\to Y^n}(\overleftarrow{P}_{0,n}, \overrightarrow{Q}_{0,n}) &= \mathbb{D}(P_{0,n}||\overrightarrow{\Pi}_{0,n}) \\ &= \sup_{\overleftarrow{S}_{0,n}\otimes\overrightarrow{R}_{0,n}:\ \overleftarrow{S}_{0,n}\in\mathcal{Q}^{\mathbf{C3}},\ \overrightarrow{R}_{0,n}\in\mathcal{Q}^{\mathbf{C4}}} \\ &\quad \int \log\left(\frac{d(\overleftarrow{S}_{0,n}\otimes\overrightarrow{R}_{0,n})}{d(\overrightarrow{\Pi}_{0,n})}\right) d(\overleftarrow{P}_{0,n}\otimes\overrightarrow{Q}_{0,n}) \end{aligned} \tag{36.9}$$

and the supremum is achieved when $\Lambda_{0,n}(x^n, y^n) \triangleq \frac{d(\overleftarrow{P}_{0,n}\otimes\overrightarrow{Q}_{0,n})}{d(\overleftarrow{S}_{0,n}\otimes\overrightarrow{R}_{0,n})} = 1-a.s.,\ n \in \mathbb{N}$.

36.6 Conclusion

In this chapter, we have discussed functional and topological properties of directed information on Polish spaces, and their applications in extremum problems related to feedback capacity, nonanticipative lossy data compression. We have also presented two variational equalities for directed information which are of interest in developing generalized BAA. These results are also important in developing tools for nonanticipative or real-time communication, and communication for control.

References

1. Marko H (1973) The bidirectional communication theory—a generalization of information theory. IEEE Trans Commun 21(12):1345–1351
2. Massey JL (1990) Causality, feedback and directed information. In: International symposium on information theory and its applications (ISITA), 27–30 Nov, pp 303–305
3. Kramer G (1998) Directed information for channels with feedback. PhD thesis, Swiss Federal Institute of Technology, Zurich, Switzerland

4. Charalambous CD, Stavrou PA, Ahmed NU (2014) Nonanticipative rate distortion function and relations to filtering theory. IEEE Trans Autom Control 59(4):937–952
5. Charalambous CD, Stavrou PA (2014) Optimization of directed information and relations to filtering theory. In: European control conference (ECC) (to appear). Strasbourg, France, pp 24–27
6. Csiszár I, Körner J (1981) Information theory: coding theorems for discrete memoryless systems. Academic Press, New York
7. Charalambous CD, Stavrou PA (2013) Directed information on abstract spaces: properties and variational equalities. IEEE Trans Inf Theo (Online). Available: http://arxiv.org/abs/1302.3971
8. Stavrou PA, Kourtellaris CK, Charalambous CD (2014) Applications of information nonanticipative rate distortion function. In: IEEE international symposium on information theory (ISIT) (to appear), Honolulu, HI, USA, 29 June–5 July 2014 (Online). Available: http://arxiv.org/abs/1401.5828v4
9. Charalambous CD, Stavrou PA (2012) Directed information on abstract spaces: properties and extremum problems. In: IEEE international symposium on information theory (ISIT), Cambridge, MA, USA, 1-â 6 July 2012, pp 518–522

Chapter 37
Information Nonanticipative Rate Distortion Function and Its Applications

Photios A. Stavrou, Christos K. Kourtellaris and Charalambos D. Charalambous

37.1 Motivation

In lossy compression source coding with fidelity [1], the sequence of real-valued symbols $X^{\infty} \triangleq \{X_0, X_1, \ldots\}$, $X_j \in \mathscr{X}_j, \forall j \geq 0$, generated by a source distribution $P_{X^{\infty}}$, is transformed by the encoder into a sequence of symbols, the compressed representation $Z^{\infty} \triangleq \{Z_0, Z_1, \ldots\}$ (taking values in a finite alphabet set), which is then transmitted over a noiseless channel. The decoder at the channel output upon observing the compressed representation symbols produces the reproduction sequence $Y^{\infty} \triangleq \{Y_0, Y_1, \ldots\}$, $Y_j \in \mathscr{Y}_j, \forall j \geq 0$. Such a compression system is called causal [2] if the reproduction symbol Y_n of the source symbol X_n, depends on the present and past source symbols $\{X_0, \ldots, X_n\}$ but not on the future source symbols $\{X_{n+1}, X_{n+2}, \ldots\}$. The cascade of the encoder–decoder called the reproduction coder is a family of measurable functions $\{f_n(X_0, \ldots, X_n) \in \mathscr{Y}_n : n = 0, 1, \ldots\}$, while the compressed representation itself may be noncausal and have variable rate [2]. Consequently, the decoder can generate the reproductions with arbitrary delay.

Zero-delay source coding is a subclass of causal coding, with the additional constraint that the compressed representation symbol Z_n, depends on the past and present source symbols $X^n \triangleq \{X_0, X_1, \ldots, X_n\}$, while the reproduction at the decoder Y_n of the present source symbol X_n, depends only on the compressed representation $Z^n \triangleq \{Z_0, Z_1, \ldots, Z_n\}$ received so far. Thus, a zero-delay coding system

P.A. Stavrou (✉) · C.K. Kourtellaris · C.D. Charalambous
Department of Electrical and Computer Engineering (ECE),
University of Cyprus,75 Kallipoleos, P.O. Box 20537, 1678 Nicosia, Cyprus
e-mail: stavrou.fotios@ucy.ac.cy

C.K. Kourtellaris
e-mail: kourtellaris.christos@ucy.ac.cy

C.D. Charalambous
e-mail: chadcha@ucy.ac.cy

J.H. van Schuppen and T. Villa (eds.), *Coordination Control of Distributed Systems*,
Lecture Notes in Control and Information Sciences 456,
DOI 10.1007/978-3-319-10407-2_37

consists of a family of encoding–decoding measurable functions $\{h_i, f_i\}_{i=0}^{\infty}$ such that $Z_i = h_i(\{X_j : j = 0, 1, \ldots, i\})$ and $Y_i = f_i(\{Z_j : j = 0, 1, \ldots, i\})$, $\forall i \geq 0$. On the other hand, the most efficient zero-delay coding system corresponds to joint source-channel coding (JSCC) using zero-delay transmission (see Chap. 38 of this book).

The optimal performance theoretically attainable (OPTA) by noncausal codes is given by the rate distortion function (RDF) of the source [1], the OPTA by causal codes is given by the minimization over causal reproduction coders of the entropy rate of the reproduction sequence [2], and it is an upper bound on the OPTA by noncausal codes, although no closed expression for the conditional distribution is given. In general, very little is known about the performance of causal, zero-delay codes, and JSCC using nonanticipative transmission, with average or excess distortion probability. Often, bounds are introduced to quantify the rate loss due to causality and zero-delay of the coding systems compared to that of noncausal coding systems.

In many delay-sensitive applications of lossy compression, limited end-to-end decoding delay is often desirable, while for real-time systems, such as, communication for control over finite rate channels [3, 4], and in general, for systems involving feedback, causal and zero-delay coding is preferable to noncausal coding.

37.2 Objectives

In this chapter, we introduce the information nonanticipative RDF, as an alternative to the classical information RDF (OPTA by noncausal codes), we identify certain limitations of the later, for delay-sensitive applications, and we proceed further to illustrate the importance of the nonanticipative RDF in

1. JSCC using nonanticipative transmission;
2. bounding the OPTA by noncausal and causal codes for general sources.

Finally, to facilitate the application of the information nonanticipative RDF in 1 and 2, we proceed further to present the expression of the optimal reproduction distribution for nonstationary source-reproduction pairs and give an example.

Chapter 35 of this book serves as an introduction to Information Theory, and Chap. 36 (see also [5]) can be used to formulate the information nonanticipative RDF on general abstract spaces and to show existence of solutions. Chapter 38 demonstrates the operational meaning of information nonanticipative RDF in JSCC using nonanticipative or real-time transmission of information.

37.3 Information Nonanticipative RDF and Motivation

Given a source distribution $P_{X^n}(dx^n)$, a sequence of reproduction distributions $\{P_{Y_i|Y^{i-1},X^i}(dy_i|y^{i-1},x^i): i=0,1,\ldots,n\}$, and a measurable distortion function

$$d_{0,n}:\mathscr{X}_{0,n}\times\mathscr{Y}_{0,n}\longmapsto[0,\infty],\quad d_{0,n}(x^n,y^n)\triangleq\sum_{i=0}^{n}\rho_i(x^i,y^i), \tag{37.1}$$

and an average fidelity set[1] $(D\geq 0)$

$$\begin{aligned}\overrightarrow{\mathscr{Q}}_{0,n}(D)\triangleq\Big\{&\overrightarrow{P}_{Y^n|X^n}(dy^n|x^n)\triangleq\otimes_{i=0}^{n}P_{Y_i|Y^{i-1},X^i}(dy_i|y^{i-1},x^i):\\ &\ell_{d_{0,n}}(\overrightarrow{P}_{Y^n|X^n})\triangleq\int d_{0,n}(x^n,y^n)(\overrightarrow{P}_{Y^n|X^n}\otimes P_{X^n})(dx^n,dy^n)\leq D(n+1)\Big\},\end{aligned} \tag{37.2}$$

the information nonanticipative RDF and its rate are defined by

$$\begin{aligned}R^{\mathrm{na}}_{0,n}(D)&=\inf_{\overrightarrow{P}_{Y^n|X^n}(\cdot|x^n)\in\overrightarrow{\mathscr{Q}}_{0,n}(D)}\int\log\Big(\frac{\overrightarrow{P}_{Y^n|X^n}(dy^n|x^n)}{P_{Y^n}(dy^n)}\Big)(\overrightarrow{P}_{Y^n|X^n}\otimes P_{X^n})(dx^n,dy^n)\\ &=\inf_{\overrightarrow{P}_{Y^n|X^n}(\cdot|x^n)\in\overrightarrow{\mathscr{Q}}_{0,n}(D)}\mathbb{I}_{X^n\rightarrow Y^n}(P_{X^n},\overrightarrow{P}_{Y^n|X^n}),\end{aligned} \tag{37.3}$$

$$R^{\mathrm{na}}(D)=\lim_{n\longrightarrow\infty}\frac{1}{n+1}R^{\mathrm{na}}_{0,n}(D). \tag{37.4}$$

provided the limit exists; if the infimum in (37.3) does not exist, we set $R^{\mathrm{na}}(D)=+\infty$. The notation $\mathbb{I}_{X^n\rightarrow Y^n}(P_{X^n},\overrightarrow{P}_{Y^n|X^n})$ indicates directed information as a functional of $\{P_{X^n},\overrightarrow{P}_{Y^n|X^n}\}$.

At this stage, it is important to recall certain limitations of the classical information RDF [1] with respect to its computation, and its applications to JSCC based on nonanticipative transmission discussed in Chap. 38 (see also [6]), and briefly outlined below, which motivated our interest in the nonanticipative RDF (37.3).

The first limitation of the classical information RDF, the OPTA by noncausal codes, is the computational complexity of finding the solution to this extremum problem. It is only known for a small class of sources, which are memoryless or Gaussian. For example, even for a Binary Symmetric Markov source with parameter p, BSMS(p), the complete characterization of the OPTA by noncausal codes is currently unknown, and only bounds are available [7].

The second limitation of the classical information RDF is the noncausality or anticipative form of the optimal reproduction distribution, which implies that for any time n, the reproduction at time $i\leq n$ of $x_i\in\mathscr{X}$ by $y_i\in\mathscr{Y}$ depends on past, present, and future source symbols (i.e., its is noncausal). That is, the optimal reproduction conditional distribution of the classical RDF has the form

---

[1] $\otimes$ denotes convolution of distributions.

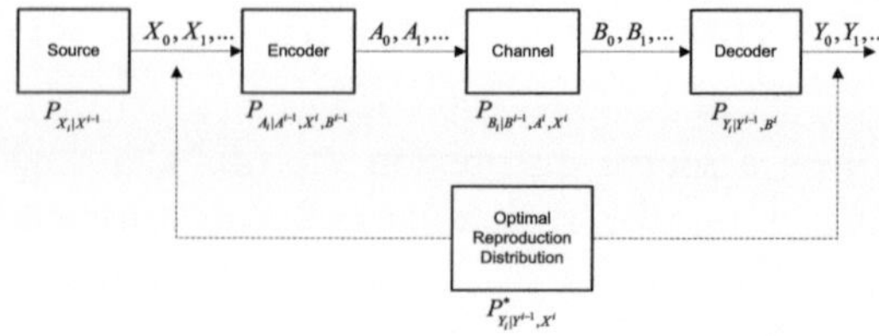

Fig. 37.1 Realization of optimal reproduction distribution

$$P^*_{Y^n|X^n}(\mathrm{d}y^n|x^n) = \otimes_{i=0}^n P^*_{Y_i|Y^{i-1},X^n}(\mathrm{d}y_i|y^{i-1}, x^n). \tag{37.5}$$

The anticipative form of the optimal reproduction distribution (37.5) implies that, in general, the classical information RDF cannot be used in JSCC using nonanticipative transmission as in Fig. 37.1). An exception is the class of independent sources. Indeed, a necessary condition for JSCC via nonanticipative transmission is the realization of the optimal reproduction distribution by an encoder-channel-decoder, which at each time instant processes symbols causally (nonanticipative). This nonanticipative feature of the reproduction distribution is a feature in the two examples of JSCC based on real-time transmission, i.e., the binary IID source with a Hamming distortion, and the IID Gaussian source with mean-square distortion [8].

37.4 Applications of Information Nonanticipative RDF

Bounds. The information nonanticipative RDF is equivalent to introducing a causal constraint on the class of reproduction conditional distributions of the classical information RDF, described by the following equivalent statements.

$$\begin{aligned} X_{i+1}^n \leftrightarrow (X^i, Y^{i-1}) \leftrightarrow Y_i \Longleftrightarrow\ & P_{Y^i|X^n}(\mathrm{d}y^i|x^n) \\ &= P_{Y^i|X^i}(\mathrm{d}y^i|x^i),\ \forall x^n,\ i = 0, 1, \ldots, n-1. \end{aligned}$$

The deterministic interpretation of this causality constraint is that at each instant of time i, Y_i depends on the past and present source symbols $\{X_0, X_1, \ldots, X_i\}$ but not on future source symbols $\{X_{i+1}, X_{i+2}, \ldots\}$, i.e., $Y_i = f_i(X_0, X_1, \ldots, X_i)$, for some measurable f_i, $\forall i \geq 0$. In the context of Neuhoff and Gilbert [2] causal codes, this MC is a probabilistic version of the definition of a causal reproduction coder.

Therefore, by recalling the definition of OPTA by causal codes in [2], denoted by $r^{c,+}(D)$, we also obtain the bounds (in view of the converse to the coding theorem)

$$\begin{aligned} r^{c,+}(D) \geq R^{\mathrm{na},+}(D) &\triangleq \limsup_{n\longrightarrow\infty} \frac{1}{n+1} R^{\mathrm{na}}_{0,n}(D) \geq R^+(D) \\ &\triangleq \limsup_{n\longrightarrow\infty} \frac{1}{n+1} R_{0,n}(D). \end{aligned}$$

where $R_{0,n}(D)$ is the OPTA by noncausal codes. Consequently, by analyzing and computing the exact expression of the nonanticipative RDF, $R^{\text{na},+}(D)$, and its reproduction distribution, we are able to obtain bounds on both $R^{+}(D)$ and $r^{c,+}(D)$, and evaluate the rate loss due to causality even for sources with memory.

Noiseless Coding Theorems using Nonanticipative Transmission. The causal conditioning dependence makes the application of standard tools from noiseless coding theorem very difficult. Nevertheless, our interest is to show achievability of $R^{\text{na}}(D)$ using a noisy coding theorem, in real time, and this can be done via uncoded transmission (i.e., the source is not matched to the channel), with respect to excess distortion probability. An extensive elaboration on the applications of bounds to specific examples, and examples of sources for which $R^{\text{na}}(D)$ is achievable is given in [6]. Chapter 38 of this book discusses JSCC based on nonanticipative transmission and presents an example of a source with memory and a channel with memory and cost constraint operating optimally in real time.

37.5 Nonstationary Optimal Reproduction Distribution of Nonanticipative RDF

The expression of reproduction conditional distribution which achieves the infimum of $R^{\text{na}}_{0,n}(D)$ requires the Gateaux differential of $\mathbb{I}_{X^n\to Y^n}(P_{X^n}, \overrightarrow{P}_{Y^n|X^n})$ in every direction of $\{P_{Y_i|Y^{i-1},X^i}(\mathrm{d}y_i|y^{i-1},x^i): i=0,1,\ldots,n\}$ (due to nonstationarity).

Theorem 37.1 *Let* $\mathbb{I}_{P_{X^n}}(P_{Y_i|Y^{i-1},X^i}: i = 0,1,\ldots,n) \triangleq \mathbb{I}_{X^n\to Y^n}(P_{X^n}, \overrightarrow{P}_{Y^n|X^n})$ *be well defined for every* $\overrightarrow{P}_{Y^n|X^n}$. *Suppose* $\{P_{Y_i|Y^{i-1},X^i}(\cdot|\cdot): i = 0,1,\ldots,n\} \to \mathbb{I}_{P_{X^n}}(P_{Y_i|Y^{i-1},X^i}(\cdot|\cdot): i = 0,1,\ldots,n)$ *is Gateaux differentiable. The Gateaux derivative at the points* $P^*_{Y_i|Y^{i-1},X^i}(\cdot|\cdot)$ *in each direction* $\delta P_{Y_i|Y^{i-1},X^i} = P_{Y_i|Y^{i-1},X^i} - P^*_{Y_i|Y^{i-1},X^i}$, $i=0,\ldots,n$, *is*

$$
\begin{aligned}
&\delta\mathbb{I}_{\mu_{0,n}}(P^*_{Y_i|Y^{i-1},X^i}, \delta P_{Y_i|Y^{i-1},X^i}: i=0,\ldots,n)\\
&=\sum_{i=0}^{n}\int \log\left(\frac{P^*_{Y_i|Y^{i-1},X^i}(dy_i|y^{i-1},x^i)}{P^*_{Y_i|Y^{i-1}}(dy_i|y^{i-1})}\right)\\
&\quad\cdot T_i(P^*_{Y_j|Y^{j-1},X^j}, \delta P_{Y_j|Y^{j-1},X^j}\ : j=0,\ldots,i)P_{X^i}(dx^i),
\end{aligned}
$$

where $T_i(P^*_{Y_j|Y^{j-1},X^j}, \delta P_{Y_j|Y^{j-1},X^j}: j=0,\ldots,i)$, $i=0,1,\ldots,n$ *are given by*

$$
\begin{aligned}
T_0(P^*_{Y_0|Y^{-1},X^0}, \delta P_{Y_0|Y^{-1},X^0}) &= \delta P_{Y_0|Y^{-1},X^0},\\
T_1(P^*_{Y_j|Y^{j-1},X^j}, \delta P_{Y_j|Y^{j-1},X^j}: j=0,1) &= \delta P_{Y_0|Y^{-1},X^0}\otimes P^*_{Y_1|Y^0,X^1} + P^*_{Y_0|Y^{-1},X^0}\otimes \delta P_{Y_1|Y^0,X^1},\\
T_i(P^*_{Y_j|Y^{j-1},X^j}, \delta P_{Y_j|Y^{j-1},X^j}: j=0,1,\ldots,i) &= \delta P_{Y_0|Y^{-1},X^0}\otimes^{i}_{j=1} P^*_{Y_j|Y^{j-1},X^j}
\end{aligned}
$$

$$+ P^*_{Y_0|Y^{-1},X^0} \otimes \delta P_{Y_1|Y^0,X^1} \otimes_{j=2}^{i} P^*_{Y_j|Y^{j-1},X^j}$$
$$+ \ldots + \otimes_{j=0}^{i-1} P^*_{Y_j|Y^{j-1},X^j} \otimes \delta P_{Y_i|Y^{i-1},X^i}.$$

The constrained problem defined by (37.3) can be reformulated using Lagrange multipliers (due to its convexity) to obtain

$$R^{\rm na}_{0,n}(D) = \sup_{s\leq 0} \inf_{\overrightarrow{P}_{Y^n|X^n}} \Big\{ \mathbb{I}_{X^n\rightarrow Y^n}(P_{X^n}, \overrightarrow{P}_{Y^n|X^n}) - s\Big(\ell_{d_{0,n}}(\overrightarrow{P}_{Y^n|X^n}) - (n+1)D\Big)\Big\}, \quad (37.6)$$

where $s \leq 0$ is the Lagrange multiplier associated with the fidelity constraint. Since $P_{Y_i|Y^{i-1},X^i}$ are probability measures, we also introduce another set of Lagrange multipliers to obtain an optimization problem without constraints, and then use (37.6) and Theorem 37.1, to obtain the following recursions.

Recursive Expressions of Optimal Nonstationary Reproduction Distribution. For $i \triangleq n-k, k=0,1,\ldots,n$, we have

$$g_{n,n}(x^n, y^n) = 0, \quad g_{i,n}(x^i, y^i) = -\int P_{X_{i+1}|X^i}(dx_{i+1}|x^i) \log\Big(\int e^{s\rho_{i+1}(x^{i+1},y^{i+1}) - g_{i+1,n}(x^{i+1},y^{i+1})} \times P^*_{Y_{i+1}|Y^i}(dy_{i+1}|y^i)\Big), \quad (37.7)$$

and the optimal (nonstationary) reproduction distributions are

$$P^*_{Y_n|Y^{n-1},X^n}(dy_n|y^{n-1},x^n) = \frac{e^{s\rho_n(x^n,y^n)} P^*_{Y_n|Y^{n-1}}(dy_n|y^{n-1})}{\int_{\mathcal{Y}_n} e^{s\rho_n(x^n,y^n)} P^*_{Y_n|Y^{n-1}}(dy_n|y^{n-1})}, \quad (37.8)$$

$$P^*_{Y_i|Y^{i-1},X^i}(dy_i|y^{i-1},x^i) = \frac{e^{s\rho_i(x^i,y^i) - g_{i,n}(x^i,y^i)} P^*_{Y_i|Y^{i-1}}(dy_i|y^{i-1})}{\int_{\mathcal{Y}_i} e^{s\rho_i(x^i,y^i) - g_{i,n}(x^i,y^i)} P^*_{Y_i|Y^{i-1}}(dy_i|y^{i-1})}. \quad (37.9)$$

The closed form expression of the nonstationary nonanticipative RDF is given by

$$R^{\rm na}_{0,n}(D) = s(n+1)D - \sum_{i=0}^{n} \int \Big(\int g_{i,n}(x^i, y^i) P^*_{Y_i|Y^{i-1},X^i}(dy_i|y^{i-1},x^i) + \log \int e^{s\rho_i(x^i,y^i) - g_{i,n}(x^i,y^i)} P^*_{Y_i|Y^{i-1}}(dy_i|y^{i-1})\Big) \overrightarrow{P}^*_{Y^{i-1}|X^{i-1}}(dy^{i-1}|x^{i-1}) P_{X^i}(dx^i).$$

The above recursions are necessary conditions for the optimality of the nonstationary nonanticipative RDF. Several additional properties, including necessary and sufficient conditions, should be derived to aid the computation of the optimal nonstationary reproduction distributions, similar to the stationary case discussed in [6].

Discussion of Recursions (37.7)–(37.9). These recursions illustrate the nonanticipation, since $g_{i,n}(\cdot,\cdot)$, $i=n-k$, $k=0,1,\ldots,n$, appearing in the exponent of the reproduction distributions (37.9) integrate out future reproduction distributions, which is a consequence of dynamic programming. The above recursions are general, while depending on the assumptions imposed on the distortion function and source they can be simplified considerably.

Controlled Sources. For controlled sources *cau*sally affected by reproduction symbols, the source conditional distributions are $\{P_{X_i|X^{i-1},Y^{i-1}} : i=0,1,\ldots,n\}$.

Stationary Case. If the joint process $\{X_i, Y_i) : i=0,1,\ldots,\}$ is stationary the terms $g_{i,n}(\cdot,\cdot)$ appearing in (37.9) cancel, and we obtain $P^*_{Y_n|Y^{n-1},X^n}(\cdot|\cdot)=P^*(\cdot|\cdot), i=0,1,\ldots$ given by (37.8). This case is investigated in [6, 9], in the context of realizable filters and JSCC based on nonanticipative transmission for general Gaussian processes.

Example: BSMS(p) and Hamming Distortion. Consider a BSMS(p), with stationary transition probabilities $\left\{P_{X_i|X_{i-1}}(x_i|x_{i-1}) : (x_i,x_{i-1})\in\{0,1\}\times\{0,1\}\right\}$ given by $P_{X_i|X_{i-1}}(0|0)=P_{X_i|X_{i-1}}(1|1)=1-p$, $P_{X_i|X_{i-1}}(1|0)=P_{X_i|X_{i-1}}(0|1)=p$, $i\in 0,1,\ldots$, and single letter Hamming distortion $\rho(x,y)=0$ if $x=y$ and $\rho(x,y)=1$ if $x\neq y$. The solution to the nonanticipative RDF is given below [6, Theorem IV.11].

$$R^{\mathrm{na}}(D)=\begin{cases} H(m)-H(D) & \text{if } D\leq \frac{1}{2},\\ 0 & \text{otherwise,}\end{cases}\qquad m=1-p-D+2pD, \tag{37.10}$$

and the optimal reproduction distribution is

$$P^*_{Y_i|X_i,Y_{i-1}}(y_i|x_i,y_{i-1})=\begin{pmatrix} \alpha & \beta & 1-\beta & 1-\alpha\\ 1-\alpha & 1-\beta & \beta & \alpha\end{pmatrix},$$

where $\alpha=\frac{(1-p)(1-D)}{1-p-D+2pD}$, $\beta=\frac{p(1-D)}{p+D-2pD}$.

For $p=\frac{1}{2}$ the BSMS($\frac{1}{2}$) reduces to an IID Bernoulli source, and $R^{\mathrm{na}}(D)=1-H(D)\equiv R(D)$, $D<\frac{1}{2}$, as expected. The graph of $R^{\mathrm{na}}(D)$ is illustrated in Fig. 37.2. The operational meaning of $R^{\mathrm{na}}(D)$ using real-time transmission is shown

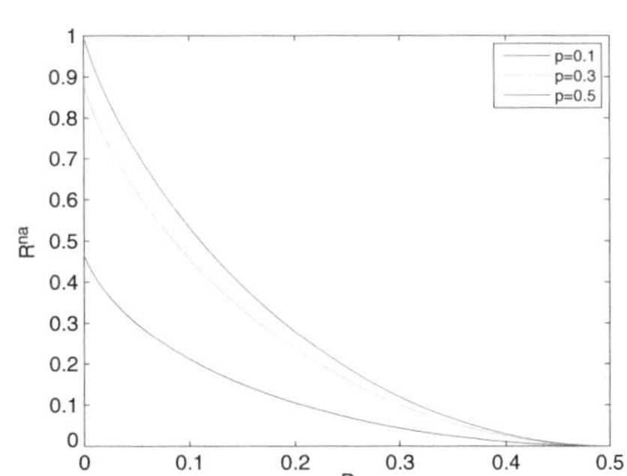

Fig. 37.2 $R^{\mathrm{na}}(D)$ for different values of parameter p

via a noisy coding theorem, with respect to the excess distortion probability, for finite "n, " while the error exponent is computed for $n \to \infty$ (see [6]).

The evaluation of the upper bound $R(D) \leq R^{\mathrm{na}}(D)$ and its comparison to the upper bound given by Berger in [7], and of the rate loss of noncausal codes with respect to causal and delayless codes is discussed in [6].

37.6 Conclusion and Open Issues

In this chapter, we discussed various applications of the information nonanticipative RDF in bounding the OPTA by noncausal and causal codes, and in designing JSCC systems using nonanticipative transmissions. The Binary Symmetric Markov Source is used as an illustrative example.

Computing the information nonanticipative RDF for nonstationary sources using the recursions (37.7)–(37.9) is an open problem, and so is the problem of designing JSCC systems operating in real time, for nonstationary sources. Further understanding of the structural properties of the nonanticipative optimal reproduction distributions, and properties of $R^{\mathrm{na}}_{0,n}(D)$ is necessary to complete the analysis.

References

1. Berger T (1971) Rate distortion theory: a mathematical basis for data compression. Prentice-Hall, Englewood Cliffs
2. Neuhoff DL, Gilbert RK (1982) Causal source codes. IEEE Trans Inf Theor 28(5):701–713
3. Charalambous CD, Farhadi A (2008) LQG optimality and separation principle for general discrete time partially observed stochastic systems over finite capacity communication channels. Automatica 44(12):3181–3188
4. Charalambous CD, Kourtellaris CK, Hadjicostis C (2011) Optimal encoder and control strategies in stochastic control subject to rate constraints for channels with memory and feedback. In: 50th IEEE CDC/ECC, 12–15 Dec 2011, pp 4522-â4527
5. Charalambous CD, Stavrou PA (2012) Directed information on abstract spaces: properties and extremum problems. In: IEEE international symposium on information theory (ISIT), Cambridge, MA, USA, 1-â6 Jul 2012, pp 518–522
6. Stavrou PA, Kourtellaris CK, Charalambous CD (2014) Information nonanticipative rate distortion function and its applications. IEEE Trans Inf Theor (submitted) [Online]. Available: http://arxiv.org/abs/1405.1593v1
7. Berger T (1977) Explicit bounds to R(D) for a binary symmetric Markov source. IEEE Trans Inf Theor 23(1):52–59
8. Berger T (2003) Living information theory. IEEE Inf Theor Soc Newslett 53(1):6–19
9. Charalambous CD, Stavrou PA, Ahmed NU (2014) Nonanticipative rate distortion function and relations to filtering theory. IEEE Trans Autom Control 59(4):357–352

Chapter 38
Nonanticipative Duality of Sources and Channels with Memory and Feedback

Christos K. Kourtellaris, Charalambos D. Charalambous and Photios A. Stavrou

38.1 Nonanticipative Model of Communication

The main objective of this chapter is to unfold a duality of sources and channels reported by Shannon (see Chap. 35). The framework is general, sources, and channels have memory and feedback, and a transmission cost is often imposed. The methodology is based on addressing the following questions.

- Question 1: What are the information structures of capacity achieving encoders and channel input distributions?
- Question 2: What are the information structures of optimal reproduction distributions of nonanticipative RDF?

The other three chapters of this book by the same authors, introduced background material on information theoretic concepts, such as, direct and converse coding theorems, extremum problems of directed information, and nonanticipative RDF and its applications, which they are utilized in this chapter.

Nonanticipative Communication Systems. Consider Fig. 38.1. The alphabet spaces of the source output, channel input, channel output, and decoder output are sequences of Polish spaces (complete separable metric spaces), $\{\mathcal{X}_i : i = 0, \ldots, n\}$, $\{\mathcal{A}_i : i = 0, \ldots, n\}$, $\{\mathcal{B}_i : i = 0, \ldots, n\}$ and $\{\mathcal{Y}_i : i = 0, \ldots, n\}$, and $\mathcal{X}^n \triangleq \times_{k=0}^n \mathcal{X}_k$, $\mathcal{A}^n \triangleq \times_{k=0}^n \mathcal{A}_k$, $\mathcal{B}^n \triangleq \times_{k=0}^n \mathcal{B}_k$, $\mathcal{Y}^n \triangleq \times_{k=0}^n \mathcal{Y}_k$. Let $x^n \triangleq \{x_0, x_1, \ldots, x_n\} \in \mathcal{X}^n$, and similarly for $a^n \in \mathcal{A}^n$, $b^n \in \mathcal{B}^n$, $y^n \in \mathcal{Y}^n$. The conditional independence of

C.K. Kourtellaris (✉)
The Department of Electrical and Computer Engineering (ECE),
University of Cyprus, 75 Kallipoleos, P.O. Box 20537, 1678 Nicosia, Cyprus
e-mail: kourtellaris.christos@ucy.ac.cy

C.D. Charalambous
e-mail: chadcha@ucy.ac.cy

P.A. Stavrou
e-mail: stavrou.fotios@ucy.ac.cy

J.H. van Schuppen and T. Villa (eds.), *Coordination Control of Distributed Systems*,
Lecture Notes in Control and Information Sciences 456,
DOI 10.1007/978-3-319-10407-2_38

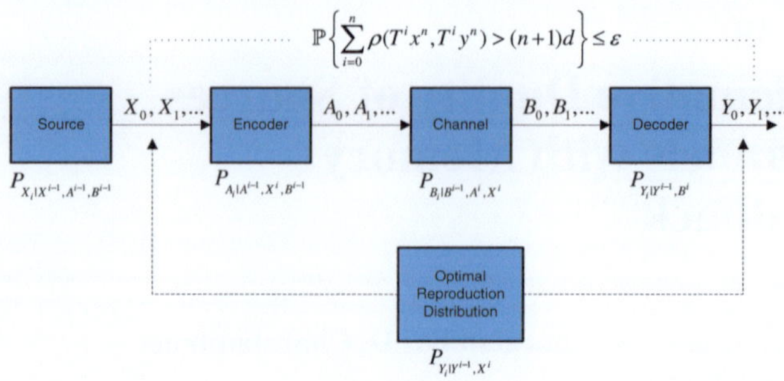

Fig. 38.1 Diagram of a nonanticipative communication system

Random Variables (RVs) X, Y given RV B is defined by the conditional distribution property $P(\mathrm{d}x, \mathrm{d}y|B = b) = P\mathrm{d}x|B = b) \otimes P(\mathrm{d}y|B = b) - a.a.b \in \mathscr{B}$ or equivalently by the Markov Chain (MC) $X \leftrightarrow B \leftrightarrow Y$. Conditional distributions $P(\mathrm{d}b|A = a)$ are represented by stochastic kernels $Q(\mathrm{d}b|a)$.
In nonanticipative communication, the outputs of subsystems appearing in Fig. 38.1 depend causally on input–outputs, as defined below.

Definition 38.1 (*Source–Encoder–Channel–Decoder*) The source, encoder, channel, and decoder in Fig. 38.1 are, respectively, defined, for $n = 0, 1, \ldots$, by

$$\overleftarrow{S}_{0,n}(\mathrm{d}x^n|a^{n-1}, b^{n-1}) \triangleq \otimes_{i=0}^{n} s_i(\mathrm{d}x_i|x^{i-1}, a^{i-1}, b^{i-1}), \quad , i = 0, \ldots, n; \quad (38.1)$$

$$\overrightarrow{P}_{0,n}(\mathrm{d}a^n|b^{n-1}, x^n) \triangleq \otimes_{i=0}^{n} p_i(\mathrm{d}a_i|a^{i-1}, b^{i-1}, x^i), \quad i = 0, \ldots, n \quad (38.2)$$

$$\overrightarrow{Q}_{0,n}(\mathrm{d}b^n|a^n, x^n) \triangleq \otimes_{i=0}^{n} q_i(\mathrm{d}b_i|b^{i-1}, a^i, x^i); \quad (38.3)$$

$$\overrightarrow{R}_{0,n}(\mathrm{d}y^n|b^n) \triangleq \otimes_{i=0}^{n} r_i(\mathrm{d}y_i|y^{i-1}, b^i). \quad (38.4)$$

Deterministic feedback encoders classes $\mathscr{E}_{[0,n]}^{DF}, \mathscr{E}_{[0,n]}^{DM}$ are measurable functions

$$\mathscr{E}_{[0,n]}^{DF} \triangleq \Big\{e_i : \mathscr{A}^{i-1} \times \mathscr{B}^{i-1} \times \mathscr{X}^i \mapsto \mathscr{A}_i, a_i = e_i(a^{i-1}, b^{i-1}, x^i) : i = 0, \ldots, n\Big\},$$

$$\mathscr{E}_{[0,n]}^{DM} \triangleq \Big\{\{e_i(\cdot)\}_{i=0}^{n} \in \mathscr{E}_{[0,n]}^{DF} : e_i(a^{i-1}, x^i, b^{i-1}) = g_i(x_i, b^{i-1}), i = 0, \ldots, n, \Big\},$$

and deterministic decoders $\{y_i = d_i(y^{i-1}, b^i) : i = 0, \ldots, n\}$ are similarly defined.

Any sequence of stochastic kernels appearing in the right-hand side of (38.1)–(38.4) uniquely defines the measures on the left, and vice-versa (see Chap. 36). Given the source, encoder, channel, and decoder, the joint measure is defined by

$$\mathbb{P}(\mathrm{d}x^n, \mathrm{d}a^n, \mathrm{d}b^n, \mathrm{d}y^n) = \otimes_{i=0}^{n} \Big(r_i(\mathrm{d}y_i|y^{i-1}, b^i) \otimes q_i(\mathrm{d}b_i|b^{i-1}, a^i, x^i) \\ \otimes p_i(\mathrm{d}a_i|a^{i-1}, b^{i-1}, x^i) \otimes s_i(\mathrm{d}x_i|x^{i-1}, a^{i-1}, b^{i-1}) \Big). \tag{38.5}$$

The definition of the measure (38.5) implies the following conditional independence:

$$Y^{i-1} \leftrightarrow (X^{i-1}, A^{i-1}, B^{i-1}) \leftrightarrow X_i, \quad i = 1, \ldots, n; \tag{38.6}$$

$$Y^{i-1} \leftrightarrow (A^{i-1}, B^{i-1}, X^i) \leftrightarrow A_i, \quad i = 0, \ldots, n; \tag{38.7}$$

$$Y^{i-1} \leftrightarrow (A^i, B^{i-1}, X^i) \leftrightarrow B_i, \quad i = 0, \ldots, n; \tag{38.8}$$

$$(A^i, X^i) \leftrightarrow (B^i, Y^{i-1}) \leftrightarrow Y_i, \quad i = 0, \ldots, n. \tag{38.9}$$

The distortion function between the source and its reproduction, $d_{0,n} : \mathscr{X}^n \times \mathscr{Y}^n \longmapsto [0, \infty]$, and cost of transmitting symbols over the channel, $c_{0,n} : \mathscr{X}^n \times \mathscr{A}^n \times \mathscr{B}^{n-1} \longmapsto [0, \infty]$ are measurable functions defined by

$$d_{0,n}(x^n, y^n) \triangleq \sum_{i=0}^{n} \rho(T^i x^n, T^i y^n), \quad c_{0,n}(x^n, a^n, b^{n-1}) \triangleq \sum_{i=0}^{n} \gamma(T^i x^n, T^i a^n, T^i b^{n-1}).$$

where for each $i \in \{0, 1, \ldots, n\}$, $T^i z^n$ is a causal mapping, i.e, a measurable function of $z^i \triangleq \{z_0, z_1, \ldots, z_i\}$, and $T^i z^{n-1}$ a measurable function of z^{i-1}.

A nonanticipative code of a source and a channel is defined as follows.

Definition 38.2 (*Nonanticipative Code*) A nonanticipative code, (n, d, ϵ, P), is a 10-tuple

$$\Big\{ \mathscr{X}_{0,n}, \mathscr{A}_{0,n}, \mathscr{B}_{0,n}, \mathscr{Y}_{0,n}, \overleftarrow{S}_{0,n}(\mathrm{d}x^n|a^{n-1}, b^{n-1}), \overrightarrow{P}_{0,n}(\mathrm{d}a^n|b^{n-1}, x^n), \\ \overrightarrow{Q}_{0,n}(\mathrm{d}b^n|a^n, x^n), \overrightarrow{R}_{0,n}(\mathrm{d}y^n|b^n), d_{0,n}, c_{0,n} \Big\}, \tag{38.10}$$

with excess distortion probability and transmission cost

$$\mathbb{P}\Big\{ d_{0,n}(X^n, Y^n) > (n+1)d \Big\} \leq \epsilon, \qquad \epsilon \in (0, 1), \qquad d \geq 0, \tag{38.11}$$

$$\frac{1}{n+1} \mathbb{E}\Big\{ c_{0,n}(X^n, A^n, B^{n-1}) \Big\} \leq P, \qquad P \geq 0. \tag{38.12}$$

A nonanticipative symbol-by-symbol (SbS) code is a code (n, d, ϵ, P), in which the encoder distribution satisfies $p_i(\mathrm{d}a_i|a^{i-1}, b^{i-1}, x^i) = p_i(\mathrm{d}a_i|b^{i-1}, x_i) - a.a.\ (a^{i-1}, b^{i-1}, x^i)$ and/or $r_i(\mathrm{d}y_i|y^{i-1}, b^i) = r_i(\mathrm{d}y_i|y^{i-1}, b_i) - a.a.(y^{i-1}, b^i), i = 0, \ldots, n$.
An uncoded transmission is a nonanticipative code with identity map encoder and decoder, i.e., $A_i = X_i, Y_i = B_i, i = 1, \ldots, n$. Nonanticipative codes embed deterministic nonanticipative codes as a special case, (i.e., $\mathscr{E}^{DF}_{[0,n]}, \mathscr{E}^{DM}_{[0,n]}$).

Definition 38.3 (*Minimum Excess Distortion*) The minimum excess distortion achievable by a nonanticipative code, (n, d, ϵ, P), is defined by

$$D^o(n, \epsilon, P) \stackrel{\triangle}{=} \inf\{d : \exists (n, d, \epsilon, P) \ \text{nonanticipative code}\}.$$

38.2 Information Structures of Extremum Problems of Capacity and Nonanticipative RDF

This section provides answers to Question 1 and Question 2.

Information Structures of Extremum Problems of Nonanticipative RDF. Given a source $\{P_{X_i|X^{i-1},Y^{i-1}}(dx_i|x^{i-1}, y^{i-1}) : i = 0, \dots, n\} \equiv \overleftarrow{P}_{X^n|Y^{n-1}}(x^n|y^{n-1})$, the optimal reproduction distribution $\{Q_{Y_i|Y^{i-1},X^i}(dy_i|y^{i-1}, x^i) : i = 0, \dots, n\} \equiv \overrightarrow{Q}_{Y^n|X^n}(dy^n|x^n)$, with fidelity $\mathscr{Q}^{na}_{0,n}(D) \stackrel{\triangle}{=} \{\overrightarrow{Q}_{Y^n|X^n} : \frac{1}{n+1}\mathbb{E}\Big(d_{0,n}(X^n, Y^n)\Big) \leq D\}$, is given by the solution of nonanticipative RDF (see Eq. 38.18)

$$R^{na}(D) \stackrel{\triangle}{=} \lim_{n\to\infty} \frac{1}{n+1} R^{na}_{0,n}(D), \quad R^{na}_{0,n}(D) = \inf_{\overrightarrow{P}_{Y^n|X^n} \in \mathscr{Q}^{na}_{0,n}(D)} I(X^n \to Y^n). \tag{38.13}$$

The following information structures of $R^{na}_{0,n}(D)$ follow directly from Chap. 37.

Theorem 38.1 (Information Structures of $R^{na}_{0,n}(D)$) *Suppose the following conditions hold.*
(a) the distortion function has the form $d_{0,n}(x^n, y^n) \stackrel{\triangle}{=} \sum_{i=0}^n \rho(x_i, y^i)$;
(b) the source is Markov with respect to $\{X_i : i = 0, \dots, n\}$.
Then, the optimal reproduction distribution has the property $P^*_{Y_i|Y^{i-1},X^i}(dy_i|y^{i-1}, x^i) = P^*_{Y_i|Y^{i-1},X_i}(dy_i|y^{i-1}, x_i)$ $-a.a.(y^{i-1}, x^i)$, *and it is given by*

$$P^*_{Y_i|Y^{i-1},X_i}(dy_i|y^{i-1}, x_i) = \frac{e^{s\rho(x_i,y^i) - g_{i,n}(x_i,y^i)} P^*_{Y_i|Y^{i-1}}(dy_i|y^{i-1})}{\int_{\mathscr{Y}_i} e^{s\rho(x_i,y^i) - g_{i,n}(x_i,y^i)} P^*_{Y_i|Y^{i-1}}(dy_i|y^{i-1})}, \ i = 0, \dots, n. \tag{38.14}$$

for some $\{g_{i,n}(\cdot, \cdot) : i = 0, \dots, n\}$, $g_{n,n} = 0$, $s < 0$, *i.e., it is Markov with respect to* $\{X_i : i = 0, \dots, n\}$. *If in addition, (c)* $\{(X_i, Y_i) : i = 0, \dots, n\}$ *is jointly stationary, then* (38.14) *is stationary, and* $\{g_{i,n}(\cdot, \cdot) = 0 : i = 0, \dots, n\}$.

The optimal stationary reproduction distribution of the nonanticipative RDF for partially observed Gaussian sources is derived in [1].

Information Structures of Extremum Problems of Capacity. From converse channel coding theorem, Chap. 35, any achievable transmission rate is bounded above by $\sum_{i=0}^{n} I(X^i, A^i; B_i|B^{i-1})$, $A_i = e_i(A^{i-1}, B^{i-1}, X^i)$, $i = 0, \ldots, n$. Define the transmission cost constraint

$$\mathscr{P}_{0,n}(P) \triangleq \Big\{p_i(\mathrm{d}a_i|a^{i-1}, b^{i-1}, x^i), i = 0, \ldots, n : \frac{1}{n+1}\sum_{i=0}^{n} \mathbb{E}\{\gamma(T^i X^n, T^i A^n, T^i B^{n-1})\} \le P\Big\}.$$

The next theorem summarizes the information structures of encoders $\mathscr{E}_{[0,n]}^{DF}$ (or distributions $\mathscr{P}_{0,n}(P)$) , maximizing this upper bound (derivations are found in [2]).

Theorem 38.2 (Information Structures of Extremum Problems of Capacity) *Suppose the source and channel are defined by* (38.1) *and* (38.3), *respectively.*
A. For any encoder of class $\mathscr{E}_{[0,n]}^{DF}$ *the following hold.*

$$I(X^n \to B^n) = \sum_{i=0}^{n} I(X^i, A^i; B_i|B^{i-1})\Big|_{\{A_j=e_j(A^{j-1},B^{j-1},X^j)\}_{j=0}^{i}} \tag{38.15}$$

$$= \Big(\sum_{i=0}^{n} I(X^i; B_i|B^{i-1}, A^i) + \sum_{i=0}^{n} I(A^i; B_i|B^{i-1})\Big)\Big|_{\{A_j=e_j(A^{j-1},B^{j-1},X^j)\}_{j=0}^{i}}$$

$$\stackrel{(a)}{=} \sum_{i=0}^{n} I(A^i; B_i|B^{i-1})\Big|_{\{A_j=e_j(A^{j-1},B^{j-1},X^j)\}_{j=0}^{i}}, \tag{38.16}$$

where (*a*) *holds if the map* $\{e_j(\cdot, b^{j-1}) : \mathscr{X}^j \mapsto \mathscr{B}_j, j = 0, \ldots, n\}$ *is one-to-one and onto and its inverse is measurable.*
B. Suppose the source and channel satisfy the conditional independence

$$s_i(x_i|x^{i-1}, a^{i-1}, b^{i-1}) = s_i(dx_i|x_{i-1}, a_{i-1}, b^{i-1}), \quad i = 0, \ldots, n, \tag{38.17}$$

$$q_i(db_i|b^{i-1}, a^i, x^i) = q_i(db_i|b^{i-1}, a_i, x_i), \quad i = 0, \ldots, n. \tag{38.18}$$

Then, the following hold.
1. For $\{e_i(\cdot)\}_{i=0}^{n} \in \mathscr{E}_{[0,n]}^{DF}$ *then* $I(X^n \to B^n) = \sum_{i=0}^{n} I(X_i, A_i; B_i|B^{i-1})$, *where*

$$I(X_i, A_i; B_i|B^{i-1}) = \mathbb{E}^e\Big\{\log\Big(\frac{q_i(dB_i|B^{i-1}, X_i, g_i(A^{i-1}, X^i, B^{i-1}))}{\nu_i^e(dB_i|B^{i-1})}\Big)\Big\}, \quad i = 0, \ldots, n; \tag{38.19}$$

2. The maximization of $I(X^n \to B^n)$ *over encoders* $\mathscr{E}_{[0,n]}^{DF}$ *occurs in* $\mathscr{E}_{[0,n]}^{DM}$ *and*

$$\sup_{\{e_j(x^j,a^{j-1},b^{j-1})\}_{j=0}^n \in \mathscr{E}^{DF}_{[0,n]}} I(X^n \to B^n) = \sup_{\{g_j(x_j,b^{j-1})\}_{j=0}^n \in \mathscr{E}^{DM}_{[0,n]}} \sum_{i=0}^n I(X_i, A_i; B_i|B^{i-1}),$$

$$I(X_i, A_i; B_i|B^{i-1}) = \mathbb{E}^g\Big\{\log\Big(\frac{q_i(dB_i|B^{i-1}, X_i, g_i(X_i, B^{i-1}))}{\nu_i^g(dB_i|B^{i-1})}\Big)\Big\}, i = 0, \ldots, n, \tag{38.20}$$

$$\nu_i^g(db_i|b^{i-1}) = \int_{\mathscr{X}_i} q_i(db_i|b^{i-1}, x_i, g_i(x_i, b^{i-1})) \otimes P_i^g(dx_i|b^{i-1}), \; i = 0, 1, \ldots, n.$$

3. If $\{\gamma(T^i x^n, T^i a^n, T^i b^{n-1}) = \overline{\gamma}(x_i, a_i, b^{i-1})\}_{i=0}^n$, *the statements 1. and 2. hold, when transmission cost* $\mathscr{E}_{0,n}(P) \triangleq \Big\{e \in \mathscr{E}^{DF}_{[0,n]} : \frac{1}{n+1}\mathbb{E}\Big\{c_{0,n}(X^n, A^n, B^{n-1})\Big\} \leq P\Big\}$, *is imposed.*
4. If the right sides of (38.17), (38.18) *are* $\{s_i(dx_i|x_{i-1}, a_{i-1}, b_{i-1}), q_i(db_i|b_{i-1}, a_i, x_i)\}_{i=0}^n$, *then the optimal encoders are* $\{e_i(a^{i-1}, x^i, b^{i-1}) = g_i^M(x_i, b_{i-1})\}_{i=0}^n$ *and*

$$I(X^n \to B^n) = \sum_{i=0}^n I(X_i, A_i; B_i|B_{i-1}) = \mathbb{E}^{g^M}\Big\{\log\Big(\frac{q_i(dB_i|B_{i-1}, X_i, g_i^M(X_i, B_{i-1}))}{\nu_i^{g^M}(dB_i|B_{i-1})}\Big)\Big\}. \tag{38.21}$$

5. If $\{\gamma(T^i x^n, T^i a^n, T^i b^{n-1}) = \overline{\gamma}(x_i, a_i, a_{i-1}, b^{i-1})\}_{i=0}^n$ *then*

$$C_{0,n}(P) \triangleq \sup_{\{p_i(da_i|a^{i-1}, b^{i-1}, x^i): i=0,\ldots,n\} \in \mathscr{P}_{0,n}(P)} \sum_{i=0}^n I(X^i, A^i; B_i|B^{i-1}). \tag{38.22}$$

$$= \sup_{\{p_i(da_i|a_{i-1}, b^{i-1}, x_i): i=0,\ldots,n\} \in \mathscr{P}_{0,n}(P)} \sum_{i=0}^n I(X_i, A_i; B_i|B^{i-1}). \tag{38.23}$$

Theorem 38.2 admits generalizations, when the right-hand sides of (38.17), (38.18), depend, on past variables of (x^{i-1}, a^{i-1}), and/or $X^i \leftrightarrow (A^i, B^{i-1}) \leftrightarrow B_i, i = 0, \ldots, n$, which implies capacity is defined via $\sup_{\{p_i(da_i|a^{i-1}, b^{i-1})\}_{i=0}^n \in \mathscr{P}_{0,n}(P)} \sum_{i=0}^n I(A^n \to B^n)$.

38.3 Achievability of Nonanticipative Code

The instantaneous definition of realization of $R^{na}_{0,n}(D)$ and its limit are the following.

Definition 38.4 (*Realizations of* $R^{na}_{0,n}(D)$ *and* $R^{na}(D)$) *(a) Instantaneous Realization.* The reproduction distribution $\{P^*_{Y_i|Y^{i-1},X^i}(\mathrm{d}y_i|y^{i-1}, x^i) : i = 0, 1, \ldots, n\}$ of a source $\{P_{X_i|X^{i-1},Y^{i-1}}(\mathrm{d}x_i|x^{i-1}, y^{i-1}) : i = 0, \ldots, n\}$, achieving the infimum of $\frac{1}{n+1} R^{na}_{0,n}(D)$, is realizable, if there exists a channel $\{P_{B_i|B^{i-1},A^i,X^i}(\mathrm{d}b_i|b^{i-1}, a^i, x^i) :$

$i=0,1,\ldots,n\}$, an encoder $\{P_{A_i|A^{i-1},B^{i-1},X^i}\ (da_i|a^{i-1},b^{i-1},x^i):i=0,1,\ldots,n\}$, and a decoder $\{P_{Y_i|Y^{i-1},B^i}\ (dy_i|y^{i-1},b^i):i=0,1,\ldots,n\}$ such that

$$P^*_{Y_i|Y^{i-1},X^i}(dy_i|y^{i-1},x^i)=P_{Y_i|Y^{i-1},X^i}(dy_i|y^{i-1},x^i),\quad i=0,\ldots,n, \tag{38.24}$$

where the right-hand side term in (38.24) is obtained from joint distribution of the source–encoder–channel–decoder. Moreover, the rate (R_n,D_n) is realizable if, the realization operates at per unit time average distortion $D_n\triangleq\frac{1}{n+1}\mathbb{E}\{d_{0,n}(X^n,Y^n)\}$, rate $R_n=\frac{1}{n+1}R^{na}_{0,n}(D_n)\equiv\frac{1}{n+1}I(X^n\rightarrow Y^n)$, and there exists a $P\in[0,\infty)$ such that the channel rate, $C_{0,n}(P)$ is finite.
(b) Limiting Realization. The reproduction distribution $\{P^*_{Y_i|Y^{i-1},X^i}(dy_i|y^{i-1},x^i):i=0,1,\ldots,\}$ of a source achieving the infimum of $R^{na}(D)$, is realizable, if there exists a channel, encoder, decoder as in *(a)*. Moreover, the rate (R,D) is realizable if, the realization operates at per unit time average distortion $D\triangleq\lim_{n\longrightarrow\infty}\frac{1}{n+1}\mathbb{E}\{d_{0,n}(X^n,Y^n)\}$, rate $R=\lim_{n\longrightarrow\infty}\frac{1}{n+1}R^{na}_{0,n}(D)\equiv\lim_{n\longrightarrow\infty}\frac{1}{n+1}I(X^n\rightarrow Y^n)$, and there exists a $P\in[0,\infty)$ such that the channel capacity with transmission cost, $\lim_{n\longrightarrow\infty}\frac{1}{n+1}C_{0,n}(P)$ is finite.

Theorem 38.3 (Achievability of nonanticipative code). *A. Instantaneous. Suppose the following conditions hold for any finite n.*

1. *$R^{na}_{0,n}(D)$ has a solution;*
2. *$C_{0,n}(P)$ has a solution;*
3. *The optimal reproduction distribution $\overrightarrow{P}^*_{Y^n|X^n}(dy^n|x^n)$ is realizable, and $(R_{n,n})$ is realizable;*
4. *For a given $D_n\in[D_{min},D_{max}]$ there exists a P such that the realization gives $R^{na}_{0,n}(D_n)=C_{0,n}(P)=I(X^n\rightarrow B^n)$.*

B. Limiting. Suppose the following conditions hold.

1. *$R^{na}(D)$ has a solution;*
2. *$C(P)$ has a solution;*
3. *The optimal reproduction distribution $\overrightarrow{P}^*_{Y^\infty|X^\infty}(dy^\infty|x^\infty)$ for $R^{na}(D)$ is stationary and realizable, and (R,D) is realizable;*
4. *For a given $D\in[D_{min},D_{max}]$ there exists a P such that the realization gives $R^{na}(D)=C(P)=\lim_{\longrightarrow\infty}\frac{1}{n+1}I(X^n\rightarrow B^n)$.*

If $\mathbb{P}^*_{X^n,Y^n}\Big\{\sum_{i=0}^n\rho(T^iX^n,T^iY^n)>(n+1)d\Big\}\leq\epsilon, d>D$, where $\mathbb{P}^*$ is taken with respect to $P^*_{Y^n,X^n}(dy^n,dx^n)=\otimes_{i=0}^n\Big(P^*_{Y_i|Y^{i-1},X^i}(dy_i|y^{i-1},x^i)\otimes P_{X_i|X^{i-1},Y^{i-1}}(dx_i|x^{i-1},y^{i-1})\Big)$ then there exists an (n,d,ϵ,P) nonanticipative code.

An analog of Theorem 38.3 can be obtained using outage capacity, instead of excess distortion probability.

38.4 Example: Duality of a Source and a Channel with Memory

The duality discussed in Chap. 35, of the binary symmetric Markov source BSMS(p) and the binary symmetric unit memory channel, called BSSC(α_1, β_1), is addressed, by constructing an encoder–decoder operating optimally (achieving channel capacity and nonanticipative RDF), processing information in real time.

Nonanticipative RDF of BSMS(p) with Hamming Distortion. By Chap. 37, for a BSMS(p, $P(x_i = 0|x_{i-1} = 0) = 1 - p$ and $P(x_i = 0|x_{i-1} = 1) = p, i = 1, \dots, n$ with a distortion $\rho(x, y) = 0$ if $x = y$ and $\rho(x, y) = 1$ if $x \neq y$, the nonanticipative RDF is

$$R^{na}(D) = \begin{cases} H(m) - H(D) = H(p) - mH(\alpha_1) - (1 - m)H(\beta_1), & \text{if } D \leq \frac{1}{2} \\ 0, & \text{otherwise} \end{cases} \tag{38.25}$$

$$P^*_{Y_i|X_i,Y_{i-1}}(y_i|x_i, y_{i-1}) = \begin{matrix} & \begin{matrix} 0,0 & 0,1 & 1,0 & 1,1 \end{matrix} \\ \begin{matrix} 0 \\ 1 \end{matrix} & \begin{pmatrix} \alpha & \beta & 1-\beta & 1-\alpha \\ 1-\alpha & 1-\beta & \beta & \alpha \end{pmatrix} \end{matrix}, \quad i = 1, \dots, n, \tag{38.26}$$

where $m = 1 - p - D + 2pD$, $\alpha = \frac{(1-p)(1-D)}{1-p-D+2pD}$, $\beta = \frac{p(1-D)}{p+D-2pD}$.
Without loss of generality, the optimal reproduction distribution is realized by fixing the channel, and identifying an encoder–decoder pair so that the end-to-end design operates in real time and optimally. It appears easier to use the form of the optimal reproduction distribution as the channel.
Capacity of Binary Unit Memory Channel with Feedback and Cost. The channel and transmission cost are defined by

$$P_{B_i|A_i,B_{i-1}}(b_i|a_i, b_{i-1}) = \begin{matrix} & \begin{matrix} 0,0 & 0,1 & 1,0 & 1,1 \end{matrix} \\ \begin{matrix} 0 \\ 1 \end{matrix} & \begin{pmatrix} \alpha_1 & \beta_1 & 1-\beta_1 & 1-\alpha_1 \\ 1-\alpha_1 & 1-\beta_1 & \beta_1 & \alpha_1 \end{pmatrix} \end{matrix}, \tag{38.27}$$

$$\frac{1}{n}\mathbb{E}\Big\{\sum_{i=1}^{n} c(A_i, B_{i-1})\Big\} \leq \kappa, \quad c(a_i, b_{i-1}) \triangleq a_i \oplus b_{i-1}, \quad i = 1, 2, \dots, n, \tag{38.28}$$

where $\kappa \in [0, \kappa_{\max}]$, $(\alpha_1, \beta_1) \in [\frac{1}{2}, 1] \times [\frac{1}{2}, 1]$. Define the state of the channel as the modulo2 addition, $s_i \triangleq a_i \oplus b_{i-1}$; for $s_i = 0$, $s_i = 1$ the channel breaks down into two symmetric channels, for $i = 1, \ldots, n$, hence the name BSSC(α_1, β_1). The transmission cost is expressed as, $\frac{1}{n}\mathbb{E}\{\overline{S_i}\} \leq \kappa$, where $\overline{(\cdot)}$ denotes compliment of $(\cdot)$.

Theorem 38.4 [2]. *The capacity of the $BSSC(\alpha_1, \beta_1)$, with or without feedback, subject to the average cost constraint $\frac{1}{n}\mathbb{E}\{\sum_{i=1}^{n} c(A_i, B_{i-1})\} = \kappa$ is*

$$C(\kappa) = H(\alpha_1\kappa + (1-\beta_1)(1-\kappa)) - \kappa H(\alpha_1) - (1-\kappa)H(\beta_1). \tag{38.29}$$

The capacity achieving channel input distributions are given, for $i = 1, \ldots, n$, by

$$\textit{Feedback:} \quad P^*_{A_i|B_{i-1}}(a_i|b_{i-1}) = \begin{pmatrix} \kappa & 1-\kappa \\ 1-\kappa & \kappa \end{pmatrix}; \tag{38.30}$$

$$\textit{No Feedback:} \quad P^*_{A_i|A_{i-1}}(a_i|a_{i-1}) = \begin{pmatrix} \dfrac{1-\kappa-\gamma}{1-2\gamma} & \dfrac{\kappa-\gamma}{1-2\gamma} \\ \dfrac{\kappa-\gamma}{1-2\gamma} & \dfrac{1-\kappa-\gamma}{1-2\gamma} \end{pmatrix}, \tag{38.31}$$

where $\gamma = \alpha_1\kappa + \beta_1(1-\kappa)$.

Optimal Coding–Decoding Realization Schemes. Theorem 38.3, B.1. B.2. hold.

Matching the Rates. The first part of Theorem 38.3, B.4. is information matching, i.e., $C(\kappa) = R^{na}(D)$, established as follows.

$$\text{For any} \quad D \in [0, \frac{1}{2}] \quad \text{if} \quad \alpha_1 = \alpha,\ \beta_1 = \beta,\ \kappa = m \quad \text{then} \quad C(\kappa) = R^{na}(D). \tag{38.32}$$

Statement (38.32) is equivalent to using the optimal reproduction distribution as the encoder–channel–decoder, and setting the average cost κ to m.
It remains to address Theorem 38.3, B.3, which is the realization of the optimal reproduction conditional distribution and nonanticipative RDF, by an encoder–channel–decoder (channel is fixed), so that the source–encoder–channel–decoder realizes the optimal reproduction distribution and end-to-end distortion is achieved, and source–encoder–channel operates at capacity equals minimum compression rate.

Realization without Feedback Encoder. Substituting $\alpha_1 = \alpha$, $\beta_1 = \beta$, $\kappa = m$ into (38.31) the capacity achieving distribution without feedback is identical to that of the source, hence letting $A_i = X_i, i = 1, \ldots,$ (encoder is an identity map), the channel capacity is achieved and the transmission cost is satisfied. Letting $Y_i = B_i, i = 1, \ldots,$ (decoder is an identity map) the fidelity of reconstruction is satisfied. Thus, uncoded transmission is optimal for this source–channel and distortion-cost pairs.

Excess Distortion Probability. Existence of an (n, d, ε, d) nonanticipative code is shown by computing the excess distortion probability via Hoeffding's inequality

[3], since $\{Z_i \triangleq (Y_i, X_i) : i = 1, \ldots, \}$ is Markov. Letting $S_n \triangleq \sum_{i=1}^{n} \rho(X_i, Y_i)$, $d \triangleq \delta + \frac{\mathbb{E}[S_n]}{n}$, $\delta > 0$, gives

$$\mathbb{P}^*_{X^n,Y^n}\Big\{S_n > nd\Big\} \leq \exp\Big(-\frac{\lambda^2(n\delta - 2\|f\|m/\lambda)^2}{2n\|f\|^2 m^2}\Big), \; n > 2\|f\|m/(\lambda\delta), \tag{38.33}$$

where $\|f\| = 1$, $m = 1$, $\lambda = \min\{p, 1-p\}\min\{\alpha, \beta, 1-\alpha, 1-\beta\}$. This bound is illustrated in Fig. 38.2. Tighter bounds are given in [2].

Realization with Feedback Encoder. The feedback encoder realization is shown in Fig. 38.3. The optimal encoder consists of an identity pre-encoder and a modulo2 encoder of the current output of the pre-encoder, $A_i = X_i$, and the previous channel output, $S_i = A_i \oplus B_{i-1}$, and the optimal decoder is an identity map $Y_i = B_i, i = 1, \ldots$. The channel and the cost constraint are transformed to their equivalent forms, $P_{B_i|S_i,B_{i-1}}(b_i|s_i, b_{i-1})$, $c(s_i)$, $i = 1, 2, \ldots,$ respectively. In this case, the $S_i : i = 1, 2, \ldots,$ in an independent processes. The performance of this realization is evaluated via the excess distortion probability. A complete analysis is found in [2]. Other choices of realizations are possible, using memoryless channels, and decoders with memory.

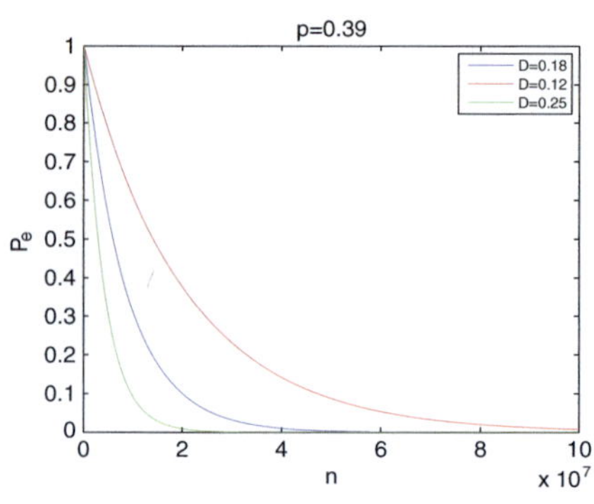

Fig. 38.2 Graph of excess distortion probability, $\delta = 0.01$

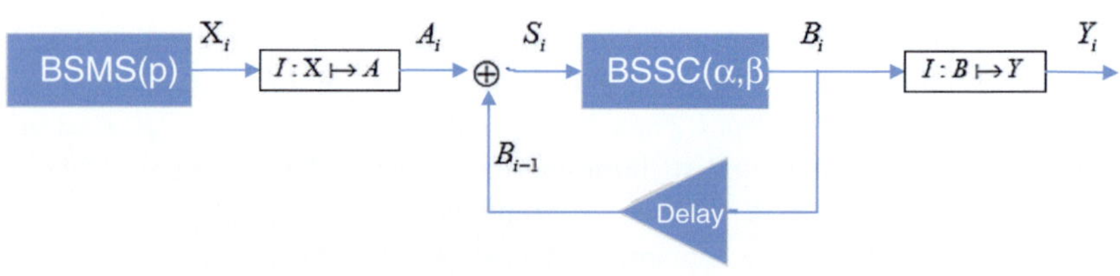

Fig. 38.3 Diagram displaying optimal feedback realization of encoder/decoder

38.5 Conclusions

A duality of a source and a channel is discussed for sources and channels with memory and feedback, based on nonanticipative transmission, with respect to excess distortion probability.

Similarly, a duality of a source and a channel, with respect to outage capacity, instead of excess distortion probability, can be carried out.

Whether such dualities suffice as a basis for nonanticipative point-to-point, network communication, and communication for control, remain, however, to be seen [4].

References

1. Charalambous CD, Stavrou PA, Ahmed NU (2014) Nonanticipative rate distortion function and relations to filtering theory. IEEE Trans Autom Control 59(4):937–952
2. Kourtellaris CK (2014) Nonanticipative information theory, Ph. D thesis, University of Cyprus, June 2014.
3. Glyn P, Ormoneit WD (2002) Hoeffdingâs inequality for uniform ergodic Markov chains. Stat Probab Lett 56:143–146
4. Gastpar M, Rimoldi B, Vetterli M (2003) To code or not to code: lossy source-channel communication revisited. IEEE Trans Info Theor 49(5):1147–1158

Part VIII
Informatics of Distributed Systems: Modeling and Verification

Computer science has different views of distributed system than control theory. The focus is more on computability, verification, and complexity.

The first two chapters deal with computational tools for hybrid automata and the underlying mathematical theory. An example of verification by formal methods is described in Chap. 40. The formal model of world automata (an extension of hybrid automata) motivated by their application in distributed systems is presented in Chap. 42. How to obtain consensus in network structured as digraphs by distributed averaging is shown in Chap. 43.

Chapter 39
An Introduction to the Verification of Hybrid Systems Using ARIADNE

Davide Bresolin, Luca Geretti, Tiziano Villa and Pieter Collins

39.1 Introduction

Hybrid systems are dynamical systems that exhibit both a discrete and a continuous behaviors. To model and specify hybrid systems in a formal way, the notion of *hybrid automata* has been introduced [1]. Intuitively, a hybrid automaton is a "finite-state automaton" with continuous variables that evolve according to dynamics characterizing each discrete state (called a *location* or *mode*). Of particular importance in the analysis of a hybrid automaton is the computation of the *reachable set*, i.e., the set of all states that can be reached under the dynamical evolution starting from a given initial state set. Many approximation techniques and tools to estimate the reachable set have been proposed in the literature. Most of the available software packages, like PhaVer [8] and SpaceEx [9], are limited to affine dynamics. Others, like HSOLVER [10] can handle non-linear dynamics, but can verify only safety properties.

To overcome the limitations of current tools, we recently proposed a development environment for hybrid systems verification, called ARIADNE [3–5], which differs

D. Bresolin (✉) · L. Geretti · T. Villa
Dipartimento di Informatica, Università di Verona, 37134 Verona, Italy
e-mail: davide.bresolin@univr.it

L. Geretti
e-mail: luca.geretti@univr.it

T. Villa
e-mail: tiziano.villa@univr.it

P. Collins
Department of Knowledge Engineering, Maastricht University,
6211 LH Maastricht, The Netherlands
e-mail: pieter.collins@maastrichtuniversity.nl

J.H. van Schuppen and T. Villa (eds.), *Coordination Control of Distributed Systems*,
Lecture Notes in Control and Information Sciences 456,
DOI 10.1007/978-3-319-10407-2_39

from existing tools by being based on a rigorous function calculus [7]. This calculus provides a rigorous mathematical semantics for the numerical analysis of dynamical systems, suitable for implementing formal verification algorithms.

39.2 The Reachability Problem for Hybrid Automata

We first give a formal definition of a hybrid automaton.

Definition 39.1 A *hybrid automaton* is a tuple $\mathscr{A} = \langle \text{Loc}, \text{Edg}, \mathbb{R}^n, Inv, Dyn, Act, Res\rangle$ such that:

1. $\langle \text{Loc}, \text{Edg}\rangle$ is a finite directed graph; the vertices, Loc, are called *locations* or *control modes*, and the directed edges, Edg, are called *control switches*;
2. each location $q \in \text{Loc}$ is labeled by the *invariant condition* Inv[q] on $\mathbb{R}^n$ and the *dynamic law* Dyn[q] on $\mathbb{R}^n \times \mathbb{R}^n \times \mathbb{R}^{\geq 0}$ such that if Inv[q](x) is true then Dyn[q](x, x, 0) is true;
3. each edge $e \in \text{Edg}$ is labeled by the *activation condition* Act[e] on $\mathbb{R}^n$ and the *reset relation* Res[e] on $\mathbb{R}^n \times \mathbb{R}^n$.

A *state* ℓ of a hybrid automaton $\mathscr{A}$ is a pair , vx, where $v \in \text{Loc}$ is a location and $x \in \mathbb{R}^n$ is an assignment of values for the continuous variables. A state vx is said to be *admissible* if Inv[v](x) holds.

A *trajectory* ξ of a hybrid automaton is a (finite or infinite) sequence $(\xi_i)_{i\geq 0}$ of continuous functions $\xi_i : [\tau_i, \tau_{i+1}] \to \text{Loc} \times \mathbb{R}^n$ such that $\text{Dyn[q]}(\xi_\text{i}(\text{s}), \xi_\text{i}(t), t - \text{s})$ holds for all $\tau_i \leq s \leq t \leq \tau_{i+1}$, and both $\text{Act[e]}(\xi_\text{i}(\tau_{\text{i}+1}))$ and $\text{Res[e]}(\xi_\text{i}(\tau_{\text{i}+1}), \xi_{i+i}(\tau_{i+1}))$ hold for some $e \in \text{Edg}$. Here, $\xi_i(t)$ represents the state of the system after i events and at time t.

Definition 39.2 Let $\mathscr{A}$ be a hybrid automaton. A state $\langle q_r r\rangle$ *reaches* a state $q_s s$ if there exists a finite trajectory $\xi = (\xi_i)_{0\leq i\leq n}$ such that $\xi_0(0) = q_r r$ and $\xi_n(\tau_{n+1}) = q_s s$. We use $ReachSet_{\mathscr{A}}(q_r r)$ to denote the set of states reachable from $q_r r$. Moreover, given a set of states $X_0 \subseteq \text{Loc} \times \mathbb{R}^n$, we use $ReachSet_{\mathscr{A}}(X_0)$ to denote the set $\cup_{q_r r\in X_0} ReachSet_{\mathscr{A}}(q_r r)$.

Checking safety properties on hybrid automata reduces to the reachability problem. Suppose we wish to verify that a safety property φ holds for a hybrid automaton H; in other words, that φ remains true for all possible executions starting from a set X_0 of initial states. Then, we only need to prove that $ReachSet_{\mathscr{A}}(X_0) \subseteq \text{Sat}(\varphi)$, where $\text{Sat}(\varphi)$ is the set of states where φ is true. Unfortunately, the reachability problem is not decidable in general [1]. Nevertheless, formal verification methods can be applied to hybrid automata: Suppose we can compute an over-approximation S to $ReachSet_{\mathscr{A}}(X_0)$, that is, a set $S \supseteq ReachSet_{\mathscr{A}}(X_0)$. Then, if S is a subset of $\text{Sat}(\varphi)$, then so is the reachable set and the automaton H respects the property. Conversely, if we can compute an under-approximation S to $ReachSet_{\mathscr{A}}(X_0)$ (that is, a set $S \subseteq ReachSet_{\mathscr{A}}(X_0)$) that turns out to contain at least one point outside $\text{Sat}(\varphi)$, we have proved that H does not respect the safety property φ.

39.3 The ARIADNE Software Package

ARIADNE's computational engine is based on a rigorous *function calculus*, where continuous functions $f : \mathbb{R}^m \mapsto \mathbb{R}^n$ are the basic building blocks used to represent and compute the evolution of hybrid automata [7]. Every component of a hybrid automaton $\mathscr{A}$ can be represented in the function calculus setting as follows:

- For every discrete location, a function $\mathrm{Dyn} : \mathbb{R}^\mathrm{n} \mapsto \mathbb{R}^\mathrm{n}$ is used to represent the continuous dynamics $\dot{x} = \mathrm{Dyn}(\mathrm{x})$.
- Invariants are represented using single-valued functions $\mathrm{Inv} : \mathbb{R}^\mathrm{n} \mapsto \mathbb{R}$ that are *negative* exactly when the invariant is true.
- Discrete transitions are represented using a function $\mathrm{Act} : \mathbb{R}^\mathrm{n} \mapsto \mathbb{R}$ that is *positive* when the guard of the transition is true (and negative otherwise), and a reset function $\mathrm{Res} : \mathbb{R}^\mathrm{n} \mapsto \mathbb{R}^\mathrm{n}$.

Additionally:

- Regions of space $R \subseteq \mathbb{R}^n$ are represented using *ImageSets*, i.e., functions mapping a box $[-1, 1]^p$ to a subset of $\mathbb{R}^n$ that approximates the desired region. [1]

In particular, given $f : \mathbb{R}^m \mapsto \mathbb{R}^n$, and a point $x \in \mathbb{R}^m$, an *approximation* of f near x can be computed. The result is a pair $\hat{f} = (T, I)$ where T is a *polynomial expansion* of f to a given degree and I an interval such that $f(z) - T(z) \in I$ for all points z near x, giving an *error bound* on the approximation. We also call this set an *enclosure*, since it encloses the exact set of points. If we express a starting set S as an enclosure, the evolution under the continuous dynamics can be obtained by numerical integration, using the *flow tube* of the continuous evolution: A function $flow : \mathbb{R}^{n+1} \mapsto \mathbb{R}^n$ such that $flow(S, t)$ is the set of points reached from S after t time units of continuous evolution; taking also discrete evolution into account, we call *reached set* the set of points reached from S after t time units.

39.3.1 Evolution of Enclosures

ARIADNE is able to compute approximations to the reachable set by "patching together" enclosures of the reached sets obtained by evolving the system for finite time intervals. Such evolution uses either an *upper semantics* or a *lower semantics*:

- upper semantics implies that, if we evolve the system for a finite time, the set of points that we obtain is a superset of the reachable set;
- lower semantics implies that, if we evolve the system for a finite time, each point that we obtain has a bounded distance to a point of the reachable set.

[1] *ImageSets* are used in the stable version of ARIADNE. The development version uses a more accurate representation based on *ConstrainedImageSets* [7].

While the computation of evolution for a finite time is straightforward, the same could not be said for the case of infinite time. To do that, we need to be able to perform the intersection and subtraction of sets. Unfortunately, these two operations cannot be performed on enclosures, which instead must be *discretised* onto a set of *cells* according to a *grid*.

Given a hybrid automaton $\mathscr{A}$, and an initial set of states S_0, ARIADNE can compute two kinds of approximations to the reachable set:

- An *outer approximation* O of the reachable set using upper semantics, for both finite and infinite time evolution. Formally, a *closed set* O such that the *closure* of $ReachSet_{\mathscr{A}}(S_0)$ is strictly contained in the *interior* of O.
- An ε-*lower approximation* L_ε of the reachable set using lower semantics, for both finite and infinite time evolution. Formally, an *open set* L_ε where for every point $x \in L_\varepsilon$ there exists a point $y \in ReachSet_{\mathscr{A}}(S_0)$ such that $|x - y| < \varepsilon$.

In the case of O, the evolution can be performed either in the *forward* or *backward* direction. On the contrary, backward evolution for L_ε, while computable, has not been implemented, since lower semantics causes backward transitions to yield very coarse results that become ineffective for reachability analysis.

39.3.2 Verification

Verification in ARIADNE relies on reachability analysis. ARIADNE currently offers two classes of verification routines: *safety* and *dominance*. The safety verification routine accepts a space region in which the reachable set should be included for all times. The dominance checking routine instead compares the reachable sets of two hybrid systems on a common subspace and decides which one is included into the other. Both routines can be applied to a specific hybrid automaton or be *parametric*. Parametric safety and parametric dominance verification identify some constants of the hybrid automaton and treat them as parameters that take values within a given interval: The routines split the parametric space and verify each subspace. This approach is able to identify optima for design parameters of a system.

All verification routines based on approximations are necessarily dependent on the coarseness of the approximation: An answer to the verification problem may be unattainable only because the accuracy is insufficient to the task. ARIADNE defines the accuracy of computation by means of some *settings*, the most important being:

- the grid used for each location, to control the granularity of the state space;
- the integration step, to control the accuracy of evaluation of the flow function.

In particular, the output of the reachability routines converges to the "best possible" approximations when the accuracy settings converge to zero. Now, since discretised evolution routines are built upon the evolution of enclosures, and verification makes use of approximations obtained using discretised evolution, it is apparent that efficiency of verification (and effectiveness, if we have a limited time to obtain a definite answer) depends on a proper choice of such accuracy settings. Since we cannot decide beforehand which values are optimal for a given system and verification task, we must resort to an *iterative refinement* procedure: if we do not obtain a result with the current accuracy "level," we repeat the calculation of the approximation for finer values of the settings.

A proper manual tuning of the accuracy settings would be quite difficult, especially if iterative refinement is considered. In other words, it is desirable to automate the choice of the accuracy settings in order to improve both usability and efficiency. ARIADNE, therefore, does not require the user to tune these settings: Instead, it extracts reasonable values after a pre-analysis of the domain, at each refinement step. This implies that the choice of the domain for the state space is the only mandatory information that must be provided together with the hybrid automaton.

39.4 The Water Tank Example

In the following, we consider an application from hydraulic control, i.e., a water tank system composed of a cylindrical *tank*, equipped with an inlet pipe at the top, an outlet pipe at the bottom, and a valve that controls the inlet flow. The outlet flow is proportional to the water level, while the inlet flow is controlled by a *valve* that receives *open* and *close* position commands from a controller. In response to a command, the valve aperture $\alpha(t)$ changes linearly in time with a rate $1/T$. The inlet flow is proportional to the inlet pressure $p(t)$ and to the valve aperture $\alpha(t) \in [0, 1]$:

$$F_{in}(t) = \alpha(t) f(p(t)).$$

Location q_1 of Fig. 39.1a represents the nominal state of the tank, when the water level is under the overflow limit, while location q_2 represents overflow. When overflow occurs, the automaton stays in location q_2 until the inlet flow $u(t)$ is less than or equal to the outlet flow $\lambda\sqrt{H}$.

The current water level $x(t)$ is measured by a sensor that outputs a signal $x_s(t)$ affected by an unknown sensor error $\delta(t)$:

$$x_s(t) = x(t) + \delta(t), \tag{39.1}$$

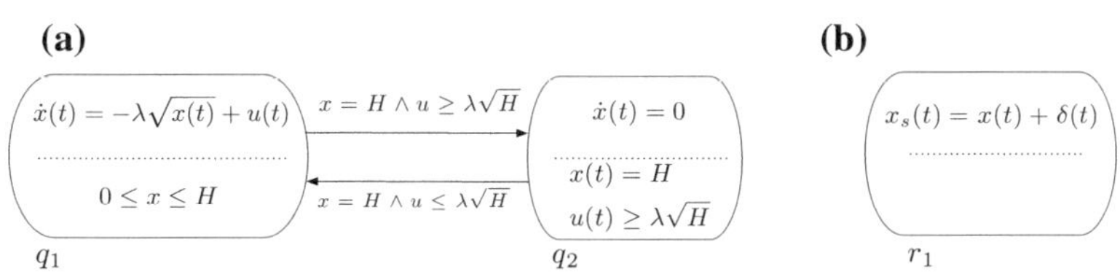

Fig. 39.1 The hybrid automata for the tank and the sensor. **a** Tank. **b** Sensor

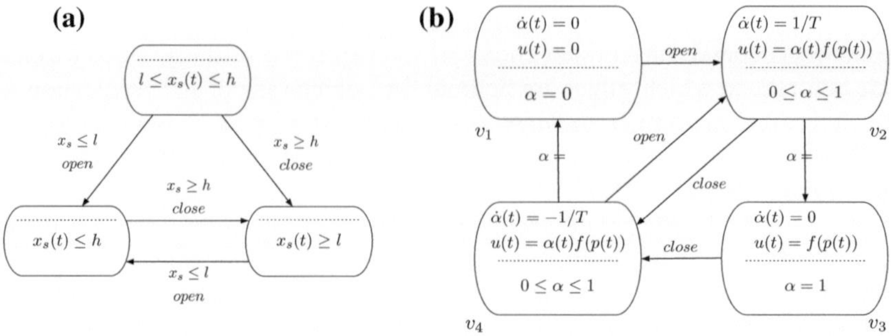

Fig. 39.2 The hybrid automata for the controller and the valve. **a** Hysteretic controller. **b** Valve

and can be modeled by the single location automaton of Fig. 39.1b, where x and δ are input variables, and x_s is the only output variable.

The automaton for the controller is depicted in Fig. 39.2a. It reads the water level $x_s(t)$ measured by the sensor and sends the position commands *open* and *close* to the valve following a simple hysteretic loop:

- when the valve is closed and the water level is decreasing, the *open* command is produced when $x_s(t) \leq l$ (location c_3);
- conversely, when the valve is opened and the water level is increasing, the *close* command is produced when $x_s(t) \geq h$ (location c_2).

where l and h represent *lower* and *upper* threshold values for the water level. Location c_1 is the initial location, corresponding to the situation in which the controller does not know whether the water level is increasing or decreasing. The automaton has no output variables, two output events *open* and *close*, and one input variable x_s.

ARIADNE allows to specify these systems separately, and then automatically compose them into a monolithic automaton for evolution and verification.

In this particular example, given an initial water level $x_i \in [6.5, 7.0]$, we want to verify whether $x \in [5.25, 8.25]$, where $l \in [5.25, 6.25]$ and $h \in [7.25, 8.25]$. This is a parametric safety verification problem, with parameters l and h. The algorithm offered by the ARIADNE library splits the parameter space with a granularity chosen by the user; then for each subspace, a verification loop is performed, in which we progressively refine the accuracy settings until a definite answer is obtained (or a user-defined time budget is hit).

It must be noticed that if we reach the minimum allowed values for the accuracy settings without getting a positive or negative answer, then an *indeterminate* result is returned. Figure 39.3 shows the verification outcomes for this problem, where the squares represent the verification subspaces. A safe result is shown in green, an unsafe result in red and an indeterminate one in yellow. It can be noticed that low values of h and high values of l are required to provide a positive answer.

Finally, in Fig. 39.4, we show the result for approximations to the reachable set as computed for two different values of the two parameters. The green region represents

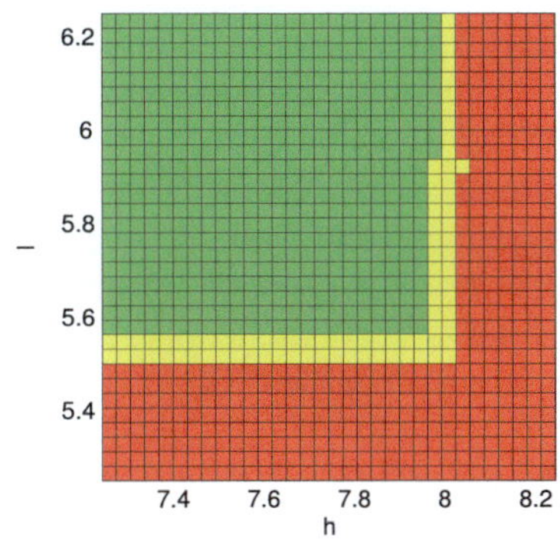

Fig. 39.3 The verification results for the contract satisfaction and dominance problems

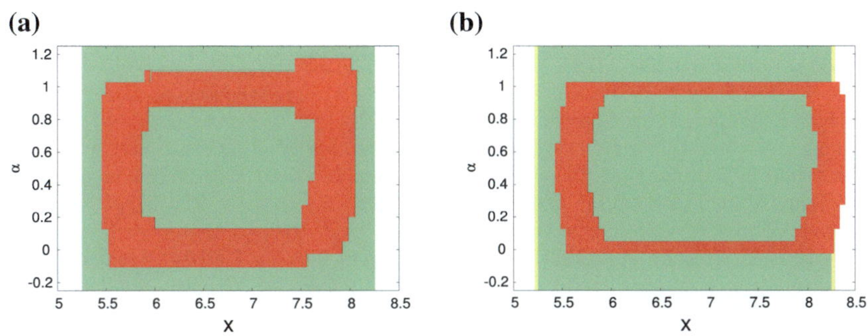

Fig. 39.4 Reachable set for two choices of the l and h thresholds, with a safe (**a**) and unsafe (**b**) result. **a** $l = [5.75, 5.781]$, $h = [7.75, 7.781]$. **b** $l = [5.75, 5.781]$, $h = [8.125, 8.156]$

the safe region, while the red region is the computed approximation. In Fig. 39.4a, the O approximation is used to verify that safety is guaranteed. In Fig. 39.4b, the L_ε approximation allows us to state that the system is unsafe. The yellow region represents the ε-tolerance on the safe region given by the approximation error. These results look rather coarse, but this is due to the fact that we need to periodically discretise the reached set onto a grid in order to decide when to stop evolving.

39.5 Conclusions and Future Work

Formal verification of hybrid systems is still in its infancy, but tools like ARIADNE show promising results also on non-trivial case studies (see Chap. 40). Further information on the framework can be found in [7] about functional calculus, [4] regarding the reachability routines, and [5] for advanced verification strategies.

ARIADNE can manage non-linear dynamics and can compute both outer and lower approximations to the reachable set. However, this high expressivity is also the main reason for some shortcomings, in particular with respect to scalability and accuracy of the approximations. The recent introduction of support functions increased substantially the size of linear hybrid systems that can be verified [9] and showed that the choice of the correct set representation is crucial. To overcome the current limitations of the tool, we are working on the following improvements: addressing scalability by means of counter-example based abstraction refinement techniques [2], handling specification properties beyond safety [6], extending the tool to synthesize switching controllers with respect to safety properties [11].

References

1. Alur R, Courcoubetis C, Henzinger TA, Ho PH (1992) Hybrid automata: an algorithmic approach to the specification and verification of hybrid systems. In: Hybrid systems, LNCS. Springer, Berlin, pp 209–229
2. Alur Rajeev, Dang Thao, Ivančić Franjo (2006) Counterexample-guided predicate abstraction of hybrid systems. Theor Comput Sci 354(2):250–271
3. Ariadne: An open tool for hybrid system analysis. http://ariadne.parades.rm.cnr.it
4. Benvenuti L, Bresolin D, Collins P, Ferrari A, Geretti L, Villa T (2012) Ariadne: Dominance checking of nonlinear hybrid automata using reachability analysis. Reachability Problems., volume 7550 of LNCSSpringer, Berlin Heidelberg, pp 79–91
5. Benvenuti L, Bresolin D, Collins P, Ferrari A, Geretti L, Villa T (2014) Assume-guarantee verification of nonlinear hybrid systems with ARIADNE. Int J Robust Nonlinear Control 24(4):699–724
6. Bresolin D (2013) Improving HyLTL model checking of hybrid systems. In: Proceedings of the 4th international symposium on games, automata, logics and formal verification (GandALF2013), vol 119 of EPTCS, pp 79–92
7. Collins P, Bresolin D, Geretti L, Villa T (2012) Computing the evolution of hybrid systems using rigorous function calculus. In: Proceedings of the 4th IFAC conference on analysis and design of Hybrid Systems (ADHS12), pp 284–290
8. Frehse G (2008) PHAVer: algorithmic verification of hybrid systems past HyTech. Int J Softw Tools Technol Transf (STTT) 10:263–279
9. Frehse G, Le Guernic C, Donzé A, Cotton S, Ray R, Lebeltel O, Ripado R, Girard A, Dang T, Maler O (2011) SpaceEx: scalable verification of hybrid systems. In: Proceedings 23rd international conference on computer aided verification (CAV 2011), volume 6806 of LNCS. Springer, Berlin, pp 379–395
10. Ratschan S, She Z (2007) Safety verification of hybrid systems by constraint propagation based abstraction refinement. ACM Trans Embed Comput Syst 6(1)
11. Tomlin CJ, Lygeros J, Sastry SS (2000) A game theoretic approach to controller design for hybrid systems. Proc IEEE 88(7):949–970

Chapter 40
Formal Verification Applied to Robotic Surgery

Davide Bresolin, Luca Geretti, Riccardo Muradore, Paolo Fiorini and Tiziano Villa

40.1 Introduction

In engineering practice, the analysis of a complex system is usually carried out via simulation, which allows the researcher to explore one of the possible system executions at a time. Formal verification instead aims at exploring all possible executions, in order to ensure proper functionality of the system in all cases or conversely to acquire information about potential fault cases.

Nowadays, formal methods are becoming a vital aspect in the design of safety-critical systems, including robotics and automation systems [6]. An area where they can greatly improve the reliability of the design process is autonomous robotic surgery (ARS). The aim of ARS is to perform simple tasks without the presence (or telepresence) of surgeons. Therefore, with ARS, basic tasks will be executed by robots, allowing the surgeons to focus only on the most difficult aspects of the intervention. This implies that the overall control architecture must respect strict safety constraints and must guarantee the successful accomplishment of the surgical task, independently of uncertainties and unmodeled subsystems.

In this work, we will show how formal verification can provide accurate and reliable answers to help the designer in the development of ARS systems by considering

D. Bresolin (✉) · L. Geretti · R. Muradore · P. Fiorini · T. Villa
Department of Computer Science, University of Verona, Strada Le Grazie 15, 37134 Verona, Italy
e-mail: davide.bresolin@univr.it

L. Geretti
e-mail: luca.geretti@univr.it

R. Muradore
e-mail: riccardo.muradore@univr.it

P. Fiorini
e-mail: paolo.fiorini@univr.it

T. Villa
e-mail: tiziano.villa@univr.it

J.H. van Schuppen and T. Villa (eds.), *Coordination Control of Distributed Systems*,
Lecture Notes in Control and Information Sciences 456,
DOI 10.1007/978-3-319-10407-2_40

the automatic execution of a simple surgical action such as puncturing. We first model the overall task as a finite sequence of atomic actions that should be accomplished to guarantee the success of the surgical action. This model takes the form of a *hybrid automaton* consisting of a discrete control part that operates in a continuous environment [1]. Then, we specify the safety constraints that the system should respect in a precise mathematical way as reachability properties of the hybrid automaton model. Finally, we compare the results that can be obtained by current state-of-the-art tools for reachability analysis of hybrid automata, and we formally prove that the sequence of subtasks planned on preoperative data can successfully accomplish the surgical operation while respecting the safety constraints.

40.2 Modeling a Surgical Task

Puncturing is the act of penetrating a biological tissue with a needle, e.g., when performing a biopsy. Together with other elementary tasks such as cutting and suturing, this action can be used to build more complex surgical tasks. To model the puncturing task in a formal way, we divided its execution into three subtasks: (i) a *free motion* phase, where the end effector of the robot approaches the patient's tissue starting from its home position; and (ii) a *perpendicular attitude* phase, where the end effector is in contact with the tissue, and the robot moves its wrist to have the tool orthogonal with the patient's surface; (iii) a *puncturing* phase, where the robot increases the force applied by the end effector until the tissue is penetrated.

We assume that the controller for each subtask stabilizes the plant, while the switching between controllers preserves the stability. Our goal is not to prove the stability of the overall system but to prove in a formal way that the *task* itself can be executed correctly. Thus, the focus of the analysis is to show the feasibility of the task rather than the stability of the plant. The test case under consideration is a typical example of a *hybrid system*, i.e., a system consisting of a discrete part that operates in a continuous environment, and for this reason, it is sensitive not only to time-driven phenomena but also to event-driven phenomena. We model the surgical task by using the well-known formalism of hybrid automata introduced in Chap. 39. In a hybrid automaton, continuous variables evolve according to dynamics specified at each location (or discrete state). In our test case, each location of the automaton corresponds to one of the subtasks identified for the surgical action.

The automaton describing a simplified version of the puncturing action is shown in Fig. 40.1a. The continuous space is given by the nine-dimensional set of variables

$$X = \{x, v_x, y, v_y, \phi, v_\phi, t, x_c, y_c\}, \tag{40.1}$$

where x and y represent the position of the end effector on a plane, ϕ is the orientation of the needle, v_x, v_y, and v_ϕ are the first derivatives of x, y, and ϕ, respectively, t is the time elapsed, and x_c and y_c are auxiliary variables that store the position of the contact point between the end effector and the patient's surface. The patient is assumed to lie on a plane orthogonal to the x coordinate, with an unknown position

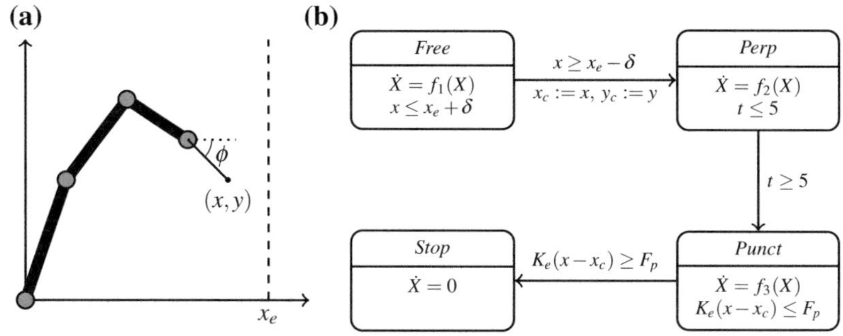

Fig. 40.1 The surgical test bench. **a** Cartesian state space. **b** Automation model

in the range $[x_e - \delta, x_e + \delta]$, where δ is the uncertainty and $x_e = 0.95$ the nominal position.

Locations *Free*, *Perp*, and *Punct* correspond to the three subtasks in which the puncturing action is divided, while location *Stop* is reached upon successful execution of the task. Transitions describe the switching from one subtask to another and are defined as follows:

- the transition from *Free* to *Perp* is activated when the end effector touches the tissue. To model the uncertainty in the position of the patient, this transition can be taken as soon as x is greater or equal to $x_e - \delta$, but the automaton is not forced to leave location *Free* if x is less or equal to $x_e + \delta$. Upon activation of the transition, the actual value of the contact point between the end effector and the patient is stored in variables x_c and y_c;
- the manipulator remains in location *Perp*, while the time is less than 5 s. At $t = 5$ s, the tool is perpendicular to the tissue, the transition is activated, and the automaton moves to *Punct*;
- the last transition becomes active as soon as the force applied by the needle, given by the expression $K_e(x - x_c)$, is greater than the threshold F_p.

In location *Stop*, the robot is assumed to remain motionless; hence, the derivatives of all variables are set to 0. To give the details of the continuous dynamics in locations *Free*, *Perp*, and *Punct*, we have to briefly recall the dynamic model of the robot and of the controller. The serial link manipulator in Cartesian or task space is depicted in Fig. 40.1b and is described by the set of nonlinear differential equations

$$\Lambda(\mathbf{x})\ddot{\mathbf{x}} + \Xi(\mathbf{x}, \dot{\mathbf{x}})\dot{\mathbf{x}} + \gamma(\mathbf{x}) = \tau + \mathbf{f}_e \tag{40.2}$$

where $\mathbf{x} = \begin{bmatrix} x & y & \phi \end{bmatrix}^{\mathbf{T}}$ is the pose (position and orientation) of the end effector (i.e., the tip of the needle). The matrices $\Lambda(\mathbf{x})$ and $\Xi(\mathbf{x}, \dot{\mathbf{x}})$ are the inertia matrix and the Coriolis/centrifugal matrix, respectively. The torque τ is the control vector, and $\mathbf{f}_e$ is the vector of generalized interaction force. We refer the reader to [8] for more details and properties of this mathematical representation.

In our example, the model of the robot is known, and hence, it is possible to implement the inverse dynamics control within the parallel force/position scheme [8]. Let $\mathbf{x} - \mathbf{x}_d$ be the tracking error between the reference trajectory $\mathbf{x}_d$ and the actual Cartesian position $\mathbf{x}$. In free motion ($\mathbf{f}_e = 0$), the robot model coupled with the position control action is described by

$$M_d \ddot{\mathbf{x}} = M_d \ddot{\mathbf{x}}_d + K_D(\dot{\mathbf{x}} - \dot{\mathbf{x}}_d) + K_P(\mathbf{x} - \mathbf{x}_d), \tag{40.3}$$

where M_d, K_D, and K_P are positive design parameters whose choice is application dependent. When the end effector is in contact with the patient, the right-hand side of the control equation also has a term associated with the interaction force $\mathbf{f}_e = K_e(\mathbf{x} - \mathbf{x}_c)$

$$M_d \ddot{\mathbf{x}} = M_d \ddot{\mathbf{x}}_d + K_D(\dot{\mathbf{x}} - \dot{\mathbf{x}}_d) + K_P(\mathbf{x} - \mathbf{x}_d) + K_f K_e(\mathbf{x} - \mathbf{x}_c). \tag{40.4}$$

where $\mathbf{x}_c$ is the contact point in Cartesian coordinate and K_e is the stiffness matrix of the tissue. The control force action $K_f K_e(\mathbf{x} - \mathbf{x}_c)$ depends on the design parameter K_f. This parameter is a sort of trade-off between the position action and the force action: It is in charge of adapting the nominal trajectory $\mathbf{x}_d$ according to the current interaction force.

When the automaton is in locations *Free* and *Perp*, the reference trajectory $\mathbf{x}_d(t)$ starts from $(0, 0, \frac{\pi}{2})$ and ends in $(1.0, 0.2, 0)$ after 5 s following the equations

$$\begin{aligned} x_d(t) &= -\tfrac{1}{2\pi}\sin(\tfrac{2\pi}{5}t) + \tfrac{1}{5}t \qquad y_d(t) = -\tfrac{1}{10\pi}\sin(\tfrac{2\pi}{5}t) + \tfrac{1}{25}t \\ \phi_d(t) &= \tfrac{1}{4}\sin(\tfrac{2\pi}{5}t) - \tfrac{\pi}{10}t + \tfrac{\pi}{2} \end{aligned} \tag{40.5}$$

In location *Punct*, the reference trajectory keeps y and ϕ constant and increases the value of x to penetrate the tissue:

$$x_d(t) = x_c + \tfrac{1}{2}(t-5) \qquad y_d(t) = y_c \qquad \phi_d(t) = 0 \tag{40.6}$$

In all locations, the dynamics of variables x, y, ϕ, t, x_c, and y_c are fixed and set to $\dot{x} = v_x$, $\dot{y} = v_y$, $\dot{\phi} = v_\phi$, $\dot{t} = 1$, $\dot{x}_c = 0$, and $\dot{y}_c = 0$. The dynamics for v_x, v_y, and v_ϕ are obtained by substituting $\mathbf{x}_d$ with (40.5) in (40.3) for location *Free*, $\mathbf{x}_d$ with (40.5) in (40.4) for location *Perp*, and $\mathbf{x}_d$ with (40.6) in (40.3) for location *Punct*.

40.3 Expressing Properties of a Surgical Task

We recall from Chap. 39 that the verification of hybrid systems is usually carried out by reachability analysis. Current tools typically accept a subset of the state space called *safe set* as specification and test whether the set of states that can be reached under dynamical evolution is included into the safe set or not. This means that they are limited to *safety properties*, that is, properties that hold for all possible executions and for all time instants. Moreover, the property cannot be specified using some specification language, but must be transformed into a safe set by the user. Nevertheless, they can still be used to verify some meaningful properties of the automatic puncturing test case:

1. *The task is feasible, and the position of the needle at the end of the task is always inside a given target region R.* In the most general case, this is not a safety property and thus cannot be verified by current tools. However, if we impose the task to be completed before a maximum time $t_{\max}$ and if we add an auxiliary variable t representing time elapse, we can rephrase the property as a safety property:

$$\text{Always}(t \geq t_{\max} \rightarrow (\text{Stop} \wedge \mathbf{x} \in R)),$$

   which corresponds to the safe set $\{(\ell, X) \mid t < t_{\max}\} \cup \{(\ell, X) \mid t \geq t_{\max} \wedge \ell = \text{Stop} \wedge \mathbf{x} \in R\}$.
2. *The force applied to the patient by the end effector during the perpendicular attitude subtask is always less than a given threshold.* This can be formally stated by the following property:

$$\text{Always}(\text{Perp} \rightarrow \|K_e(\mathbf{x} - \mathbf{x}_c)\| < f_{\max}),$$

   which corresponds to the safe set $\{(\ell, X) \mid \ell \neq \text{Perp}\} \cup \{(\ell, X) \mid \ell = \text{Perp} \wedge \|K_e(\mathbf{x} - \mathbf{x}_c)\| < f_{\max}\}$.

In particular, we study the dependence of the satisfaction of these properties on the value of two design parameters, namely the uncertainty on the position of the patient δ for Property 1 and the control parameter K_f for Property 2.

40.4 Verification of a Surgical Task

As described in Chap. 39, approximation techniques can be exploited to verify hybrid automata with complex dynamics. Among others, the most relevant state-of-the-art tools that are currently developed and publicly available are PHAVER [4], SPACEEX [5], HSOLVER [7], and ARIADNE [3]. In the following, we briefly describe the four tools. We refer the reader to the specific literature for a comprehensive description of their algorithms and state space representation choices. Table 40.1 shows whether they are able to verify the two properties of the robotic surgery case study or not and summarizes their differences under the following criteria:

- **Class of system they can verify**: Do they support nonlinear dynamics? Can the system be specified as a composition of smaller components?
- **Soundness of the results**: Is the verification result guaranteed to be mathematically correct?

Table 40.1 Comparison of PHAVER, SPACEEX, HSOLVER, and ARIADNE

	PHAVER	SPACEEX	HSOLVER	ARIADNE
State space representation	Polyhedra	Support functions	Predicate abstraction	Image sets
Nonlinear dynamics	No	No	**YES**	**YES**
Composition of automata	**YES**	**YES**	No	**YES**
Rigorous results	**YES**	No	**YES**	**YES**
User-definable accuracy	**YES**	**YES**	No	**YES**
Graphical output of results	**YES**	**YES**	No	**YES**
Maximum no. of variables*	~10	~100	~10	~10
Property 1	No answer	**True if** $\delta \leq 0.025$	No answer	**True if** $\delta \leq 0.0375$
Property 2	No answer	**True if** $K_f \geq 0.14375$	No answer	**True if** $K_f \geq 0.1625$

*These numbers of variables were reached in some cases reported in the literature. A bold item signifies that the tool of the column exhibits the activity of the corresponding row

- **Accuracy control**: Is it possible to choose the quality of the approximations?
- **Output**: Is it possible to obtain a graphical output of the results, or is only a YES/NO answer provided?
- **Scalability**: What is the maximum size of a system that they can verify?

PHAVER [4] is one of the first tools that enabled verification of hybrid automata with complex dynamics: It handles affine dynamics and guards and supports the composition of hybrid automata. The state space is represented using polytopes. Results are formally sound by means of an exact and robust arithmetic with unlimited precision. Scalability is limited: Systems with more than 10 continuous variables are usually out of the capabilities of the tool.

SPACEEX [5] improves upon PHAVER with particular regard to scalability (systems with 100 variables have been analyzed with this tool). It combines polyhedra and supports function representations of the state space of systems with piecewise affine, nondeterministic dynamics. Differently from PHAVER, the result of SPACEEX is not guaranteed to be numerically sound. This means that when the tool finds the system safe, we can only conclude that more sophisticated methods are necessary to find bugs for that system.

HSOLVER [7] uses constraint propagation and abstraction refinement techniques to discretize the state space of the system and verify safety properties. HSOLVER supports systems with complex nonlinear dynamics and guards, but not the composition of automata. Because of the particular state space representation, it cannot provide a graphical output of the reachable set, but only a safe/possibly unsafe answer to the verification problem.

ARIADNE [3] uses rigorous numerical methods for working with real numbers, functions, and sets in the Euclidean space to verify hybrid systems with nonlinear dynamics, guards, and reset functions. It supports composition to build complex systems from simpler components and can compute both upper approximations and lower approximations of the reachable set. By combining outer and lower approximations, ARIADNE can provide both positive and negative answers to safety properties and other more complex verification problems. Its high expressivity, however, affects the performance and scalability of the tool, which is currently limited to systems with 10 continuous variables.

We tested the capabilities of the four tools with respect to Property 1 and Property 2. It turns out that the puncturing test case is outside the capabilities of PHAVER and HSOLVER, which cannot give an answer in reasonable time. SPACEEX and ARIADNE, on the other hand, are able to verify both properties.

In the first step of this verification example, we determine the values of δ for which Property 1 is satisfied. In this particular case, the truth of the property is monotonic with respect to the value of the parameter, that is, if Property 1 is satisfied by some δ, then it is also satisfied by all values $\delta' \leq \delta$. Hence, we can use bisection to find the greatest value of the parameter for which the property is still true. We start from a range of possible values $[\delta_{\min}, \delta_{\max}] = [0, 0.05]$, and we fix the initial value of δ to $\frac{\delta_{\min}+\delta_{\max}}{2}$. This range is then refined: If for the current value of δ the system respects Property 1, then δ is a safe value for the uncertainty and the next range for δ is $[\delta, \delta_{\max}]$. Otherwise, δ is a possibly unsafe value, and the next range for δ is $[\delta_{\min}, \delta]$. The procedure stops when $|\delta_{\min} - \delta_{\max}|$ is under a threshold of "acceptable precision" equal to 10^{-2}. Table 40.1 shows that the two tools report different values for the smallest safe δ, with ARIADNE being more accurate.

As a second verification example, we focus our attention on another design parameter: the proportional gain constant K_f of the force loop (40.4). We are now interested in finding under what values of K_f Property 2 is respected. We fix $\delta = 0.025$, a value for which both tools report the system as safe, and we proceed in the very same way as the previous analysis. We start from an initial range of possible values for K_f, and we refine it by bisection. We set the approximation threshold to be 10^{-2}, and the final safe values for the parameter turn out to be $K_f \geq 0.1625$ when using ARIADNE and $K_f \geq 0.14375$ when using SPACEEX. Notice that for this second property, the most accurate result is the one obtained by SPACEEX.

Figure 40.2 shows the approximations of the reachable set computed by ARIADNE (first row) and SPACEEX (second row) for $\delta = 0.025$ and $K_f = 0.1625$, projected on the (t, x), (t, y), and (t, ϕ) planes, respectively. We would like to point out that both tools allow the user to set an arbitrary level of accuracy for the approximation. Hence, by tuning the accuracy settings, it is possible to obtain coarser or finer results, depending on the chosen trade-off between accuracy and computation time. Due to the complexity of the algorithms and of the state space representations, the accuracy is controlled by setting a number of parameters, which are different between the two tools. It is not possible to set the same accuracy level for ARIADNE and SPACEEX. Hence, for our experiments, we chose to match the computation time for the two tools: some minutes on a 2.4 GHz Intel Core 2 Duo machine with 4 GB of RAM.

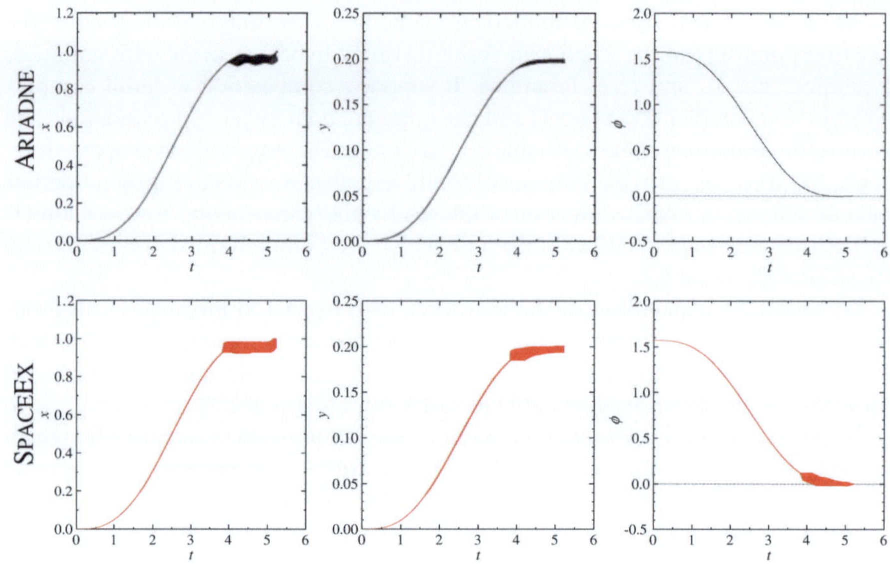

Fig. 40.2 Approximation of the reachable set computed by ARIADNE and SPACEEX

These experiments prove that formal verification can be useful within the control design cycle to prove the correctness of task execution and to guide the designers in the tuning of the control parameters. However, the existing tools still in general lack scalability and robustness to be truly practical for control engineers. This motivates the need for further research efforts to improve these features, possibly by exploiting abstraction techniques that can simplify the model of the system to make it tractable without loosing information on the properties of interest [2].

References

1. Alur R, Courcoubetis C, Henzinger TA, Ho PH (1992) Hybrid automata: an algorithmic approach to the specification and verification of hybrid systems. In: Hybrid systems, LNCS, vol 736. Springer, Berlin, pp 209–229
2. Alur R, Dang T, Ivančić F (2006) Counterexample-guided predicate abstraction of hybrid systems. Theor Comput Sci 354(2):250–271
3. Benvenuti L, Bresolin D, Collins P, Ferrari A, Geretti L, Villa T (2014) Assume-guarantee verification of nonlinear hybrid systems with ARIADNE. Int J Robust Nonlinear Control 24:699–724
4. Frehse G (2008) PHAVer: algorithmic verification of hybrid systems past HyTech. Int J Softw Tools Technol Transfer (STTT) 10:263–279
5. Frehse G, Le Guernic C, Donzé A, Cotton S, Ray R, Lebeltel O, Ripado R, Girard A, Dang T, Maler O (2011) SpaceEx: scalable verification of hybrid systems. In: Proceddings of 23rd international conference on computer aided verification (CAV 2011), LNCS, vol 6806. Springer, Berlin/Heidelberg, pp 379–395

6. Muradore R, Bresolin D, Geretti L, Fiorini P, Villa T (2011) Robotic surgery: formal verification of plans. Robot Autom Mag IEEE 18(3):24–32
7. Ratschan S, She Z (2007) Safety verification of hybrid systems by constraint propagation based abstraction refinement. ACM Trans Embed Comput Syst 6(1):8
8. Siciliano B, Sciavicco L, Villani L, Oriolo G (2008) Robotics: modelling, planning and control. Springer, Berlin

Chapter 41
Computing Reachable Sets of Differential Inclusions

Sanja Živanović Gonzalez and Pieter Collins

41.1 Introduction

In this essay, we compute over-approximations to reachable sets of *differential inclusions*,

$$\dot{x}(t) \in F(x(t)), \quad x(0) \in X_0, \tag{41.1}$$

where $F : \mathbb{R}^n \rightrightarrows \mathbb{R}^n$ is a continuous set-valued map with compact and convex values. A function $x : [0, \infty) \to \mathbb{R}^n$ is a solution of (41.1) if it is absolutely continuous and satisfies $\dot{x}(t) \in F(x(t))$ for almost every t.

In applications, differential inclusions arise as *nondeterministic noisy systems*

$$\dot{x}(t) = f(x(t), v(t)), \quad x(0) \in X_0, \quad v(t) \in V. \tag{41.2}$$

It is known (see [1]) that if the set V is compact and separable, f is continuous, and $f(x, V)$ is convex for all x, then the solution sets of (41.1) and (41.2) are equivalent for measurable functions $v(\cdot)$. Nondeterministic noisy systems are appropriate models for systems with bounded disturbances for which a stochastic description is unknown or inappropriate, such as for systems with nontrivial internal dynamics.

The reachable set of a differential inclusion at time t is defined as

$$\text{Reach}(X_0, t) = \{x(t) \mid x(\cdot) \text{ is a solution of } (41.1)\}. \tag{41.3}$$

S. Živanović Gonzalez (✉)
Department of Mathematics and Computer Science, Barry University,
11300 NE 2nd Ave, Miami Shores, FL, USA
e-mail: szivanovic@barry.edu

P. Collins
Department of Knowledge Engineering,
Maastricht University, 6211 LH Maastricht, The Netherlands
e-mail: pieter.collins@maastrichtuniversity.nl

J.H. van Schuppen and T. Villa (eds.), *Coordination Control of Distributed Systems*,
Lecture Notes in Control and Information Sciences 456,
DOI 10.1007/978-3-319-10407-2_41

Safety properties of a system can be formulated in terms of the reachable sets. In order to verify safety properties, we must rigorously compute *over-approximations* to the reachable sets.

Several different techniques and various types of numerical methods have been proposed as approximations to the solution set of a differential inclusion. For example, ellipsoidal calculus was used in [2], a Lohner-type algorithm in [3], grid-based methods in [4], and [5] optimal control in [6] and discrete approximations in [7–10]. However, these algorithms either do not give rigorous over-approximations, or are approximations of low order (e.g. Euler approximations with a first-order single-step truncation error). Essentially, the only algorithms mentioned above that could give arbitrary accurate error estimates are the ones that use grids. However, higher-order discretization of a state space greatly affects efficiency of the algorithm. It was noted in [5] that if one is trying to obtain higher-order error estimates on the solution set of differential inclusions, then grid methods should be avoided.

Our approach to computing over-approximations to reachable sets relies on higher-order methods for computing reachable sets over small time steps. We approximate the infinite-dimensional space of possible inputs to (41.2) by a finite-dimensional space and derive estimates for the error between the exact reachable set and the finite-dimensional approximations.

41.2 Motivation

Suppose we are given a coordinated system which comprises multiple autonomous underwater vehicles (AUVs). Obviously, vehicles must avoid collisions, and we would like to be able to automatically verify this property in the closed-loop system.

Due to the complexity of the system, in practice, it would be infeasible to verify the system as a whole. One possible way to verify the whole system would be to simplify models of each component, verify that the simplified model is an abstraction of a physical model, and verify the whole system using the simplified component models. Further, we should verify the system under a full spectrum of possible operating environments. There are then three ways in which differential inclusions arise for this problem.

The vehicles are subject to unknown disturbances due to water currents, and since these have their own dynamics but may be assumed to be bounded, nondeterministic noisy systems are an appropriate model class.

Each vehicle has its own complicated (nonlinear) dynamics, which we may simplify by a lower-order system with approximately the same behavior. In order that the simplified model is an abstraction of a physical model, we need to include the errors of the approximation. If the physical model is given by a differential equation $\dot{x}(t) = f(x(t))$, the reduced state by $z(t) = g(x(t))$ and a simplified model by $\dot{\tilde{z}}(t) = h(\tilde{z}(t))$, then we can show that $\dot{\tilde{z}} - \dot{z} = h(g(x)) - g'(x)f(x)$, giving a bound on the error.

As we would like to be able to treat vehicles independently, we need to use abstract models of the other vehicles. For example, we may treat other vehicles as subject only to physical constraints of the dynamics. This again yields a nondeterministic noisy system model for the component.

41.3 Single-Step Approximation

In this section, we outline an approach for computing accurate over-approximations to the reachable set over small time steps.

For $x_k \in \mathbb{R}^n$ and $v_k(\cdot) \in L^\infty([t_k, t_{k+1}])$, define $\phi(x_k, v(\cdot))$ to be the point $x(t_{k+1})$ which is the value at time t_{k+1} of the solution of (41.2) with $x(t_k) = x_k$. Then, the set of points attainable at time t_{k+1} from the set X_k at time t_k is given by

$$\text{Reach}(X_k, t_k, t_{k+1}) = \{\phi(x_k, v(\cdot)) \mid x_k \in X_k \text{ and } v(\cdot) \in L^\infty([t_k, t_{k+1}]; V)\}.$$

Since the space of bounded measurable functions is infinite-dimensional, we aim to approximate this set by restricting the disturbances to a finite-dimensional space.

Consider a set of approximating functions $W_k \subset C([t_k, t_{k+1}]; \mathbb{R}^m)$. The approximate system is then

$$\dot{y}(t) = f(y(t), w_k(t)), \quad y(t_k) = x(t_k), \quad w_k(\cdot) \in W_k, \tag{41.4}$$

for $t \in [t_k, t_{k+1}]$. We then need to find an error bound ϵ_k such that

$$\forall x_k \in X_k,\ \forall v_k \in L^\infty([t_k, t_{k+1}]; V),\ \exists w_k \in W \text{ s.t. } \|\phi(x_k, v_k(\cdot)) - \phi(x_k, w_k(\cdot))\| \leq \epsilon_k. \tag{41.5}$$

We parameterise W_k as $\{w_k(a_k, \cdot) \mid a_k \in A \subset \mathbb{R}^p\}$. Setting $\tilde{\phi}(x_k, a_k) = \phi(x_k, w(a_k, \cdot))$, i.e., $\tilde{\phi}$ is also the solution at $t = t_{k+1}$ of $\dot{x}(t) = f(x_k, w(a_k, \cdot))$ with $x(t_k) = x_k$, we obtain the over-approximation

$$\text{Reach}(X_k, t_k, t_{k+1}) \subset \{\tilde{\phi}(x_k, a_k) \pm \tilde{\varepsilon}_k \mid x_k \in X_k \text{ and } a_k \in A\}.$$

In practice, the local error $\tilde{\varepsilon}_k$ for a time step consists of an analytical error ϵ_k given by (41.5) and a numerical error which is a remainder of a polynomial model $\hat{\phi}(x_k, a_k) \pm \hat{\varepsilon}_k$ representing the solution $\tilde{\phi}(x_k, a_k)$ of $\dot{x}(t) = f(x(t), w_k(a_k, t))$. In other words, $\tilde{\varepsilon}_k = \epsilon_k + \hat{\varepsilon}_k$.

We would like to choose approximating functions $w_k = w(a_k, \cdot) : [t_k, t_{k+1}] \to \mathbb{R}$ such that the solution of (41.4) is an approximation of high order to the solution of (41.2). In Sect. 41.5 we shall see that the choice

$$w_k(a, t) = a_0 + a_1(t - t_{k+1/2})/h_k$$

where $t_{k+1/2} = t_k + h_k/2 = (t_k + t_{k+1})/2$ gives good error bounds for input-affine systems. For more information on the choice of functions w_k and formulae for the single-step errors of order $O(h)$, $O(h^2)$, $O(h^3)$ in the case of and generalizations of it to some other differential inclusions, see [11, 12].

41.4 Algorithm for Computing the Reachable Set

In this section, we present an algorithm for computation of the solution set of (41.2), using the single-step computation presented in Sect. 41.3.

Let $[0, T]$ be an interval of existence of (41.2). Let $0 = t_0, t_1, \ldots, t_{n-1}, t_n = T$ be a partition of $[0, T]$, and let $h_k = t_{k+1} - t_k$. Let R_k be an over-approximation of the set $\mathrm{Reach}(x_0, t_k)$. Then, to compute an over-approximation R_{k+1} of $\mathrm{Reach}(x_0, t_{k+1})$:

1. Compute the flow $\tilde{\phi}_k(x_k, a_k)$ of $\dot{x}(t) = f(x(t), w_k(a_k, t))$, $x(t_k) = x_k$, for $t \in [t_k, t_{k+1}]$, $x_k \in R_k$, and $a_k \in A$.
2. Compute the uniform error bound ε_k.
3. Compute the set R_{k+1} which over-approximates $R(x_0, t_{k+1})$ as $R_{k+1} \supset \{\tilde{\phi}(x_k, a_k) \pm \epsilon_k \mid x_k \in R_k,\ a_k \in A\}$.
4. Reduce the number of parameters a_k (if necessary).
5. Split the new obtained domain (if necessary).

Step 1 of the algorithm produces an approximated flow in the form $\tilde{\phi}_k(x_k, a_k) \approx \phi(x_k, w(a_k, \cdot))$ which is guaranteed to be valid for all $x_k \in R_k$. In practice, we cannot represent $\tilde{\phi}$ exactly and instead use Taylor (polynomial) model approximation (see [13, 14]) with guaranteed error bound $\hat{\phi}$. In Step 2, we add the uniform error bound ε_k to make sure an over-approximation is achieved. In Step 3, we compute a new approximating set by applying the approximated flow to the initial set of points. Steps 4 and 5 are crucial for the efficiency and the accuracy of the algorithm. It is important to notice that the number of parameters (a_k initially) grows over the time steps, and Step 4 is vital in reducing the complexity of the representation of the R_k. If the approximating set could become too large, it may be hard to compute "good" approximations to the flow and/or the error valid over the entirety of R_k. Step 5 allows the flow to be computed over smaller domains, but should be used sparingly, since it doubles the number of future flow step computations required.

41.5 Input-Affine Systems

In this section, we restrict attention to the input-affine system

$$\dot{x}(t) = f(x(t)) + \sum_{i=1}^{m} g_i(x(t))v_i(t); \quad x(t_0) = x_0. \tag{41.6}$$

For some $r \geq 1$ depending on the desired order of approximation, we assume that

- $f : \mathbb{R}^n \to \mathbb{R}^n$ is a C^r function,
- each $g_i : \mathbb{R}^n \to \mathbb{R}^n$ is a C^r function,
- $v_i(\cdot)$ is a measurable function such that $v_i(t) \in [-V_i, +V_i]$ for some $V_i > 0$.

Then, the Eq. (41.4) becomes

$$\dot{y}(t) = f(y(t)) + \sum_{i=1}^{m} g_i(y(t))w_i(a_k, t); \quad y(t_k) = y_k, \ t \in [t_k, t_{k+1}]. \tag{41.7}$$

In what follows, we assume that we have a bound B on the solutions of (41.6) and (41.7) for all $t \in [0, T]$. We take constants $K, K_i, L, L_i, H, \Lambda$ such that

$$\begin{aligned} &\|f(z(t))\| \leq K, \quad \|g_i(z(t))\| \leq K_i, \quad \lambda(Df(\cdot)) \leq \Lambda, \\ &\|Df(z(t))\| \leq L, \quad \|Dg_i(z(t))\| \leq L_i, \\ &\|D^2 f(z(t))\| \leq H, \quad \|D^2 g_i(z(t))\| \leq H_i, \end{aligned} \tag{41.8}$$

for each $i = 1, \ldots, m$, and for all $t \in [0, T]$, and $z(\cdot) \in B$. We also set

$$K' = \sum_{i=1}^{m} V_i K_i, \quad L' = \sum_{i=1}^{m} V_i L_i, \quad H' = \sum_{i=1}^{m} V_i H_i.$$

Here, Df denotes the Jacobian matrix, $D^2 f$ denotes the Hessian matrix, and $\lambda(\cdot)$ denotes the logarithmic norm of a matrix (see [15]).

Below we present formulas for obtaining uniform single-step error bound of $O(h^2) + O(h^3)$ and in some special cases pure $O(h^3)$ error—Step 2 of the algorithm presented in Sect. 41.4. For proofs, detail error derivation and other orders of error please refer to [12]. In what follows, we write $h_k = t_{k+1} - t_k$, $t_{k+1/2} = t_k + h_k/2 = (t_k + t_{k+1})/2$.

As previously mentioned, the $w_i(a, \cdot)$ are real-valued finitely parametrized functions. If they are taken to be affine, $w_i(t) = a_{i,0} + a_{i,1}(t - t_{k+1/2})/h_k$, and satisfying

$$\int_{t_k}^{t_{k+1}} v_i(t) - w_i(t)\, \mathrm{d}t = 0; \quad \int_{t_k}^{t_{k+1}} (t - t_{k+1/2})\ (v_i(t) - w_i(t))\, \mathrm{d}t = 0, \tag{41.9}$$

then

$$a_{i,0} = \frac{1}{h_k} \int_{t_k}^{t_{k+1}} v_i(t)\mathrm{d}t; \quad a_{i,1} = \frac{12}{h_k^2} \int_{t_k}^{t_{k+1}} v_i(t)\,(t - t_{k+1/2})\,\mathrm{d}t. \tag{41.10}$$

It is easy to see that

$$|a_{i,0}| \leq V_i, \ |a_{i,1}| \leq 3\,V_i, \ |w_i(t)| \leq 5V_i/2, \text{ and } |\dot{w}(t)| \leq 3V_i/2h_k. \qquad (41.11)$$

Theorem 41.1 *For any $k \geq 0$, and all $i = 1, \ldots, m$, if*

- *$f(\cdot)$ is a C^2 vector function,*
- *$g_i(\cdot)$ are non-constant C^2 functions, and*
- *the w_i satisfy* (41.9),

then an error of $O(h^2) + O(h^3)$ is obtained. Moreover, if the w_i are affine functions, $w_i(t) = a_{i,0} + a_{i,1}(t - t_{k+1/2})/h_k$, then a formula for calculating the error is given by

$$\begin{aligned}
&\left(1 - L(h_k/2) - h_k L'\right) \|x(t_{k+1}) - y(t_{k+1})\| \leq (h_k^2/4)L' \left(11K + (69/2)K'\right)\\
&\quad + (7h_k^3/8)\,K' \left[(4H' + H)\,(K + (5/2)K') + L^2 + \left((9/2)L + 5L'\right)L'\right] \frac{e^{\Lambda h_k} - 1}{\Lambda h_k}\\
&\quad + (7h_k^3/48)\left(H\,K' + L\,L'\right)\left(K + K'\right).
\end{aligned}$$

Corollary 41.1 *For any $k \geq 0$,*

- *if the system has additive noise, i.e. $\dot{x}(t) = f(x(t)) + v(t)$,*
- *$f(\cdot)$ is a C^2 function, and*
- *$w_i(t)$ are real-valued functions defined on $[t_k, t_{k+1}]$ which satisfy Eqs.* (41.9),

then an error of $O(h^3)$ is obtained. Moreover, for $w_i(t) = a_{i,0} + a_{i,1}(t - t_{k+1/2})/h_k$, the formula for the local error is given by:

$$\begin{aligned}
&\left(1 - (h_k/2)L\right)\|x(t_{k+1}) - y(t_{k+1})\|\\
&\quad \leq \frac{7}{48}\,h_k^3\,K'\,H\,(K + K') + \frac{7}{8}\,h_k^3\,K'\left(L^2 + H\,(K + 5K'/2)\right)\frac{e^{\Lambda h_k} - 1}{\Lambda\,h_k}.
\end{aligned} \qquad (41.12)$$

The formula for the error in the additive noise case is simplified because $L' = H' = 0$. If we write $||v(t)|| = K'$, then the result follows directly from Theorem 41.1.

We also achieve an $O(h^3)$ method in the case of a single-input system:

Proposition 41.1 *For any $k \geq 0$, if*

- *the input-affine system has single input, i.e., $m = 1$ in* (41.6)
- *$f(\cdot)$ and $g(\cdot)$ are C^2 functions, and*
- *$w(t)$ is a real valued function defined on $[t_k, t_{k+1}]$ which satisfies Eqs.* (41.9),

then an error of $O(h^3)$ is obtained. Moreover, for $w(t) = a_0 + a_1(t - t_{k+1/2})$, the formula for the local error is given by

$$(1 - (h_k/2)L - h_k L')\,\|x(t_{k+1}) - y(t_{k+1})\|$$
$$\le \frac{7\,h_k^3}{8}\,K'\Big((H + 10\,H')(K + (5/2)K') + L^2 + (25/2)\,L\,L' + 25\,(L')^2\Big)\,\frac{e^{\Lambda h_k} - 1}{\Lambda\,h_k}$$
$$+ \frac{h_k^3}{48}\,(K + K')\,\Big((7/6)(H\,K' + L\,L') + 28\,(H'\,K + L\,L') + 29\,(H'\,K' + (L')^2)\Big).$$

41.6 Computational Results

The algorithm is being tested using the tool Ariadne [16] for reachability analysis of hybrid systems. It is based on set-valued (interval) analysis of real systems. We tested several linear and nonlinear systems, and while the algorithm appears to be efficient, further investigation of steps 4 and 5 is necessary in order to obtain better efficiency for higher-dimensional nonlinear systems.

In order to give a better explanation of the algorithm, we demonstrate how to compute reachable sets for a simple perturbed Van der Pol oscillator,

$$\dot{x} = y; \quad \dot{y} = -x + 2\,(1 - x^2)\,y + v,$$

where v represents additive noise. Let $D = [0, 2] \times [-1, 3]$ be the region of computation and let $v(\cdot) \in [-0.08, 0.08]$, i.e., $A = 0.08$. Then $K = 20$, $L = 31$, $\Lambda = 27$, $H = 12$, and

$$\varepsilon = \|x(t_{k+1}) - y(t_{k+1})\| \le 11.24\,h^3 + 168.17\,h^3\,\frac{e^{27\,h} - 1}{27h}.$$

If the set of initial points is $X_0 = [0.1, 0.105] \times [1.5, 1.505]$ and time step is $h = 0.001$, then analytical single-step error is $\epsilon = 1.817092608 \times 10^{-7}$.

In Figs. 41.1 and 41.2, we show the solution set of the perturbed Van der Pol oscillator where the time interval of computation is $[0, 1.5]$. In Fig. 41.1, splitting of

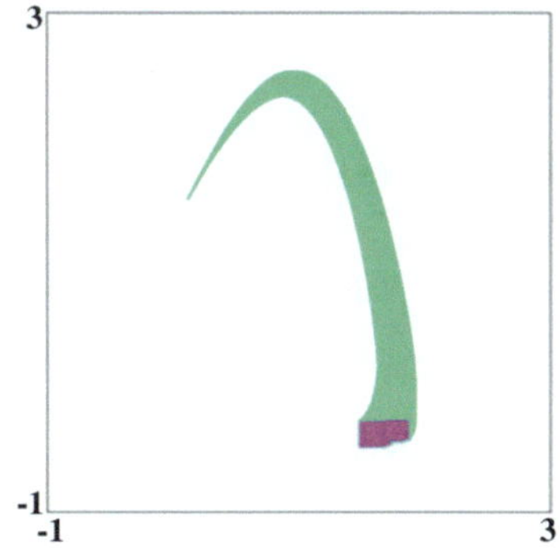

Fig. 41.1 Evolution of the Perturbed Van der Pol oscillator: splitting performed at $t_1 = 0.6$ and $t_2 = 1.2$

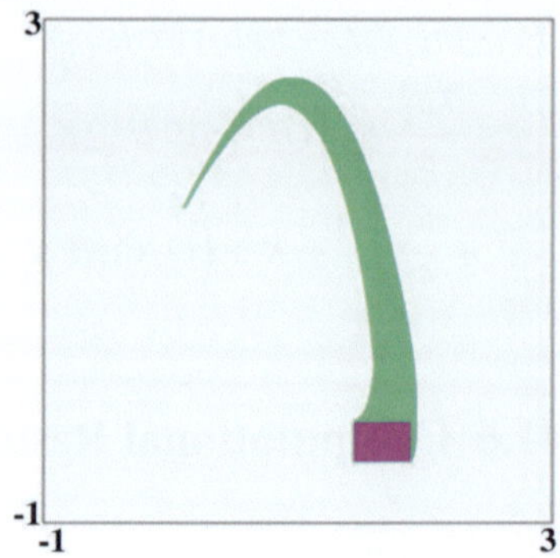

Fig. 41.2 Evolution of the Perturbed Van der Pol oscillator: no splitting performed

the domain was performed at $t_1 = 0.6$ and $t_2 = 1.2$. At t_1, the set was divided in half along x-axis, and at t_2, the set was divided in half along y-axis. In Fig. 41.2, there was no splitting performed. As predicted, the reachable set at $T = 1.5$ was smaller when splitting was performed.

41.7 Conclusions and Further Research

In this essay, we have given an algorithm for the computation of the reachable sets of a differential inclusion based on higher-order single-step methods. This algorithm is an important basis of more advanced safety verification algorithms for distributed systems.

There are several directions that we can take on from here in order to expand proposed theory. An important extension is to consider more general classes of disturbance $v(\cdot)$, such as lying in a general convex set rather than just a coordinate-aligned box. Another direction is to consider different forms of approximate inputs $w_k(t)$ which may yield better error bounds, such as step functions. Finally, it is important to consider more advanced methods of reducing the number of parameters in the enclosures R_k.

References

1. Aubin JP, Cellina A (1984) Differential inclusions: set-valued maps and viability theory. Grundlehren der Mathematischen Wissenschaften, 264. Springer, Berlin
2. Kurzhanski A, Valyi I (1997) Ellipsoidal calculus for estimation and control (English summary). Systems and control: foundations and applications. BirkhÃ user Boston Inc, Boston, MA; International Institute for Applied Systems Analysis, Laxenburg
3. Zgliczynski P, Kapela TA (2009) Lohner algorithm for perturbation of ODEs and differential inclusions. Discrete Contin Dyn Syst Ser B 11(2):365–385

4. Puri A, Borkar V, Varaiya P (1995) ϵ-approximation of differential inclusions. In: Proceedings of the 34th IEEE conference on decision and control, pp 2892–2897
5. Beyn W-J, Rieger J (2007) Numerical fixed grid methods for differential inclusions (English summary). Computing 81(1):91–106
6. Baier R, Gerdts M (2009) A computational method for non-convex reachable sets using optimal control. In: Proceedings of the European control conference 2009, Budapest, Hungary, pp 97–102
7. Dontchev A, Lempio F (1992) Difference methods for differential inclusions: a survey. SIAM Rev 34(2):263–294
8. Dontchev T (2002) Euler approximation of nonconvex discontinuous differential inclusions (English summary). An Stiint Univ Ovidius Constanta Ser Mat 10(1):73–86
9. Dontchev AL, Farkhi EM (1989) Error estimates for discretized differential inclusion. Computing 41(4):349–358
10. Grammel G (2003) Towards fully discretized differential inclusions. Set-Valued Anal 11(1):1–8
11. Živanović Gonzalez S, Collins P (2010) Numerical solutions to noisy systems. In: Proceedings of the 49th IEEE conference on decision and control
12. Živanović Gonzalez S, Collins P (2010) Higher order methods for differential inclusions. CWI technical report
13. Makino K, Berz M (2003) Taylor models and other validated functional inclusion methods. Int J Pure Appl Math 4(4):379–456
14. Revol N, Makino K, Berz M (2005) Taylor models and floating-point arithmetic: proof that arithmetic operations are validated in COSY. J Log Algebr Program 64(1):135–154
15. Söderlind G (2006) The logarithmic norm. History and modern theory (English summary). BIT 46(3):631–652
16. Benvenuti L, Bresolin D, Casagrande A, Collins P, Ferrari A, Mazzi E, Sangiovanni-Vincentelli A, Villa T (2008) Reachability computation for hybrid systems with Ariadne. In: Proceedings of the 17th IFAC World Congress

Chapter 42
Modeling Objects Moving in a Complex Environment with World Automata

Marta Capiluppi and Roberto Segala

42.1 Motivation and Main Idea

In this chapter, we focus on automata-based representations of hybrid systems [1, 2] and address the problem of representing faithfully situations where a hybrid automaton exists within an environment and derives information about other automata by observing the environment itself, rather than by using any form of direct communication. We call this kind of communication *implicit*. The main sources of motivation for these studies are real applications presented in the European project C4C, such as agents performing a *search* mission, e.g., UAVs [3] or autonomous underwater vehicles [4], but also road traffic problems [5, 6] and autonomous straddle carriers in harbors [7]. In all these situations, we have a collection of agents that need to communicate and coordinate to achieve a common goal; hence, distributed systems. Moreover, the agents move within an environment that changes dynamically and detect each other's presence not necessarily via direct communication but rather by observations of the environmental changes. Current techniques for modeling similar cases are not able to represent in a realistic and natural way the ability of each agent to take autonomous decisions, based only on its perception of the surrounding environment. This implicit communication is needed when direct exchange of information is not possible, unreliable or forbidden. Instead of providing the system with an omniscient machine, we decided to copy the nature and exploit sensors embedded in each agent. Since we need to satisfy compositionality properties, due to the multi-agent nature of the case studies, we start from the Hybrid I/O Automata (HIOAs) [8] representation and add some features to model their interaction with the environment. We will show how to apply our framework to the following example

M. Capiluppi (✉) · R. Segala
Dipartimento di Informatica, Università di Verona, Strada le Grazie 15, 37134 Verona, Italy
e-mail: marta.capiluppi@univr.it

R. Segala
e-mail: roberto.segala@univr.it

J.H. van Schuppen and T. Villa (eds.), *Coordination Control of Distributed Systems*,
Lecture Notes in Control and Information Sciences 456,
DOI 10.1007/978-3-319-10407-2_42

Fig. 42.1 Characteristics of the scenario. **a** A sandy area with cars. **b** Level of the ground. **c** A car moving from a point to another in the sandy area

describing a typical problem of coordination of agents, even though simplified to enlighten only the main challenges the designer has to face in finding a suitable model for this situation.

Example 42.1 Consider a sandy area where two cars move, as in Fig. 42.1a. We assume an underlying metric space $\mathbb{R}^2$. When a car takes a certain position, its pressure provokes a depression of the ground (Fig. 42.1b). Hence, the car changes the characteristics of the environment in a permanent way, since sand retains the shape. The other car is, then, able to *see* where a car has moved (Fig. 42.1c). We aim at avoiding collisions between the two cars.

42.2 World Automata

Here, we recall the Hybrid I/O Automata (HIOAs) [8] model.

First, we need to recall some notation. We write $f \lceil P$ for the restriction of function f to set P, that is, the function g with $\mathrm{dom}(g) = \mathrm{dom}(f) \cap P$ such that $g(c) = f(c)$ for each $c \in \mathrm{dom}(g)$. If f is a function whose range is a set of functions and P is a set, then we write $f \downarrow P$ for the function g with $\mathrm{dom}(g) = \mathrm{dom}(f)$ such that $g(c) = f(c) \lceil P$ for each $c \in \mathrm{dom}(g)$.

For each variable v, we assume both a *(static) type*, $type(v)$, which gives the set of values it may take on, and a *dynamic type*, $dtype(v)$, which gives the set of trajectories it may follow. A *valuation* **v** for a set of variables V is a function that associates with each variable $v \in V$ a value in $type(v)$. We write vals(V) for the set of valuations for V. Let J be a left-closed interval of T (the time axis) with left endpoint equal to 0. Then a *J-trajectory* for V is a function $\tau : J \rightarrow \text{vals(V)}$, such that for each $v \in V$, $\tau \downarrow v \in dtype(v)$. A *trajectory* for V is a J-trajectory for V, for any J. Trajectory τ is a *prefix* of trajectory τ', denoted by $\tau \leq \tau'$, if τ can be obtained by restricting τ' to a subset of its domain. We define $\tau \trianglerighteq t \triangleq (\tau \lceil [t, \infty)) - t$. The concatenation $^\frown$ of two trajectories is obtained by taking the union of the first trajectory and the function obtained by shifting the domain of the second trajectory until the start time agrees with the limit time of the first trajectory; the last valuation of the first trajectory, which may not be the same as the first valuation of the second trajectory, is the one that appears in the concatenation. Prefix, suffix, and concatenation operations return trajectories. For more detail, the interested reader can refer to [8].

Definition 42.1 Hybrid I/O Automaton (HIOA)
A HIOA $\mathscr{A}$ is a tuple $((U, X, Y), (I, H, O), Q, \Theta, D, \mathscr{T})$ where

- (U, X, Y) are disjoint sets of *input, internal, and output variables*, respectively. Let V denote the set $U \cup X \cup Y$ of *variables.*
- (I, H, O) are disjoint sets of *input, hidden, and output actions*, respectively. Let A denote the set $I \cup H \cup O$ of *actions.*
- $Q \subseteq$ vals(X) is the set of *states.*
- $\Theta \subseteq Q$ is a nonempty set of *initial states.*
- $D \subseteq Q \times A \times Q$ is the *discrete transition relation.*
- $\mathscr{T}$ is a set of trajectories on V s.t. $\tau(t) \lceil X \in Q, \forall \tau \in \mathscr{T}$ and that satisfy the following axioms

T1 *(Prefix closure)* For every $\tau \in \mathscr{T}$ and every $\tau' \leq \tau, \tau' \in \mathscr{T}$.
T2 *(Suffix closure)* For every $\tau \in \mathscr{T}$ and every $t \in \mathrm{dom}(\tau), \tau \trianglerighteq t \in \mathscr{T}$.
T3 *(Concatenation closure)* Let $\tau_0, \tau_1, \tau_2, \ldots$ be a sequence of trajectories in $\mathscr{T}$ such that, for each nonfinal index i, τ_i is closed and $\tau_i.lstate = \tau_{i+1}.fstate$. Then $\tau_0 \frown \tau_1 \frown \tau_2 \ldots \in \mathscr{T}$.

Now, we apply this model to the agents of Example 42.1.

Example 42.2 To represent HIOAs, we use a variant of the TIOA language [9], with some extensions for hybrid systems [10]. The HIOA of a car is reported in Fig. 42.2. It has an output variable K representing the ground pressure provided by the car and an output variable P representing the car position. The input variables are: the level of the ground *groundlevel* as a Boolean saying if the level is low (1) or high (0); the *collisionrisk* saying if another car is in collision risk (1) or not (0). A function f is defined by giving the surface of the ground occupied by the car area starting from its position p_T and its orientation angle ϕ. We can imagine that f returns a rectangle centered in p_T with orientation ϕ. The pressure variable is updated with a function z depending on the mass m and area of the car. The velocity vel of the car is 0 if collisionrisk is true. Similarly, the car slows down when groundlevel is true. For the sake of simplicity, we used a Boolean variable to represent the ground

Fig. 42.2 HIOA representing a car

type Rad = $\mathbb{R}|2\pi$
hioa Car
variables
  input collisionrisk: Bool, groundlevel: Bool
  internal ϕ: Rad, p_T: Real^2, m: Real, *vel*:Real
  output P: Real^2, K: Real
trajectories
  $K(t) = z(m, f(\phi, p_T));$

$$vel(t) = \begin{cases} 0 & \text{if } collisionrisk \\ 0.5 & \text{if } groundlevel \\ 1 & \text{otherwise.} \end{cases};$$

  $P(t) = p_T(t).$

level changes, but any other function can be used, such as more general and complex diffusion equations.

The representation of Fig. 42.2 is one of the possible ways of modeling the agents of Example 42.1. If we imagine to have replicas of the automaton representing a car, we obtain a model that needs another machine able to collect all the positions of the cars at any instant of time, in order to check possible conflicts and avoid collisions. This implies that all the agents need to communicate their data to the machine at any time and that the machine needs to send to the cars some signals to avoid collisions and all the positions of the agents involved in the conflict.

To avoid this expensive (and sometimes impossible) exchange of data, we extend the HIOA modeling framework by specializing some variables of the automaton and calling them *world variables*. The name is due to the fact that we want them to represent the connection between the agents and the surrounding world. Moreover, world variables represent the changes in the environment as might be perceived by the agents. Hence, the set of variables V is partitioned in a set W of world variables and a set S of standard automaton variables. The set W is partitioned in sets (U_w, X_w, Y_w) of *world input, internal, and output variables*, respectively, such as $U_w \subseteq U$, $X_w \subseteq X$, $Y_w \subseteq Y$. To avoid confusion, we will add to automaton variables the subscript a: U_a, X_a, Y_a. The main difference between world and automaton variables is that the values of world variables are function of time and space, not only of time as in standard automaton variables. Hence, world variables values (and trajectories) will depend both on the instant of time and the position in the underlying space. An automaton $\mathscr{A}$ will use its world inputs U_w to receive stimuli from the world it lives in. Analogously, it will use its world outputs Y_w to give stimuli to the world it lives in. Finally, internal world variables X_w are used to represent the world characteristics of $\mathscr{A}$. To keep the theory consistent with previous descriptions of automata, all the X variables represent persistent characteristics of the system.

Example 42.3 We now represent the car in Fig. 42.2 using a HIOA with world variables, extending the TIOA language to include them. Note that world variables are always described using their trajectories in time and space, i.e., they are described for any instant of time t and any point in space p. Each car is represented by the automaton in Fig. 42.3. It has an output world variable k representing the ground pressure provided by the car and an output world variable representing the car color ξ. The input world variables are: the level of the ground g and its color c. Each car perceives the ground level through a Boolean variable g saying if the ground is low (1) or high (0). We used the Boolean representation for the sake of simplicity. Of course, any other function, such as diffusion equations, may be used. Each car can check the color of the ground at each point of the area by the variable c, which represents a kind of colored map of the area. We assume that the color variable ξ takes the value black for all the points inside the car area given by f and white outside. The pressure variable k is updated with a function h depending on the mass m and area of the car, associating to each point in the area of the car the value of its pressure in time, and to each point outside the area of the car a 0 value. Two actions *collision, level* represent the possibility that another car is in the neighborhood and

type Rad = $\mathbb{R}|2\pi$
hioaw Car
world variables
  input g: Bool, c: Color;
  output k: Real, ξ: Color;
automaton variables
  internal ϕ: Rad, p_T: Real^2, m: Real, vel:Real, r:Real, stop: Bool, slow: Bool;
actions
  hidden collision, level;
transitions
  hidden collision
  pre $\exists p^* \in q(p_T, r, f(\phi, p_T))$ s.t. $c(t, p^*) =$ black
  eff stop = true;
  hidden level
  pre $\exists p^* \in q(p_T, r, f(\phi, p_T))$ s.t. $g(t, p^*) =$ true
  eff slow = true;
trajectories

$$\xi(t,p) = \begin{cases} \text{black if } p \in f(\phi, p_T) \\ \text{white otherwise} \end{cases};$$

$$k(t,p) = h(m, f(\phi, p_T));$$

$$vel(t) = \begin{cases} 0 & \text{if stop} \\ 0.5 & \text{if slow} \\ 1 & \text{otherwise.} \end{cases}$$

Fig. 42.3 HIOA with world variables (here called **hioaw**), representing a car

that the level of the ground in the neighborhood is low, respectively. Action *collision* activates a Boolean variable *stop* if there is any black point p^* in the neighborhood of the car, which is calculated by the function q returning a circle of radius r (bigger than the semi-diagonal of the rectangle representing the area of the car) and centered in p_T, but excluding the area of the car given by function f. Action *level* activates a Boolean variable *slow* if there is any point p^* in the neighborhood of the car for which the ground level variable g is true, i.e., the level of the ground is low. Hence, the velocity vel of the car is 0 if *stop* is true. Similarly, the car slows down when *slow* is true. All the presented equations describing the car dynamics are very simple, but the description of the motion is out of the scope of this paper. Indeed, they can be substituted by other equations.

The reader can notice that in Fig. 42.2, the position of the car is explicit in variable P, which is an output that must be collected by the supervisor at each instant of time to check where the automaton is in the space. In the new automaton of Fig. 42.3, the position is embedded in the world variables and does not need to be explicitly put in an automaton variable. Indeed both color and pressure world variables carry the information about the position of the automaton in the space, due to their nature.

What is missing in the above description is the interaction between agents and environment. How does the environment communicates to the agents the changes it is undergoing? To keep the compositionality theory consistent, we represent also the environment with the same model of the agents. This choice introduces a hierarchy of automata that can be both and contemporarily agents living in an environment and environments for other agents. We call the extended HIOAs *World Automata* (WAs). We renamed the automata because, even if they are an extension of the HIOAs, they will need slightly different operators to prove compositionality results.

Since we now have a hierarchy of WAs, we need to distinguish between the variables used by an automaton to communicate with the world in which it lives and the variables it uses to communicate with the world it creates. We could simply partition the sets of variables into variables used to communicate with the outside world and variables used to communicate with the inside world. Nevertheless, this method will hide the hierarchy of automata. On the contrary, we want to keep the hierarchy in order to be able to always retrieve automata at different levels of depth. For this reason, we equip variables with levels, and we assume that variables at different levels are distinct. Hence, we introduce a level function described as $l : S \to \mathbb{N}$ and extracting the level of a variable in any set S. Basically, if we consider only variables of level 0, then we have an ordinary HIOA equipped with some input and output world variables that are used to interact with its external world. The world variables of level 1 describe the world provided by an automaton. The world variables of level 2 describe the world provided by automata of level 1 and so on.

Example 42.4 In our example, the environment is given by a sandy area. At level 1, WA called Environment in Fig. 42.4 stores in an internal world variable gx the level of the ground and in another one its color cx, it receives as information from its local world the ground pressure in the input world variable k and the cars color ξ, while it gives as output to its local world the level of the ground g and its color c again. The

Fig. 42.4 WA representing the sandy area

type Color = {white, black}
worldautomaton Environment
world variables LEVEL 1
  internal gx: Bool, cx: Color
  input k: Real, ξ: Color
  output g: Bool, c: Color
automaton variables LEVEL 1
  internal ground: Bool
actions
  hidden pressure;
transitions
  hidden pressure
  pre $\exists t^* \leq t$ s.t. $k(t^*, p) \neq 0$
  eff ground = true;
trajectories
  $cx(t, p) = \xi(t, p);$
  $gx(t, p) = \begin{cases} 0 \text{ if ground} \\ 1 \text{ otherwise} \end{cases};$
  $c(t, p) = cx(t, p);$
  $g(t, p) = gx(t, p).$

color variable cx of the area is updated with the color variable ξ of the cars, while the ground level gx depends on the pressure k of the cars on the ground. An action "pressure" arises at the instants $t*$ of time in which the input world variable k is different from 0. Action "pressure" has the effect of setting Boolean variable *ground* to 1. The cars WAs have the same structure of the automaton in Fig. 42.3, but all the variables (both world and automaton ones) are of level 0. It is now easy to understand how the two automata of Fig. 42.3 (plus levels) and Fig. 42.4 interact using world variables to exchange information on the dynamical evolution of the scenario.

42.3 Operations on WAs

In our framework, parallel composition models the interaction and communication of two or more agents living in the same world, i.e., of two WAs at the same level, with the environment, i.e., the world outside. The composition rules are the same as for HIOAs, i.e., automata are synchronized on common actions and shared variables, except for output world variables. It might be that the two composing automata have some output world variables acting on the same input of the world, i.e., some output world variables with the same name, such that the intersection of the output world variables at level 0 is not empty. For all those variables, it is necessary to sum the trajectories, i.e., to sum the effect that the outputs have on the environment. The presence of levels of variables slightly changes the parallel composition policy, since we want to compose automata without losing the original hierarchy of the components. We need to impose that variables at levels not supposed to interact are not synchronized: it suffices to use disjoint sets of world variables at levels other than 0, renaming them when needed.

Since WAs create a hierarchy of automata, a second operator, called Inplacement, is introduced to represent the interaction of a WA $\mathscr{A}_2$ inside another WA $\mathscr{A}_1$ with the world created by $\mathscr{A}_1$. The result is equivalent to the composition of $\mathscr{A}_2$ with the automata of level 1 of $\mathscr{A}_1$, although there are some important differences. This operation shows how the hierarchical communication works between the automata inside a world and the automaton representing their external world. Note that, with this operator, we do not want to describe the action of a WA *moving* inside another WA, but we want to describe the static behavior of an automaton *inside* another.

42.4 Further Reading

The interested reader can find more details on the presented modeling framework and results on composability for the operators, in [11] for world variables and parallel composition, and in [12] for hierarchy and inplacement. An application of world automata to a real control case study can be found in [7]. Other examples and case studies are available in [13].

References

1. Alur R, Dill D (1990) Automata for modeling real-time systems (Lecture notes in computer science), vol 443, pp 322–335
2. Alur R, Courcoubetis C, Henziger T, Ho P (1993) Hybrid automata: an algorithmic approach to the specification and verification of hybrid systems (Lecture notes in computer science), vol 736, pp 209–229
3. Jin Y, Liao Y, Polycarpou MM (2006) Balancing search and target response in cooperative unmanned aerial vehicle (UAV) teams. IEEE Trans Syst Man Cybern 38(3)
4. De Sousa JB, Johansson KH, Speranzon A, Silva J (2005) A control architecture for multiple submarines in coordinated search missions. In: 16th IFAC World Congress on automatic control
5. Varaiya P (1993) Smart cars on smart roads: problems of control. IEEE Trans Autom Control 38(2):195–207
6. Vrancken JLM, van Schuppen JH, Soares MS, Ottenhof F (2009) A hierarchical model and implementation architecture for road traffic control. In: Proceedings of the 2009 IEEE international conference on systems, Man and Cybernetics
7. Marinica EN, Capiluppi M, Rogge JA, Segala R, Boel RK (2012) Distributed collision avoidance for autonomous vehicles: world automata representation. In: 4th IFAC conference on analysis and design of hybrid systems (ADHS)
8. Lynch N, Segala R, Vaandrager F (2003) Hybrid I/O automata. Inf Comput 185:105–157
9. Kaynar DK, Lynch N, Segala R, Vaandrager F (2010) The theory of timed I/O automata (Synthesis lectures on computer science). Morgan & Claypool Publishers
10. Mitra S, Wang Y, Lynch N, Feron E (2003) Safety verification of model helicopter controller using hybrid input/output automata. In: Maler O, Pnueli A (eds). In: Proceedings of hybrid systems: computation and control, Prague, the Czech Republic (Lecture notes in computer science), vol 2623. Springer, Berlin, pp 343–358
11. Capiluppi M, Segala R (2012) Modelling implicit communication in multi-agent systems with hybrid input/output automata. In: Third international symposium on games, automata, logics and formal verification (GandALF 2012), Electronic proceedings in theoretical computer science (EPTCS)
12. Capiluppi M, Segala S (2013) World automata: a compositional approach to model implicit communication in hierarchical hybrid systems. In: Third workshop on hybrid autonomous systems (HAS) 2013, Electronic proceedings in theoretical computer science (EPTCS)
13. Capiluppi M, Segala R (2012) Hybrid automata with worlds: a compositional approach to modeling objects that move in a complex environment. Technical report RR 87/2012. University of Verona, Department of Computer Science

Chapter 43
Distributed Average Consensus in Digraphs

Christoforos N. Hadjicostis and Alejandro D. Domínguez-García

43.1 Motivation

The design of protocols and algorithms for distributed computation and control/decision tasks has attracted significant attention by the computer science, communication, and control communities (e.g., [14–16]). Given a set of interconnected nodes (which could be sensors in a sensor network, routers in a communication network, or unmanned vehicles in a multi-agent system), motivational applications range from (i) averaging individual measurements (e.g., when each node provides a local measurement of a global quantity); (ii) transmitting data from one or multiple sources to one or multiple sinks; (iii) coordinating node speed or direction; or (iv) electing a leader.

In a typical consensus problem, each node possesses an initial value and the nodes need to follow a strategy to distributively calculate the same function of these initial values. The consensus problem has received extensive attention from the computer science community (see [15] for an introduction) and the control community [16, 20], due to its applicability to topics such as cooperative control, multi-agent systems, and modeling of flocking behavior in biological and physical systems (see, e.g., [12, 19, 21] and references therein). When the value to which the nodes agree is the average of the initial values, we say that the nodes reach *average consensus*. *Asymptotic average consensus* is reached if (typically, after executing an iterative strategy) the nodes asymptotically converge to the average of their initial values. There are a number of different characteristics that are important in average-consensus problems, including distributivity constraints (e.g., the need for each node to rely solely

---

C.N. Hadjicostis (✉)
Department of Electrical and Computer Engineering, University of Cyprus, Nicosia, Cyprus
e-mail: chadjic@UCY.AC.CY

C.N. Hadjicostis · A.D. Domínguez-García
Department of Electrical and Computer Engineering, University of Illinois at Urbana–Champaign, Urbana IL 61801, USA
e-mail: aledan@ILLINOIS.EDU

J.H. van Schuppen and T. Villa (eds.), *Coordination Control of Distributed Systems*,
Lecture Notes in Control and Information Sciences 456,
DOI 10.1007/978-3-319-10407-2_43

on locally available information), time-varying interconnections, computational and communication complexity, convergence to the average or close to it (in case the algorithm is asymptotic), speed of convergence, and others.

This chapter studies on the average-consensus problem when the interconnection topology is described by a fixed *directed* graph (digraph) and focuses primarily on establishing convergence; more details about speed of convergence and computational/communication complexity (as well as pertinent examples) can be found in the references provided. The presence of an asymmetric information structure makes the average-consensus task in digraphs particularly challenging due to the fact that nodes cannot (easily) provide acknowledgements or more generally inform the nodes that guide their decisions about the actions they are taking. Asymmetric topologies can arise in a variety of realistic scenarios (e.g., if nodes transmit at different power strengths or if interference levels are not uniform at each node).

43.2 Problem Statement

The exchange of information between components (nodes) of a distributed system can be conveniently described by a directed graph (digraph) $\mathscr{G} = \{\mathscr{V}, \mathscr{E}\}$, where $\mathscr{V} = \{1, 2, \ldots, n\}$ is the vertex set (each vertex corresponds to a component/node), and $\mathscr{E} \subseteq \mathscr{V} \times \mathscr{V}$ is the set of directed edges, where $(j, i) \in \mathscr{E}$ if node j can receive information from node i. By convention, we assume no self-loops in $\mathscr{G}$ (i.e., $(j, j) \notin \mathscr{E}$ for all $j \in \mathscr{V}$). The graph is undirected if and only if whenever $(j, i) \in \mathscr{E}$, then also $(i, j) \in \mathscr{E}$, i.e., if node j can receive information from node i, then node i can also receive information from node j. All nodes that can transmit information to node j are said to be in-neighbors of node j and are represented by the set $\mathscr{N}_j^- = \{i \in \mathscr{V} \mid (j, i) \in \mathscr{E}\}$. The number of in-neighbors of j is called the in-degree of j and is denoted by $\mathscr{D}_j^-$ (i.e., $\mathscr{D}_j^- = |\mathscr{N}_j^-|$). Nodes that receive information from node j are said to be the out-neighbors of node j and are represented by the set $\mathscr{N}_j^+ = \{l \in \mathscr{V} \mid (l, j) \in \mathscr{E}\}$. The number of out-neighbors of j is called the out-degree of j and is denoted by $\mathscr{D}_j^+$ (i.e., $\mathscr{D}_j^+ = |\mathscr{N}_j^+|$).

Let V_j be the initial value of node j. In the average-consensus problem, the objective is to have all the nodes calculate the average of these initial values, which we denote by $\mu = \frac{\sum_{\ell=1}^{n} V_\ell}{n}$. Depending on the assumptions, nodes may or may not know n, and they will presumably require several rounds of message exchanges in order to obtain μ (perhaps obtaining the values of V_ℓ, $\ell = 1, 2, \ldots, n$, in the process). In the algorithms, we present in this chapter, in order to obtain μ, each node j maintains some value $\pi_j[k]$ at round k and performs a linear iteration of the form

$$\pi_j[k] = p_{jj}[k]\pi_j[k-1] + \sum_{i \in \mathscr{N}_j^-} p_{ji}[k]\pi_i[k-1] . \tag{43.1}$$

In other words, each node j updates its value to be a linear combination of its own previous value and the values of its in-neighbors using its own self-weight ($p_{jj}[k]$) and the weights ($p_{ji}[k]$, $i \in \mathcal{N}_j^-$) on its incoming links. If we let $\pi[k] = [\pi_1[k], \pi_2[k], \ldots, \pi_j[k], \ldots, \pi_n[k]]'$ (where $'$ denotes vector transposition), then for analysis purposes (43.1) can be written in matrix form as

$$\pi[k] = P[k]\pi[k-1], \tag{43.2}$$

where the weight matrix $P[k] = [p_{ji}[k]]$ (with $p_{ji}[k]$ as the entry at the jth row ith column of matrix $P[k]$).

In (43.1), the $p_{ji}[k]$'s are a set of (time-varying) weights that need to be chosen (along with the initial conditions $\pi[0]$) so that all $\pi_j[k]$ converge for large k to μ. Node j can only choose its self-weight and the weights on its out-going links, i.e., node j can choose values for $\{p_{lj}[k] \mid l = 1, 2, \ldots, n\}$, with the constraint that $p_{lj}[k] = 0$ for all l such that $l \notin \mathcal{N}_j^+$. It is assumed that each node can observe but cannot control the (likely different) values on each of its incoming links and cannot necessarily identify the sender node associated with each value. These assumptions hold naturally for most interconnection topologies that form a digraph (in fact, in many practical situations additional information may be available at each node).

Remark 43.1 For the case when the weights $p_{lj}[k]$'s are fixed for all $k \geq 0$ (i.e., $p_{lj}[0] = p_{lj}[1] = \ldots = p_{lj}[k] = \ldots := p_{lj}$ for $(l, j) \in \mathcal{E}$), as stated in [19, 21] in various forms, the necessary and sufficient conditions for the iteration in (43.2) (with $P[k] = P$, $P(l, j) = p_{lj}$, and $\pi[0] = [V_1, V_2, \ldots, V_j, \ldots, V_n]'$) to asymptotically reach average consensus are: (i) P has a simple eigenvalue at 1, with left eigenvector $[1, 1, 1, \ldots, 1]$ and right eigenvector $[1, 1, 1, \ldots, 1]'$, and (ii) all other eigenvalues of P have magnitude strictly less than 1. If one focuses on nonnegative weights, these conditions are equivalent to the weight matrix P being a primitive doubly stochastic matrix. This is easily achievable in an undirected graph: assuming nodes know the total number of nodes n or an upper bound $n' \geq n$, each node j can easily choose fixed (nonnegative) weights on its out-going links so that $\sum_l p_{lj} = \sum_i p_{ji} = 1, \forall j$, by setting $p_{jj} = 1 - \frac{\mathcal{D}_j}{n'}$, $p_{lj} = \frac{1}{n'}$ if $(l, j) \in \mathcal{E}$, and $p_{lj} = 0$ if $(l, j) \notin \mathcal{E}$, where $\mathcal{D}_j = \mathcal{D}_j^+ = \mathcal{D}_j^-$. As long as the undirected graph is connected, this choice results in a primitive doubly stochastic (symmetric) weight matrix P. Another simple choice that results in a primitive doubly stochastic (symmetric) weight matrix P in connected undirected graphs are the *Metropolis* weights in [22] where $p_{lj} = \frac{1}{1+\max(\mathcal{D}_l, \mathcal{D}_j)}$ if $(l, j) \in \mathcal{E}$, $p_{lj} = 0$ if $(l, j) \notin \mathcal{E}$, and $p_{jj} = 1 - \sum_i p_{ji}$. In a digraph, however, a given node j may not necessarily have $\mathcal{D}_j^+ = \mathcal{D}_j^-$, and it is not as straightforward for nodes to determine appropriate weights so that $\sum_l p_{lj} = \sum_i p_{ji} = 1, \forall j$. [Strategies to obtain matrices with appropriately scaled rows and columns (including doubly stochastic matrices as a special case) have been studied under the umbrella of *matrix scaling* (see, for example, [17]) *without*, however, paying attention to distributivity constraints.]

43.3 Distributed Strategies for Average Consensus in Digraphs

In this section, we describe two strategies that can be used by nodes in a given strongly connected digraph to distributively reach average consensus. The first method relies on an embedded distributed weight adjustment algorithm, whereas the second method relies on a ratio-consensus approach that simultaneously runs two (coupled) iterations, and has each node take the ratio of its two iteration values. Both algorithms require (at the base level) that each node knows its out-degree; this requirement is rather mild, as in most protocols for ad hoc network discovery, each node not only knows which nodes it receives information from, but it also knows which nodes it transmits information to. We also discuss how broadcasting can be accommodated by ensuring that each node sends identical information to each of its out-neighbors.

43.3.1 Distributed Weight Adjustment

This section summarizes a *weight balancing* algorithm that enables the nodes in a strongly connected digraph to asymptotically reach weights that form a primitive doubly stochastic matrix. This algorithm has been presented in [6], which also showed that the weight adaptation can be combined with iteration (43.1) to ensure that the nodes asymptotically reach average consensus. We next describe these ideas starting from the distributed weight adjustment algorithm.

For distributed weight adjustment, each node iteratively updates its self-weight and the weights on its out-going links based on the sum of the weights on its incoming links. More specifically, at each iteration k, node j maintains a parameter $\delta_j[k]$, which determines its self-weight and the weights on its out-going links as $p_{jj}[k] = 1-\delta_j[k]$ and $p_{lj}[k] = c_{lj}\delta_j[k]$, where c_{lj} are constants chosen by node j at initialization so that: (i) $\sum_{l,l\neq j} c_{lj} = 1$, and (ii) $c_{lj} > 0$ if node l can receive information from node j (i.e., if $(l, j) \in \mathscr{E}$), and $c_{lj} = 0$ if node l cannot receive information from node j (i.e., if $(l, j) \notin \mathscr{E}$). For notational convenience, we will take $c_{jj} = 0$ for $j = 1, 2, \dots, n$ so that $\sum_{l,l\neq j} c_{lj} = \sum_l c_{lj} = 1$. [Note that for broadcasting to be possible, node j should choose identical c_{lj} which implies that $c_{lj} = \frac{1}{\mathscr{D}_j^+}$ for $(l, j) \in \mathscr{E}$.]

Initially, node j sets $\delta_j[0] = \frac{\mathscr{D}_j^+}{1+\mathscr{D}_j^+}$ (which implies that $p_{jj}[0] = \frac{1}{1+\mathscr{D}_j^+}$ and $p_{lj}[0] = c_{lj}\frac{\mathscr{D}_j^+}{1+\mathscr{D}_j^+}$). At each iteration k, $k \geq 1$, node j gathers the weights $p_{ji}[k]$, $(j, i) \in \mathscr{E}$, on its incoming edges and updates the value of $\delta[k]$ as

$$\delta_j[k] = \begin{cases} \delta_j[k-1]\rho_j[k-1], & \text{if } \rho_j[k-1] \leq 1, \\ 1 - \frac{1}{\rho_j[k-1]}\big(1 - \delta_j[k-1]\big), & \text{if } \rho_j[k-1] > 1, \end{cases} \tag{43.3}$$

where $\rho_j[k-1] = \sum_i p_{ji}[k-1]$. Based on $\delta_j[k]$, node j updates its self-weight and the weights on its out-going links as $p_{jj}[k] = 1 - \delta_j[k]$ and $p_{lj}[k] = c_{lj}\delta_j[k]$, $l \neq j$.

Since the initial value of $\delta_j[0]$ and the update of $\delta_j[k]$ ensure $0 \leq \delta_j[k] \leq 1$ for all k, it can be easily verified that the $n \times n$ matrix $P[k] = [p_{ji}[k]]$ (with $p_{ji}[k]$ as its $(j,i)^{th}$ entry) will be a nonnegative column-stochastic matrix (i.e., $p_{lj}[k] \geq 0$ and $\sum_{l=1}^{n} p_{lj}[k] = 1$, $j = 1, 2, \ldots, n$). In fact, one can also check that, as long as the diagonal entries of $P[k]$ are strictly smaller than one and at least one diagonal entry is strictly positive, then $P[k]$ will be primitive. This is due to the fact that $P[k]$ corresponds to a graph that is strongly connected [18].

Note that matrix $P[k]$ can also be written as

$$P[k] = \overline{P}\Delta[k] + (I - \Delta[k]), \tag{43.4}$$

where I is the $n \times n$ identity matrix, $\Delta[k] = \text{diag}(\delta_1[k], \delta_2[k], \ldots, \delta_n[k])$ is a diagonal matrix with entries $\delta_j[k]$ on the diagonal, and $\overline{P} = [c_{lj}]$ is the constant weight matrix chosen at initialization, i.e.,

$$\overline{P} = \begin{bmatrix} 0 & c_{12} & c_{13} & \ldots & c_{1n} \\ c_{21} & 0 & c_{23} & \ldots & c_{2n} \\ c_{31} & c_{32} & 0 & \ldots & c_{3n} \\ \vdots & \vdots & \vdots & \ddots & \vdots \\ c_{n1} & c_{n2} & c_{n3} & \ldots & 0 \end{bmatrix}. \tag{43.5}$$

The weight update described by the update of the $\delta_j[k]$'s in (43.3) and the parameterization in (43.4) can be shown to drive the weights, as k goes to infinity, to a limiting weight matrix that is doubly stochastic and primitive. In other words, the sequence of matrices $P[0], P[1], P[2], \ldots, P[k], \ldots$ converges to a doubly stochastic and primitive matrix P_{ss} as long as the given digraph is strongly connected [6].

The weight adaptation algorithm can be combined with iteration (43.1) to ensure that the nodes reach asymptotic average consensus. More specifically, for any $k \geq 1$, the nodes update their values according to

$$\pi[k] = P[k]\pi[k-1], \tag{43.6}$$

where $P[k]$ is determined by (43.3) and the parameterization in (43.4). Note that, at each iteration k, each node j is supposed to perform two tasks: an update of $\delta_j[k]$ (and, thus, of its self-weight and the weights on its out-going links), followed by an update of its value $\pi_j[k]$. Then, taking into account the fact that the sequence $P[0], P[1], \ldots, P[k], \ldots$ consists of column-stochastic (and primitive) matrices, the steady-state solution of (43.6), with initial conditions $\pi_j[0] = V_j$, $\forall j$, denoted by π^{ss}, is such that

$$\pi_j^{ss} = \mu, \ \forall j = 1, 2, \ldots, n. \tag{43.7}$$

The following theorem summarizes the discussions in this section; its proof can be found in [6].

Theorem 43.1 *Consider a strongly connected digraph $\mathscr{G} = \{\mathscr{V}, \mathscr{E}\}$ with $\mathscr{V} = \{1, 2, \ldots, n\}$ where each node $j \in \mathscr{V}$ has some initial value $\pi_j[0] = V_j$. Suppose that nodes set $\delta_j[0] = \frac{\mathscr{D}_j^+}{1+\mathscr{D}_j^+}$, for $j \in \mathscr{V}$, and update their values, for $k \geq 1$, according to (43.6) where $P[k] = \overline{P}\Delta[k] + (I - \Delta[k])$, and $\Delta[k] = diag\big(\delta_1[k], \delta_2[k], \ldots, \delta_j[k], \ldots, \delta_n[k]\big)$ with $\delta_j[k]$, $\forall j \in \mathscr{V}$, updated according to (43.3). Then, $\lim_{k\to\infty} P[k] = P_{ss}$ where P_{ss} is doubly stochastic and primitive, and $\lim_{k\to\infty} \pi[k] = \pi^{ss}$ where the steady-state vector π^{ss} satisfies $\pi_j^{ss} = \mu$, $\forall j = 1, 2, \ldots, n$.*

43.3.2 Ratio Consensus

This section summarizes the so-called *ratio-consensus* algorithm [5], a distributed algorithm that enables the nodes of a multi-component system to reach average consensus. For gossiping-type algorithms, an equivalent approach was also proposed in [2], which is a generalization of the algorithm proposed in [13]; another recent application of ratio consensus appears in [1], where a modified distributed Kalman consensus utilizes ratio consensus to obtain an unbiased estimate for static or dynamic communication networks. However, the idea of ratio consensus can be traced back much earlier; see the discussion on weak convergence in [18, pp. 88–89]. The ratio-consensus algorithm performs two iterative computations in parallel and allows each node to asymptotically obtain the exact average of the values the nodes posses as the ratio of the two state variables that each node maintains. The following theorem (see, for example, [5]) summarizes the basic version of ratio consensus.

Theorem 43.2 *Consider a strongly connected digraph $\mathscr{G} = \{\mathscr{V}, \mathscr{E}\}$ with $\mathscr{V} = \{1, 2, \ldots, n\}$ where each node $j \in \mathscr{V}$ has some initial value V_j. Each node j maintains, at iteration k, state variables $y_j[k]$ and $z_j[k]$ and updates them as follows:*

$$y_j[k+1] = \sum_{i \in \mathscr{N}_j^- \cup \{j\}} y_i[k] \,/\, (1 + \mathscr{D}_i^+)\,, \qquad k \geq 0, \tag{43.8}$$

$$z_j[k+1] = \sum_{i \in \mathscr{N}_j^- \cup \{j\}} z_i[k] \,/\, (1 + \mathscr{D}_i^+)\,, \qquad k \geq 0, \tag{43.9}$$

where $y_j[0] = V_j$, and $z_j[0] = 1$, for all $j \in \mathscr{V}$. Let $\pi_j[k] = \frac{y_j[k]}{z_j[k]}$; then, we have $\lim_{k\to\infty} \pi_j[k] = \frac{\sum_\ell y_\ell[0]}{\sum_\ell z_\ell[0]} = \frac{\sum_\ell V_\ell}{n} = \mu$, $\forall j \in \mathscr{V}$.

Remark 43.2 Letting $y[k] = [y_1[k], y_2[k], \ldots, y_n[k]]'$ and $z[k] = [z_1[k], z_2[k], \ldots, z_n[k]]'$, we can write the iterations in the above theorem in matrix form as

$y[k+1] = P_c y[k]$, and $z[k+1] = P_c z[k]$, where the initial conditions are the same as in the theorem and P_c is a matrix with entries $P_c(l, j) = \frac{1}{1+\mathscr{D}_j^+}$ for all $l \in \mathscr{N}_j^+$ (zero otherwise). Notice that P_c is a column-stochastic matrix that is also primitive as long as the underlying digraph is strongly connected. The ratio-consensus algorithm has each node j calculate at each time step k the ratio $\pi_j[k] = \frac{y_j[k]}{z_j[k]}$. As long as P_c is primitive column stochastic, it can be shown that the ratio $\pi_j[k]$ asymptotically converges to $\frac{\sum_\ell y_\ell[0]}{\sum_\ell z_\ell[0]}$. Thus, by appropriately choosing the initial conditions of ratio consensus, we can compute arbitrary weighted linear combinations of the initial values of the nodes. For example, if $y_j[0] = c_j V_j$ and $z_j[0] = c'_j$, then $\lim_{k\to\infty} \pi_j[k] = \frac{\sum_l y_l[0]}{\sum_l z_l[0]} = \frac{\sum_l c_l V_l}{\sum_l c'_l}$, $\forall j \in \mathscr{V}$. Note that, in general, none of the sequence $y_j[k]$, $z_j[k]$, $\pi_j[k]$ for $k \geq 0$ are monotone.

43.3.3 Other Approaches

The work in [9] proposed a broadcast-based gossip algorithm that relies on each node knowing its out-degree and performing an iteration that involves two variables (that are coupled); the authors show that average consensus is reached but a formal proof of convergence is still open. The authors of [3] use a similar approach (using so-called "surplus variables") and prove convergence for certain small-gain values of the coupling coefficients. Techniques that rely on surplus-like compensation methods to reach average consensus in the presence of unreliable communication links have received attention relatively recently in [4, 8]. Extensions of ratio consensus to handle packet drops and delays in digraphs appear in [7, 11]. Along similar lines (but not motivated by randomization induced by packet drops), the authors of [10] consider randomized discrete-time consensus systems defined over digraphs that preserve the average "on average," by providing an upper bound on the mean square deviation of the consensus value from the exact average.

43.4 Conclusion

In this chapter, we provided an introduction to the problem of average consensus in distributed control systems (with underlying possibly asymmetric communication topologies) and described two algorithms that can be used to asymptotically obtain the average.

References

1. Alighanbari M, How JP (2006) An unbiased Kalman consensus algorithm. In: American Control Conference (ACC 2006)
2. Benezit F, Blondel V, Thiran P, Tsitsiklis JN, Vetterli M (2010) Weighted gossip: Distributed averaging using non-doubly stochastic matrices. In: Proceedings of IEEE international symposium on information theory, pp 1753–1757
3. Cai K, Ishii H (2012) Average consensus on general strongly connected digraphs. Automatica 48(1):2750–2761
4. Chen Y, Tron R, Terzis A, Vidal R (2010) Corrective consensus: converging to the exact average. In: Proceedings of IEEE conference decision and control, pp 1221–1228
5. Domínguez-García AD, Hadjicostis CN (2010) Coordination and control of distributed energy resources for provision of ancillary services. In: Proceedings of IEEE SmartGridComm, pp 537–542
6. Domínguez-García AD, Hadjicostis CN (2013) Distributed matrix scaling and application to average consensus in directed graphs. IEEE Trans Autom Control 58(3):667–681
7. Domínguez-García AD, Hadjicostis CN, Vaidya NH (2012) Resilient networked control of distributed energy resources. IEEE J Sel Areas Commun 30(6):1137–1148
8. Fagnani F, Zampieri S (2009) Average consensus with packet drop communication. SIAM J Control Optim 48(1):102–133
9. Franceschelli M, Giua A, Seatzu C (2011) Distributed averaging in sensor networks based on broadcast gossip algorithms. IEEE Sens J 11(3):808–817
10. Frasca P, Hendrickx JM (2013) On the mean square error of randomized averaging algorithms. Automatica 49(8):2496–2501
11. Hadjicostis CN, Charalambous T (2014) Average consensus in the presence of delays in directed graph topologies. IEEE Trans Autom Control 59(3):763–768
12. Jadbabaie A, Lin J, Morse AS (2003) Coordination of groups of mobile autonomous agents using nearest neighbor rules. IEEE Trans Autom Control 48(6):988–1001
13. Kempe D, Dobra A, Gehrke J (2003) Gossip-based computation of aggregate information. In Proceedings of the 44th annual IEEE symposium on foundations of computer science, FOCS 2003, pp 482–491
14. Koetter R, Médard M (2003) An algebraic approach to network coding. IEEE/ACM Trans Netw 11(5):782–795
15. Lynch N (1996) Distributed algorithms. Morgan Kaufmann Publishers, San Mateo
16. Olfati-Saber R, Fax JA, Murray RM (2007) Consensus and cooperation in networked multi-agent systems. Proc IEEE 95(1):215–233
17. Parlett BN, Landis TL (1982) Methods for scaling to doubly stochastic form. Linear Algebra Appl 48:53–79
18. Seneta E (2006) Non-negative Matrices and Markov Chains, Revised printing edition edn. Springer, New York
19. Sundaram S, Hadjicostis CN (2008) Distributed function calculation and consensus using linear iterative strategies. IEEE J Sel Areas Commun 26(4):650–660
20. Tsitsiklis JN (1984) Problems in decentralized decision making and computation. PhD thesis, Massachusetts Institute of Technology, Cambridge, MA
21. Xiao L, Boyd S (2004) Fast linear iterations for distributed averaging. Syst Control Lett 53(1):65–78
22. Xiao L, Boyd S, Lall S (2005) A scheme for robust distributed sensor fusion based on average consensus. In: Proceedings of the fourth international symposium on information processing in sensor networks (IPSN 2005), pp 63–70

Part IX
Research Program

The final chapter highlights the main lines of future research to carry on the results of the C4C project and to address the growing challenges of system design in engineering, informatics, and mathematics.

Chapter 44
Research Program for Control of Distributed and of Multilevel Systems

Jan H. van Schuppen

44.1 Introduction

The reader of this book after reading many chapters has hopefully become convinced that control engineering demands from the control sciences and other research areas much more research development on control of distributed and of multilevel systems. This chapter summarizes several research issues which need to be pursued in the opinion of the author. The chapter is placed at the end of the book because it addresses research which seems of interest after the completion of the C4C Project. Its scope is in general broader than that of the C4C Project though it is motivated by the research of the project.

The term *research program* refers to the description of research issues to be pursued by one or several research groups. The description provided in the remainder of the chapter will often take the form of questions and directions rather than specific results to be achieved. Research is a pursuit into the unknown which can therefore not be prescribed. The scope in general is rather broad and the horizon is often 10 years or longer. This chapter uses the description of the term research program formulated above. A research program consists of research topics which in turn consist of research issues.

The reader will increase his/her understanding of this chapter if he/she reads the chapter in combination with the introductory chapters of the various parts of the book.

J.H. van Schuppen (✉)
Van Schuppen Control Research, Gouden Leeuw 143, 1103 KB Amsterdam, Netherlands
e-mail: jan.h.van.schuppen@xs4all.nl

J.H. van Schuppen and T. Villa (eds.), *Coordination Control of Distributed Systems*,
Lecture Notes in Control and Information Sciences 456,
DOI 10.1007/978-3-319-10407-2_44

44.2 Research Program

Control of distributed and of multilevel systems requires the attention of mathematicians and of engineers for the following list of major research topics. In the remainder of this chapter, these research topics are described in more detail.

The research topics are as follows:

1. Integration of control, communication, and informatics in multilevel-distributed systems.
2. Control of multilevel systems.
3. Coordination control of distributed systems.
4. Distributed control with communication.
5. Communication for control.
6. Informatics of distributed and of multilevel systems.

At a higher level of theory formation, the following *research areas* require attention:

1. System theory of distributed and of multilevel systems.
2. Complexity of system structure and of control, communication, and informatics issues.
3. Cognition for modeling and control.
4. An algebraic approach to multilevel and distributed systems, and computational algebra for decentralized control problems.

In the remainder of this chapter, these research topics are described in more detail.

44.3 Research Topics

44.3.1 Integration of Control, Communication, and Informatics

A research topic is integration. The main objective of a technological system is that it meets the operational specifications understood in a broad sense. These specifications include the control objectives, the effectiveness of the communication network, the correctness of the algorithms, and the adequacy of the computational resources used. Therefore, the combined objectives call for an integrated design. Such an integrated design will avoid the case in which a technological system is a combination of subsystems each of which is designed for a particular local optimum. This research issue has received little attention so far.

The research issue is thus to formulate a framework for such integration of control, communication, and informatics. A way to develop such a framework is to study many particular instances and to formulate principles for the trade-offs in the integration. The case studies described in this book offer an entry into the vast collection of examples.

44.3.2 Modeling of Multilevel-Distributed Systems

Modeling of multilevel systems is not yet a consolidated research topic. The term *multilevel system* is now preferred by the author over the term hierarchical system though both terms are used in the literature. The two terms are not quite identical; a hierarchical system is a special case of a multilevel system though no attempt to clarify this distinction will be made in this chapter. Modeling requires more research into abstraction of dynamic systems for multilevel systems, primarily of nonlinear and of hybrid systems. More experience is required though there is already quite a literature on abstraction of discrete-event systems and on abstraction of linear systems by perturbation techniques.

Modeling of the interaction of two or more subsystems is needed as this problem has not been sufficiently addressed in the literature. The extent to which the systems interact is important to the control synthesis to be carried out. System identification of distributed systems is of interest, in particular identification of the interconnections.

The above research problems are motivated by problems of control engineering as may be found in other chapters of the book. But an additional motivation is provided by the research area of biochemical reaction networks for the modeling of the chemical behavior of cells in plants and animals.

44.3.3 Control of Multilevel Systems

A major research issue is to develop control synthesis of multilevel systems. In the literature, one finds several forms of control of multilevel systems. For the discussion, restrict attention to a system with two or more levels and with two or more subsystems at each level, except for the top level.

Control synthesis at each level could be carried out by abstracting all subsystems of a layer at the next higher layer and then developing a controller for these abstracted systems. A controllability condition has to be formulated which is equivalent to the closed-loop system of two levels meeting its control objectives in which this condition is distributed over two subsequent levels.

The relations between the subsystems at different levels require further investigation. The final goal is to formulate a condition for control synthesis to achieve the control objectives for the entire multilevel system in terms of controllability conditions for the subsystems and for their interactions.

44.3.4 Coordination Control of Distributed Systems

Research is required into the many different forms of coordination control of distributed systems. The actual cooperation or interaction of subsystems is so far insufficiently formulated, and one expects new concepts to emerge. The following research issues require attention: modeling of the interaction of the subsystems, analysis of the communication needs between the coordinator and the subsystems, and the control synthesis of the coordinator for its control of the subsystems. See also Chap. 12.

44.3.5 Distributed Control with Communication Between Controllers

This research topic concerns control of distributed system in which the local controllers may communicate with each other outside the control system. A major class of examples is the control of a communication networks where each node interacts with its nearest neighbors both inside the control system and outside.

Control synthesis techniques are needed for this type of control because it is practically useful and theoretically extremely difficult. A sufficient condition for optimality of the subclass of control laws based on nearest-neighbor state communication or of nearest-neighbor partial-state communication will be useful for theory development. Think of sufficient and necessary conditions for optimality of this subclass. Even if a control law in a particular subclass is not optimal, how much does one loose in performance?

Research may focus on the following questions: What is to be communicated? When is information to be communicated? To whom is information to be communicated? Cost functions which account for the cost of communication have to be considered. Principles for the trade-off between control and communication may emerge for control with communication between controllers.

In addition, the usefulness of periodic communication in time may be of interest because this seems quite practical for control engineering. See also Chap. 22.

44.3.6 Decentralized and Distributed Control

The *concept of a system for decentralized control* is best formulated and explored. The author favors a view in which the state is local to the controller considered though extended with a model for the other controllers. See Chap. 5 for an approach in this direction. The more classical view of a central system remains of interest for specific engineering control problems.

The *concept of a system for multilevel systems* has also to be formulated. What model should a subsystem adopt for its parent and for its children? Is a degree of abstraction acceptable for such a model and if so how can the abstraction be formulated?

Equilibrium versus minimal element. A major research issue is to formulate sufficient and necessary conditions when a person-by-person equilibrium tuple of control laws is also a minimal tuple. For team problems, a sufficient condition is known; for control problems of distributed control, this is an open problem though research on it is in progress. See also Chap. 18.

Common and private information. What information to exchange between controllers depends on the concepts of common, private, and dependent information. In addition, the relation of the information with the control objectives plays a role. The research issue is to formulate concepts and to develop theory for their use in distributed control with communication. See also Chap. 26.

Signaling. In decentralized control, any controller can send signals to other controllers via the plant, called *signaling*. Which information should a controller send? The concepts of common and of private information seem appropriate here. Control synthesis for signaling between controllers of decentralized systems is basically nonexistent. This control issue requires investigation of communication over communication channels with memory and with partial-state feedback. This issue requires the interaction of control, information theory, and of communication theory. See also Chap. 20.

44.3.7 Communication and Control

A research issue is to gain a better understanding of information measures and control along the lines of those derived in the C4C Project. Thus, inputs can be used to improve the quality of the observations of a subsystem in terms of information measures, or inputs can be used to increase a control performance measure. How is one to strike a balance between these partly conflicting control objectives?

Another issue is the control of communication networks. The backpressure algorithm of control of communication networks is a form of distributed control with communication see 30. Guidelines on how to determine the structure of communication laws and of control laws in case of communication networks will be highly valuable for the actual control of these network and for control of the computer networks referred to as clouds. Research of the last decade has developed a number of variants of this form of control which could yield a better understanding of the control of distributed and of multilevel systems. The control problems generated by the technological development of clouds are a rich source for control theory.

Another research issue is the choice of the quantification of vector-valued signals for information transmission on the control performance.

44.3.8 Informatics of Distributed and of Multilevel Systems

A research issue is to formulate theoretical computer science models for the verification of system properties. The existing algebraic frameworks need further development to deal with distributed and with multilevel systems. Possibly, concepts of control of decentralized systems can be usefully explored. Existing algebraic frameworks include bisimulation relations, co-algebraic framework, and λ-calculus. A direction is to further develop an algebraic framework for decentralized, distributed, and multilevel control.

Issues of safety and security of discrete-event systems in particular of distributed systems have to be investigated.

Complexity of control laws and of control problems requires further study. Complexity of multilevel systems is likely to be a concept useful for the practice of system design. The research issues are to find concepts and to formulate theory. In particular, one expects complexity measures of multilevel systems which relate to the number of levels and to the complexity of all subsystems. In addition, the complexity will be related to the way the subsystems of adjacent levels interact.

Complexity of hybrid multilevel systems has to be considered because in large multilevel systems, both discrete and continuous subsystems are used. A promising approach seems to be computable analysis for subsets of the real numbers, see [4]. This theory has consequences for the selection of subclasses of systems for which control synthesis questions are properly solvable.

44.3.9 Economics and Distributed Systems

Can concepts and theory of economics be used to advance control of distributed and of multilevel systems? A starting point for this investigation may be the paper of Arrow and Hurwicz [3]. In that paper, the interaction of the subsystems at two adjacent levels of a multilevel system can be deduced from the cost function. In control of multilevel systems, the relation between the cost function and the structure of the control laws needs exploration. This may also lead to a better understanding of what is best communicated between systems of different levels in multilevel control systems. It may also lead to novel design principles for multilevel systems.

44.4 Research Areas

44.4.1 System Theory of Decentralized/Distributed Systems

System theory addresses research issues of representations of dynamic systems in a broad sense.

Research issues of interest are the interaction of two or more subsystems in a distributed system and the interactions of the subsystems at two or more levels of a multilevel system. In either case, a form of abstraction of the subsystems seems useful.

The main problem is to formulate a concept of state for distributed and for decentralized systems in which there are two or more controllers. A viewpoint is to make the state local so each controller has a model for the distributed system including a model for the control laws of the other controllers. This allows for each controller having its own model of other controllers and a particular belief about the models and probability distributions for the state of other controllers. See Chap. 5.

Another issue is nonsequential control in which the order of arrival of observations of outputs and of inputs is itself random. See [1, 2] for this research issue.

44.4.2 Complexity

The research issues are motivated by the observation that multilevel systems are quite effective in handling complexity of large systems and by the question: How the structure of a multilevel system can improve the functioning of such systems? The need is thus for a complexity study of multilevel systems in all its aspects.

44.4.3 Cognition

The modeling of an engineering phenomenon is one of the most time-consuming parts of control projects. Experienced researchers state that it takes about 2/3 to 70 % of the efforts of a project. For future control applications, it will therefore be of interest to develop procedures to automate the modeling and the subsequent control synthesis.

Cognition is a research area of informatics which has as objective to make computer programs which can for an unknown phenomenon carry out modeling, identification, and control design. Needed are then concepts of cognition, algorithms, and theory to obtain a model of the engineering system to be controlled and a satisfactory control laws. The envisioned approach goes much deeper than the current artificial intelligence methods. The cognition approach may not work for all problems, and a characterization is needed for the subclass for which it is effective and economical.

44.4.4 Algebra and Logic

The main mathematical basis of control and system theory is algebra, besides analysis and the theory of dynamical systems. The relations between systems and algorithms for system and control theoretic questions find their basis in algebraic concepts. For distributed and for multilevel systems, possibly novel concepts have to be developed. The finiteness of algorithms can often be obtained by imposing algebraic conditions. The algebra of particular function spaces needs further development for control and system theory. Logic can be useful for theory of control as is demonstrated by informatics.

44.5 Further Reading

For research programs on control theory in a broad sense, see [7]. For a research road maps in moving objects, see [6]. There is a vast literature on other approaches to control of distributed and of multilevel systems which the reader will find in a multitude of journals of engineering and of mathematics.

The author has taken inspiration for control from the book by Norbert Wiener on cybernetics, see [8]. The broadness of the issues considered in that book and their relations are fundaments for the development of control theory. A history of algebra and its structures may be found in [5].

References

1. Andersland MS, Teneketzis D (1992) Information structures, causality, and nonsequential stochastic control I: design-independent properties. SIAM J Control Opt 30:1447–1475
2. Andersland MS, Teneketzis D (1994) Information structures, causality, and nonsequential stochastic control: II design-dependent policies. SIAM J Control Opt 32:1726–1751
3. Arrow KJ, Hurwicz L (1963) Decentralization and computation in resource allocation. In: Pfouts RW (ed) Essays in economics and econometrics. University of North Carolina Press, Chapel Hill, pp 34–104
4. Collins P (2005) Continuity and computability of reachable sets. Theor Comput Sci 341:162–195
5. Corry L (1996) Modern algebra and the rise of mathematical structures. Number 17 in Historical Studies. Birkhäuser Verlag, Basel
6. Marrón PJ, Karnouskos S, Minder D (eds) (2009) Research road map on cooperating objects
7. Murray RM (ed) (2003) Control in an information rich world: report of the panel on future directions in control, dynamics, and systems. SIAM, Philadelphia
8. Wiener N (1948) Cybernetics: or control and communication in the animal and the machine. Wiley, New York

Index

Symbols
σ-algebra
common information σ-algebra, 217
conditionally independent σ-algebras, 217
independent private information σ-algebra, 217
observation σ-algebra, 217
private information σ-algebra, 217
sufficient σ-algebra, 217

A
Agreement problem, 222
Algebra, 391
Approximation
ϵ-lower approximation, 342
upper approximation, 342
Architecture
central control architecture, 90
control architecture, 89
network system architecture, 84
system architecture, 81
Ariadne, 339
Automaton, 339
hybrid automaton, 339, 340
control modes, 340
control switches, 340
invariant condition, 340
dynamic law, 340
activation condition, 340
reset relation, 340
state, 340
trajectory, 340
reachable set, 339
hybrid i/o automaton (HIOA), 367
HIOA of a car, 370
HIOA with world variables of a car, 372
world automaton (WA), 372
WA of a sandy area, 373
operations on world automata, 373
inplacement, 373
parallel composition, 373

B
Backpressure scheduling, 251
cluster-based backpressure scheduling, 252
enhanced layered backpressure scheduling, 257
layered backpressure scheduling, 251, 255
shortest-path backpressure scheduling, 251
Bilateral teleoperation system, 286
Blocking, 58, 132
nonblocking, 132

C
Caching, 261
cooperative caching, 261
Case study
c4c case study, 3
Channel
channel capacity, 300, 312
communication channel, 294
feedback channel, 309
feedforward channel, 309
memoryless channel, 294
Code
channel code, 300

J.H. van Schuppen and T. Villa (eds.), *Coordination Control of Distributed Systems*,
Lecture Notes in Control and Information Sciences 456,
DOI 10.1007/978-3-319-10407-2

lossless codes, 295
lossy codes, 295
nonanticipative code, 327
nonanticipative symbol-by-symbol code, 327
noncausal code, 318
realization of nonanticipative code, 330, 333
Coding, 288
causal coding, 292
joint source–channel coding (JSCC), 292, 303
noncausal coding, 292
Cognition, 391
Collision
collision avoidance, 56
Communication
asynchronous communication, 137
communication and control, 389
communication for control, 295
communication for problem, 291
communication law, 226
communication task, 166
nonanticipative communication, 293, 294, 325
real-time communication, 209
synchronuous communication, 137
Communication network
ad hoc communication network, 251
cache control of communication networks, 259
control centrality of a communication network, 267
delay-based network, 275
differentiated services packet network, 273
influential nodes of a communication network, 265
lossy networks, 274
pub–sub network, 259
VANET (vehicular ad hoc network), 266
wireless ad hoc communication network, 251
Complexity, 232, 391
complexity of hierarchical control, 232
complexity of road networks, 232
Conditional independence
conditional independence and the Markov chain condition, 294, 297, 320, 327
conditional independence of σ algebras, 217
conditional independence of linear subspaces, 115
Consensus
asymptotic average consensus, 375
distributed average consensus, 375
distributed strategies for average consensus, 378
distributed weight adjustment, 378
ratio-consensus algorithm, 380
Container
automated container terminal, 54
Continuity
lower semicontinuity, 311
upper-semi-continuity, 311
Control
control agent, 40, 41
control requirements, 61
coordination challenges, 27
coordination control, 61, 93, 99, 124, 138, 387
leader-follower coordination control, 35, 41
decentralized control, 20, 89, 155, 170, 389
distributed control, 91, 389
distributed control with direct communication between controllers, 91, 185, 388
hierarchical control, 231
model predictive control (MPC), 9, 273
coordinated model predictive control (CMPC), 69
multilevel control, 93, 387
traffic control, 53
zone-control, 53
Control law
nearest-neighbor state-feedback control law, 74, 187, 211
set of control laws depending on common conditional distribution and private information, 206
static-output nearest-neighbor feedback control law, 187
time-varying control laws based on partial observations, 170
Control objective, 102, 183
Control problem
decentralized control problem, 170
LQ optimal control problem, 123
problem of distributed control with direct communication between controllers, 185
supervisory control problem, 64, 133

Controllability, 119, 123
Coordinated discrete-event system, 138
Coordinated linear system, 100, 108, 117
Coordination
  coordinated control, 92
  motion coordination, 9, 19
Coordinator, 99, 137
  coordinator for nonblockingness, 143
  supervisory coordinator, 1, 61

D
Data
  data rate theorem, 192
Deadlock, 57
  deadlock avoidance, 56
Decoder, 294, 295
Decoder differential, 295
Differential
  Gateau differential, 321
Differential games, 22
Distortion
  Hamming distortion, 323
  IID Bernoulli source-Hamming distortion, 303
  IID Gaussian source-mean-square-error distortion, 303
  minimum excess distortion, 327
Distribution
  reproduction distribution, 320
  reproduction conditional distribution, 321
Duality
  duality of a source and a channel, 302, 332

E
Encoder, 294
Entropy, 292
  conditional entry, 296
  relative entropy, 310
  topological entropy, 192
Equality
  variational equality, 313
Equation
  Bellman-Hamilton-Jacobi equation, 16
Equilibrium
  controller-by-controller equilibrium, 209
  person-by-person equilibrium, 149
Equivalence
  feedback equivalence of a linear system, 118
Event, 130
  event set, 130
    controllable event, 131
    uncontrollable event, 131

F
Flow, 341
  flow tube, 341
Function
  information nonanticipate rate distortion function, 318
  exponential function, 158
  Hamiltonian function, 161
  nonanticipate rate distortion function, 327
  partial transition function, 130
  rate distortion function (RDF), 292, 302

G
Game
  game theory, 147
Gaussian random variable
  Canonical variable decomposition of a tuple of Gaussian random variables, 218
  covariance matrix of a Gaussian random variable, 218
  mean value of a Gaussian random variable, 218
Gaussian system
  Gaussian system with a common–private information decomposition, 220
  Gaussian system with two observation streams, 224
Generator, 130
  controlled generator, 131
  minimal generator, 140
  nonblocking generator, 142
Girsanov's theorem, 158
  Girsanov's measure transformation, 159
Graph, 69
  directed graph, 376

I
Inclusion
  computation of the reachable set, 360
  differential inclusion, 357
  input-affine systems, 360
  reachable set, 357
  single-step overapproximation, 359
Inequality

Fano's inequality, 298
Information
common information, 209, 219, 389
correlated information, 219
directed information, 292, 296, 310
independent private information, 219
mutual information, 292, 296
private information, 209, 219
sufficient information, 219
Information structure, 165, 199
r-step delayed-sharing information structure, 200
classical information structure, 200, 210
common information structure, 200, 210
information structure with perfect recall, 199
local-private information structure, 202
multi-step delayed-sharing information structure, 200, 211
nearest-neighbor-sharing information structure, 40, 201, 210
nonclassical information structure, 157, 201, 211
nonsequential information structure, 199
overlapping information structure, 200, 210
partially nested information structure, 199
private information structure, 201
sequential information structure, 199
strictly classical information structure, 200
Integration
Integration of Control, Communication, and Informatics, 386

L
Languages
$L_m(G)$-closed language, 132
conditionally closed language, 141
conditionally controllable language, 140
conditionally decomposable language, 138
controllable language, 132
languaged generated by a generator, 131
marked language, 131
prefix-closed language, 143
regular language, 131
supremal conditionally controllable language, 143
supremal controllable sublanguage, 134
synchronously nonconflicting languages, 139
Location, 339

M
Machines
complex machines, 6, 61
Manufacturing
manufacturing network, 174
Markov chain condition, 297
Matrix
communication matrix, 225
feedback matrix, 118
Measure
μ-power centrality index, 262
centrality measure, 262
information measure, 295
Mode, 339

N
Network
network system architecture, 84

O
Oscillator
Van der Pol oscillator, 363

P
Performance
no loss in performance, 207
optimal performance theoretically attainable (OPTA), 292
performance bounds in communication, 320
Platoon, 38
Prediction
adaptive prediction, 47
coordinated–distributed prediction, 50
prediction of traffic flow, 45
Principle
stochastic maximum principle, 161
Process
stationary ergodic stochastic process, 299
stochastic process, 299
Projection, 134
locally control consistent projection, 143
Property
asymptotic equipartition property, 295, 298
observer property, 140

Q
QHM (traffic management in structure of a road traffic network), 244
Quality of service (QoS), 251

R
Riccati
 algebraic Riccati equation, 123

S
Semantics
 lower semantics, 341
 upper semantics, 341
Set
 fidelity set, 302
 reachable set, 339, 357
 set of control patterns, 131
Shannon, C.E, 291
Signaling, 167, 209, 389
Simulation
 co-simulation framework, 283
 interconnection of simulation tools, 284
 SystemC Network Simulation Library, 285
Source
 binary symmetric Markov source (BSMS), 319, 324
 memoryless source, 295
 information source, 294
Space
 Polish space, 309
 reference probability space, 159
Stability
 information stability, 299
State
 initial state, 130
 marked state, 130
 state set, 130
Strategy
 admissable regular strategy, 156
 admissable relaxed strategy, 157
 person-by-person optimal relaxed strategy, 157
 team optimal relaxed strategy, 157
Subspaces
 (A, B)-invariant subspace, 118
 A-invariant subspace, 116
 conditionally linearly independent subspaces, 115
 coordinator subspace, 115
 linearly independent subspaces, 115
Supervisor, 131
 nonblocking supervisor, 132
System, 31
 asynchronous subsystem, 86
 bilateral teleoperation system, 286
 child subsystem, 83
 closed-loop discrete-event system, 131
 coordinated linear control system, 117
 coordinated linear system, 100, 108, 116
 coordinated system, 85
 decentralized system, 85
 discrete-event system, 130, 208
 discrete-time nonlinear stochastic system, 198
 distributed discrete-event system, 134
 distributed system, 84
 dynamic system structure, 27
 hybrid system, 339
 input-affine system, 360
 line-structured deterministic linear system, 186
 mammillary system, 85
 multilevel system, 84, 387
 networked control system, 83, 274
 nonanticipate communication system, 325
 observer-controller decentralized control system, 169
 parent subsystem, 83
 real-time communication system, 294
 stochastic dynamic system, 156
 synchronous subsystem, 86
 system specification, 31
System theory, 390
 system theory decentralized and distributed systems, 390

T
Team, 145
 locally finite team decision rule, 150
 optimal team decision rule, 157
 stationary team decision rule, 150
 team cost, 149
 team decision rule, 148
 team member, 148
 team objective, 148
 team optimality conditions, 161
 team problem, 149
 team theory, 147
 team theory setting, 148
Traffic
 road traffic management, 239
 road traffic network, 5, 40, 45, 53

traffic control, 53
traffic intersection control, 35
traffic flow prediction, 45
urban traffic, 35

V
Variable
random variables
jointly independent and identically distributed random variables, 298
Vehicle
aerial vehicle, 5
automated guided vehicles, 6, 53
autonomous underwater vehicle (AUV), 4
underwater vehicle, 5, 9
uninhabited aerial vehicle (UAV), 5
vehicle systems, 5, 9, 27, 53
Verification
ariadne, 339
reach set or reached set, 340
ϵ-lower approximation, 342
enclosure, 341
flow tube, 341
function calculus, 341
image sets, 341
lower semantics, 341
outer approximation, 342
upper semantics, 341
(parametric) dominance verification, 342
(parametric) safety verification, 342
formal verification, 347
vehicle formation control, 10
vehicle motion coordination, 9
watertank example, 343
surgical task, 347
modeling of a surgical task, 348
properties of a surgical task, 350
verification of a surgical task, 351

W
Witsenhausen, H., 204, 210
Witsenhausen's counter example, 161, 167, 202
Word, 130
empty word, 130
prefix of a word, 131

MIX
Papier aus verantwortungsvollen Quellen
Paper from responsible sources
FSC® C105338

If you have any concerns about our products,
you can contact us on
ProductSafety@springernature.com

In case Publisher is established outside the EU,
the EU authorized representative is:
Springer Nature Customer Service Center GmbH
Europaplatz 3, 69115 Heidelberg, Germany

Printed by Libri Plureos GmbH
in Hamburg, Germany